Signals and Systems: A Fresh Look

Chi-Tsong Chen
Stony Brook University

Also by the same author:

- *Linear System Theory and Design*, Oxford Univ. Press, [1st ed., (431 pp), 1970; 2nd ed., (662 pp), 1980; 3rd ed., (332 pp), 1999].

- *Analog and Digital Control System Design: Transfer-Function, State-Space, and Algebraic Methods*, Oxford Univ. Press, 1993.

- *Digital Signal Processing: Spectral Computation and Filter Design*, Oxford Univ. Press, 2001.

- *Signals and Systems*, 3rd ed., Oxford Univ. Press, 2004.

The following are widely recognized:

- There is a gap between what is taught at universities and what is used in industry.

- It is more important to teach how to learn than what to learn.

These were the guiding principles in developing this book. It gives an overview of the subject area of signals and systems, discussing the role of signals in designing systems and various mathematical descriptions for the small class of systems studied. It then focuses on topics which are most relevant and useful in practice. It also gives reasons for not stressing many conventional topics. Its presentation strives to cultivate readers' ability to think critically and to develop ideas logically.

Preface

Presently, there are over thirty texts on continuous-time (CT) and discrete-time (DT) signals and systems.[1] They typically cover CT and DT Fourier series; CT and DT Fourier transforms; discrete Fourier transform (DFT); two- and one-sided Laplace and z-transforms; convolutions and differential (difference) equations; Fourier and Laplace analysis of signals and systems; and some applications in communication, control, and filter design. About one-third of these texts also discuss state-space equations and their analytical solutions. Many texts emphasize convolutions and Fourier analysis.

Feedback from graduates that what they learned in university is not used in industry prompted me to ponder what to teach in signals and systems. Typical courses on signals and systems are intended for sophomores and juniors, and aim to provide the necessary background in signals and systems for follow-up courses in control, communication, microelectronic circuits, filter design, and digital signal procession. A survey of texts reveals that the important topics needed in these follow-up courses are as follows:

- *Signals*: Frequency spectra, which are used for discussing bandwidth, selecting sampling and carrier frequencies, and specifying systems, especially, filters to be designed.

- *Systems*: Rational transfer functions, which are used in circuit analysis and in design of filters and control systems. In filter design, we find transfer functions whose frequency responses meet specifications based on the frequency spectra of signals to be processed.

- *Implementation*: State-space (ss) equations, which are most convenient for computer computation, real-time processing, and op-amp circuit or specialized hardware implementation.

These topics will be the focus of this text. For signals, we develop frequency spectra[2] and their bandwidths and computer computation. We use the simplest real-world signals (sounds generated by a tuning fork and the middle C key of a piano) as examples. For systems, we develop four mathematical descriptions: convolutions, differential (difference) equations, ss equations, and transfer functions. The first three are in the time domain and the last one is in the transform domain. We give reasons for downplaying the first two and emphasizing the last two descriptions. We discuss the role of signals in designing systems and the following three domains:

- *Transform domain*, where most design methods are developed.

- *Time domain*, where all processing is carried out.

- *Frequency domain*, where design specifications are given.

We also discuss the relationship between ss equations (an internal description) and transfer functions (an external description).

Because of our familiarity with CT physical phenomena and examples, this text studies first the CT case and then the DT case with one exception. The exception is to

---

[1] See the references at the end of this book.

[2] The frequency spectrum is not defined or not stressed in most texts.

use DT systems with finite memory to introduce some system concepts because simple numerical examples can be easily developed whereas there is no CT counterpart. This text stresses basic concepts and ideas and downplays analytical calculation because all computation in this text can be carried out numerically using MATLAB. We start from scratch and take nothing for granted. For example, we discuss time and its representation by a real number line. We give the reason for defining frequency using a spinning wheel rather than using $\sin \omega t$ or $\cos \omega t$. We also give the reason that we cannot define the frequency of DT sinusoids directly and must define it using CT sinusoids. We make the distinction between amplitudes and magnitudes. Even though mathematics is essential in engineering, what is more important, in our view, is its methodology (critical thinking and logical development) than its various topics and calculational methods. Thus we skip many conventional topics and discuss, at a more thorough level, only those needed in this course. It is hoped that by so doing, the reader may gain the ability and habit of critical thinking and logical development.

In the table of contents, we box those sections and subsections which are unique in this text. They discuss some basic issues and questions in signals and systems which are not discussed in other texts. We discuss some of them below:

1. Even though all texts on signals and systems claim to study linear time-invariant (LTI) systems, they actually study only a very small subset of such systems which have the "lumpedness" property. What is true for LTI lumped systems may not be true for general TLI systems. Thus, it is important to know the limited applicability of what we study.

2. Even though most texts start with differential equations and convolutions, this text uses a simple RLC circuit to demonstrate that the state-space (ss) description is easier to develop than the aforementioned descriptions. Moreover, once an ss equation is developed, we can discuss directly (without discussing its analytical solution) its computer computation, real-time processing, and op-amp circuit implementation. Thus ss equations should be an important part of a text on signals and systems.

3. We introduce the concept of coprimeness (no common roots) for rational transfer functions. Without it, the poles and zeros defined in many texts are not necessarily correct. The concept is also needed in discussing whether or not a system has redundant components.

4. We discuss the relationship between the Fourier series and Fourier transform which is dual to the sampling theorem. We give reasons for stressing only the Fourier transform in signal analysis and for skipping Fourier analysis of systems.

5. This text discusses model reduction which is widely used in practice and yet not discussed in other texts. The discussion shows the roles of a system's frequency response and a signal's frequency spectrum. It explains why the same transfer functions can be used to design seismometers and accelerometers.

A great deal of thought was put into the selection of the topics discussed in this text.[3] It is hoped that the rationale presented is convincing and compelling and that this new text will become a standard in teaching signals and systems, just as my book *Linear system theory and design* has been a standard in teaching linear systems since 1970.

In addition to electrical and computer engineering programs, this text is suitable for mechanical, bioengineering, and any program which involves analysis and design of systems. This text contains more material than can be covered in one semester. When teaching a one-semester course on signals and systems at Stony Brook, I skip Chapter 5, Chapter 7 after Section 7.4, and Chapter 11, and cover the rest. The topics in Chapter 3 are discussed where they are needed. Clearly other arrangements are also possible.

Many people helped me in writing this book. Ms. Jiseon Kim plotted all the figures in the text except those generated by MATLAB. Mr. Anthony Oliver performed many op-amp circuit experiments for me. Dr. Michael Gilberti scrutinized the entire book, picked up many errors, and made several valuable suggestions. Professor Robert Funnell of McGill University suggested some improvements. I consulted Professors Amen Zemanian and John Murray whenever I had any questions or doubts. I thank them all.

C. T. Chen
December, 2011

---

[3]This text is different from Reference [C8] in structure and emphasis. It compares four mathematical descriptions, and discusses three domains and the role of signals in system design. Thus it is not a minor revision of [C8]; it is a new text.

Table of Contents

Chapter 1

Introduction

1.1 Signals and systems

This book studies signals and systems. Roughly speaking, anything that carries information can be considered a *signal*. Any physical device or computer program can be considered a *system* if the application of a signal to the device or program generates a new signal. In this section, we introduce some examples of signals and systems that arise in daily life. They will be formally defined in subsequent chapters.

Speech is one of the most common signals. When a speaker utters a sound, it generates an acoustic wave, a longitudinal vibration (compression and expansion) of air. The wave travels through the air and is funneled through the auditory canal to the eardrum of the listener. The middle ear is constructed for efficient transmission of vibration. An intricate physiologic mechanism of the inner ear transduces the mechanical vibration into nerve impulses which are then transmitted to the brain to give the listener an auditory perception. Clearly, the human ears can be considered a system. Such a system is very complicated and its study belongs to the domains of anatomy and physiology. It is outside the scope of this text.

If we speak into a telephone, the sound is first transformed into an electrical signal. To be more specific, the microphone inside the handset transforms the acoustic wave into an electrical signal. The microphone is called a *transducer*. A transducer is a device that transforms a signal from one form to another. The loudspeaker in the handset is also a transducer which transform an electrical signal into an acoustic wave. A transducer is a special type of system. Figure 1.1(a) shows the transduced signal of the phrase "signals and systems" uttered by this author. The signal lasts about 2.3 seconds. In order to see better the signal, we plot in Figure 1.1(b) its segment from 1.375 to 1.38 second. We will discuss in Chapter 3 how the two plots are generated.

We show in Figure 1.2 a different signal. It is generated from a 128-Hz (cycles per second) tuning fork. After it is struck, the tuning fork will generate the signal shown in Figure 1.2(a). The signal lasts roughly 13 seconds. We plot in Figure 1.2(b) a small segment of this signal and in Figure 1.2(c) an even smaller segment. The plot in Figure 1.2(c) appears to be a sinusoidal function; it repeats itself 14 times in the time interval of $1.0097 - 0.9 = 0.1097$ second. Thus, its period is $P = 0.1097/14$ in second

1

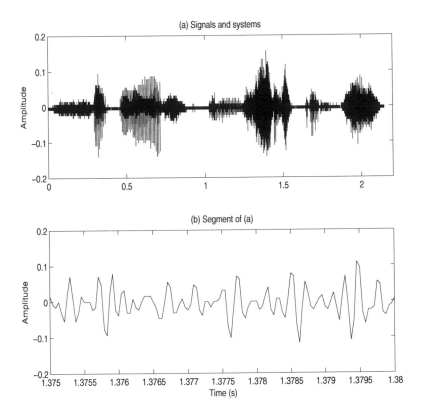

Figure 1.1: (a) Transduced signal of the sound "Signals and systems". (b) Segment of (a).

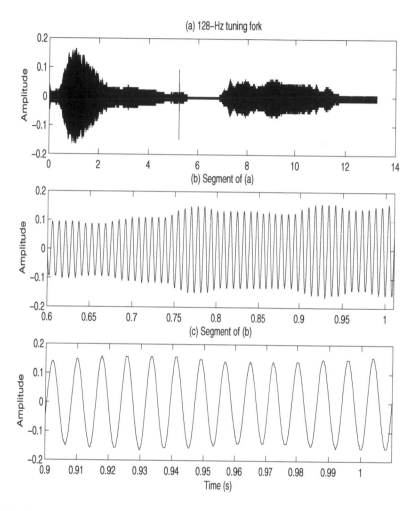

Figure 1.2: (a) Transduced signal of a 128-Hz tuning fork. (b) Segment of (a). (c) Segment of (b).

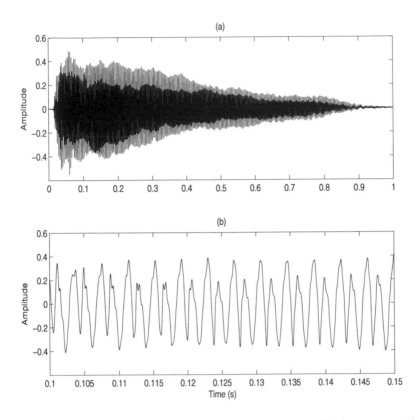

Figure 1.3: (a) Transduced signal of a middle-C sound. (b) Segment of (a).

and its frequency is $f = 1/P = 14/0.1097 = 127.62$ in Hz. (These concepts will be introduced from scratch in Chapter 4.) It is close to the specified 128 Hz. This example demonstrates an important fact that a physical tuning fork does not generate a pure sinusoidal function for all time as one may think, and even over the short time segment in which it does generate a sinusoidal function, its frequency may not be exactly the value specified. For the tuning fork we use, the percentage difference in frequency is $(128 - 127.62)/128 = 0.003$, or about 0.3%. Such a tuning fork is considered to be of high quality. It is not uncommon for a tuning fork to generate a frequency which differs by more than 1% from its specified frequency.

Figure 1.3(a) shows the transduced signal of the sound generated by hitting the middle-C key of a piano. It lasts about one second. According to Wikipedia, the theoretical frequency of middle-C sound is 261.6 Hz. However the waveform shown in Figure 1.3(b) is quite erratic. It is not a sinusoid and cannot have a single frequency. Then what does 261.6 Hz mean? This will be discussed in Sections 5.5.2.

An electrocardiogram (EKG or ECG) records electrical voltages (potentials) generated by a human heart. The heart contracts and pumps blood to the lungs for oxygenation and then pumps the oxygenated blood into circulation. The signal to induce cardiac contraction is the spread of electrical currents through the heart muscle.

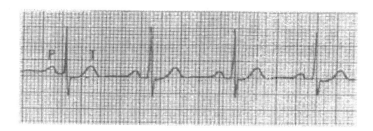

Figure 1.4: EKG graph.

An EKG records the potential differences (voltages) between a number of spots on a person's body. A typical EKG has 12 leads, called electrodes, and may be used to generate many cardiographs. We show in Figure 1.4 only one graph, the voltage between an electrode placed at the fourth intercostal space to the right of the sternum and an electrode placed on the right arm. It is the normal pattern of a healthy person. Deviation from this pattern may reveal some abnormality of the heart. From the graph, we can also determine the heart rate of the patient. Standard EKG paper has one millimeter (mm) square as the basic grid, with a horizontal 1 mm representing 0.04 second and a vertical 1 mm representing 0.1 millivolt. The cardiac cycle in Figure 1.4 repeats itself roughly every 21 mm or $21 \times 0.04 = 0.84$ second. Thus, the heart rate (the number of heart beats in one minute) is $60/0.84 = 71$.

We show in Figure 1.5 some plots which appear in many daily newspapers. Figure 1.5(a) shows the temperature at Central Park in New York city over a 24-hour period. Figure 1.5(b) shows the total number of shares traded each day on the New York Stock Exchange. Figure 1.5(c) shows the range and the closing price of *Standard & Poor's 500-stock Index* each day over three months. Figure 1.5(d) shows the closing price of the index over six months and its 90-day moving average. We often encounter signals of these types in practice.

The signals in the preceding figures are all plotted against time, which is called an *independent variable*. A photograph is also a signal and must be plotted against two independent variables, one for each spatial dimension. Thus a signal may have one or more independent variables. The more independent variables in a signal, the more complicated the signal. In this text, we study signals that have only one independent variable. We also assume the independent variable to be time.

To transform an acoustic wave into an electrical signal requires a transducer. To obtain the signal in Figure 1.4 requires a voltmeter, an amplifier and a recorder. To obtain the temperature in Figure 1.5(a) requires a temperature sensor. All transducers and sensors are systems. There are other types of systems such as amplifiers, filters, and motors. The computer program which generates the 90-day moving average in Figure 1.5(d) is also a system.

1.2 Physics, mathematics, and engineering

Even though engineering is based on mathematics and physics, it is very different from them. Physics is concerned with discovering physical laws that govern natural

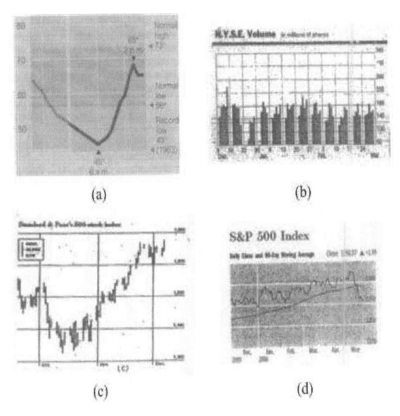

Figure 1.5: (a) Temperature. (b) Total number of shares traded each day in the New York Stock Exchange. (c) Price range and closing price of S&P 500-stock index. (d) Closing price of S&P 500-stock index and its 90-day moving average.

phenomena. The validity of a law is determined by whether it is consistent with what is observed and whether it can predict what will happen in an experiment or in a natural phenomenon. The more accurately it describes the physical world and the more widely it is applicable, the better a physical law is. Newton's laws of motion and universal gravitation (1687) can be used to explain falling apples and planetary orbits. However they assume space and time to be uniform and absolute, and cannot be used to explain phenomena that involve speeds close to the speed of light. In these situations, Newton's laws must be superseded by Einstein's special theory of relativity (1905). This theory hypothesizes that time is not absolute and depends on relative speeds of observers. The theory also leads to the famous equation $E = mc^2$, which relates energy (E), mass (m), and the speed of light (c). Einstein also developed the general theory of relativity (1915) which predicted the bending of light by the sun by an amount twice of what was computed from Newton's law. This prediction was confirmed in 1919 during a solar eclipse and established once and for all the paramount importance of Einstein's theory of relativity.

Although the *effects* of magnets and static electric charges were recognized in ancient time, it took more than eighteen centuries before human could generate and manipulate them. Alessandro Volta discovered that electric charges could be generated and stored using two dissimilar metals immersed in a salt solution. This led to the development of the battery in the 1800s and a way of generating a continuous flow of charges, called current, in a conducting wire. Soon afterwards, it was discovered that current will induce a magnetic field around a wire. This made possible man-made magnetic fields. Michael Faraday demonstrated in 1821 that passing a current through a wire which is placed in a magnetic field will cause the wire to move, which led to the development of motors. Conversely, a moving wire in a magnetic field will induce a current in the wire, which led to the development of generators. These effects were unified by the physicist James Maxwell (1831-1879) using a set of four equations, called *Maxwell's equations*.[1] These equations showed that the electric and magnetic fields, called electromagnetic (EM) waves, could travel through space. The traveling speed was computed from Maxwell's equations as 3×10^8 meters per second, same as the speed of light. Thus Maxwell argued that light is an EM wave. This was generally accepted only after EM waves were experimentally generated and detected by Heinrich Hertz (1857-1894) and were shown to have all of the properties of light such as reflection, refraction, and interference. All EM waves propagate at the speed of light but can have different frequencies or wavelengths. They are classified according to their frequency ranges as radio waves, microwaves, infrared, visible light, ultraviolet, x-rays, and so forth. Light as an EM wave certainly has the properties of wave. It turns out that light also has the properties of particles and can be looked upon as a stream of *photons*.

Physicists were also interested in the basic structure of matter. By the early 1900s, it was recognized that all matter is built from atoms. Every atom has a nucleus consisting of neutrons and positively charged protons and a number of negatively charged

---

[1] The original Maxwell's formulation consisted of twenty equations which were too complicated to be useful for engineers. The twenty equations were condensed to the now celebrated four equations in 1883 by Oliver Heaviside (1850-1925). Heaviside was a self-taught electrical engineer and was keenly interested in applying mathematics into practical use. He also developed resistive operators to study electrical circuits. See Section 9.10.1 and References [C2, N1].

electrons circling around the nucleus. Although electrons in the form of static electrical charges were recognized in ancient times, their properties (the amount of charge and mass of each electron) were experimentally measured only in 1897. Furthermore, it was experimentally verified that the electron, just as light, has the dual properties of wave and particle. These atomic phenomena were outside the reach of Newton's laws and Einstein's theory of relativity but could be explained using quantum mechanics developed in 1926. By now it is known that all matters are built from two types of particles: quarks and leptons. They interact with each other through gravity, electromagnetic interactions, and strong and weak nuclear forces. In summary, physics tries to develop physical laws to describe natural phenomena and to uncover the basic structure of matter.

Mathematics is indispensable in physics. Even though physical laws were inspired by concepts and measurements, mathematics is needed to provide the necessary tools to make the concepts precise and to derive consequences and implications. For example, Einstein had some ideas about the general theory of relativity in 1907, but it took him eight years to find the necessary mathematics to make it complete. Without Maxwell's equations, EM theory could not have been developed. Mathematics, however, has developed into its own subject area. It started with the counting of, perhaps, cows and the measurement of farm lots. It then developed into a discipline which has little to do with the real world. It now starts with some basic entities, such as points and lines, which are abstract concepts (no physical point has zero area and no physical line has zero width). One then selects a number of assumptions or axioms, and then develops logical results. Note that different axioms may lead to different mathematical branches such as Euclidean geometry, non-Euclidean geometry, and Riemann geometry. Once a result is proved correct, the result will stand forever and can withstand any challenge. For example, the Pythagorean theorem (the square of the hypotenuse of a right triangle equals the sum of the squares of both sides) was first proved about 5 B.C. and is still valid today. In 1637, Pierre de Fermat claimed that no positive integer solutions exist in $a^n + b^n = c^n$, for any integer n larger than 2. Even though many special cases such as $n = 3, 4, 6, 8, 9, \ldots$ had been established, nobody was able to prove it for all integers $n > 2$ for over three hundred years. Thus the claim remained a conjecture. It was finally proven as a theorem in 1994 (see Reference [S3]). Thus, the bottom line in mathematics is absolute *correctness*, whereas the bottom line in physics is *truthfulness* to the physical world.

Engineering is a very broad discipline. Some involve in designing and constructing skyscrapers and bridges; some in developing useful materials; some in designing and building electronics devices and physical systems; and many others. Sending the two exploration rovers to Mars (launched in Mid 2003 and arrived in early 2004) was also an engineering task. In this ambitious national project, budgetary concerns were secondary. For most engineering products, such as motors, CD players and cell phones, cost is critical. To be commercially successful, such products must be reliable, small in size, high in performance and competitive in price. Furthermore they may require a great deal of marketing. The initial product design and development may be based on physics and mathematics. Once a working model is developed, the model must go through repetitive modification, improvement, and testing. Physics and mathematics usually play only marginal roles in this cycle. Engineering ingenuity and creativity play more prominent roles.

Engineering often involves tradeoffs or compromises between performance and cost, and between conflicting specifications. Thus, there are often similar products with a wide range in price. The concept of tradeoffs is less eminent in mathematics and physics; whereas it is an unavoidable part of engineering.

For low-velocity phenomena, Newton's laws are valid and can never be improved. Maxwell's equations describe electromagnetic phenomena and waves and have been used for over one hundred years. Once all elementary particles are found and a Theory of Everything is developed, some people anticipate the death of (pure) physics and science (see Reference [L3]). An engineering product, however, can always be improved. For example, after Faraday demonstrated in 1821 that electrical energy could be converted into mechanical energy and vice versa, the race to develop electromagnetic machines (generators and motors) began. Now, there are various types and power ranges of motors. They are used to drive trains, to move the rovers on Mars and to point their unidirectional antennas toward the earth. Vast numbers of toys require motors. Motors are needed in every CD player to spin the disc and to position the reading head. Currently, miniaturized motors on the order of millimeters or even micrometers in size are being developed. Another example is the field of integrated circuits. Discrete transistors were invented in 1947. It was discovered in 1959 that a number of transistors could be fabricated as a single chip. A chip may contain hundreds of transistors in the 1960s, tens of thousands in the 1970s, and hundreds of thousands in the 1980s. It may contain several billion transistors as of 2007. Indeed, technology is open-ended and flourishing.

1.3 Electrical and computer engineering

The field of electrical engineering programs first emerged in the US in the 1880s. It was mainly concerned with the subject areas of *communication* (telegraph and telephone) and *power engineering*. The development of practical electric lights in 1880 by Thomas Edison required the generation and distribution of power. Alternating current (ac) and direct current (dc) generators and motors were developed and underwent steady improvement. Vacuum tubes first appeared in the early 1900s. Because they could be used to amplify signals, long distance telephony became possible. Vacuum tubes could also be used to generate sinusoidal signals and to carry out modulation and demodulation. This led to radio broadcasting. The design of associated devices such as transmitters and receivers spurred the creation of the subject areas of *circuit analysis* and *electronics*. Because of the large number of electrical engineers needed for the construction of infrastructure (power plants, transmission lines, and telephone lines) and for the design, testing, and maintenance of devices, engineering colleges taught mostly the aforementioned subject areas and the industrial practice of the time. This mode of teaching remained unchanged until after World War II (1945).

During World War II, many physicists and mathematicians were called upon to participate in the development of guided missiles, radars for detection and tracking of incoming airplanes, computers for computing ballistic trajectories, and many other war-related projects. These activities led to the new subject areas *computer, control and systems, microwave technology, telecommunications*, and *pulse technology*. After the war, additional subject areas appeared because of the advent of transistors, lasers, integrated circuits and microprocessors. Subsequently many electrical engineering (EE) programs changed their name to electrical and computer engineering (ECE). To show the diversity of ECE, we list some publications of IEEE (Institute of Electrical and Electronics Engineering). For each letter, we list the number of journals starting with the letter and some journal titles

- **A**: (11) Antenna and Propagation, Automatic Control, Automation.

- **B**: (2) Biomedical Engineering, Broadcasting.

- **C**: (31) Circuits and Systems, Communication, Computers, Consumer Electronics.

- **D**: (6) Device and Material Reliability, Display Technology.

- **E**: (28) Electronic Devices, Energy Conversion, Engineering Management,

- **F**: (1) Fuzzy Systems.

- **G**: (3) Geoscience and Remote Sensing.

- **I**: (17) Image Processing, Instrumentation and Measurement, Intelligent Systems, Internet Computing.

- **K**: (1) Knowledge and Data Engineering.

- **L**: (3) Lightwave Technology.

- **M**: (14) Medical Imaging, Microwave, Mobile Computing, Microelectromechanical Systems, Multimedia.

- **N**: (8) Nanotechnology, Neural Networks.

- **O**: (2) Optoelectronics.

- **P**: (16) Parallel and Distributed Systems, Photonics Technology, Power Electronics, Power Systems.

- **Q**: (1) Quantum Electronics.

- **R**: (4) Reliability, Robotics.

- **S**: (19) Sensors, Signal Processing, Software Engineering, Speech and Audio Processing.

- **T**: (1) Technology and Society.

- **U**: (1) Ultrasonics, Ferroelectrics and Frequency Control.

- **V**: (5) Vehicular Technology, Vary Large Scale Integration (VLSI) Systems, Visualization and Computer Graphics.

- **W**: (1) Wireless Communications.

IEEE alone publishes 175 journals on various subjects. Indeed, the subject areas covered in ECE programs are many and diversified.

Prior to World War II, there were some master's degree programs in electrical engineering, but the doctorate programs were very limited. Most faculty members did not hold a Ph.D. degree and their research and publications were minimal. During the war, electrical engineers discovered that they lacked the mathematical and research training needed to explore new fields. This motivated the overhaul of electrical engineering education after the war. Now, most engineering colleges have Ph.D. programs and every faculty member is required to have a Ph.D.. Moreover, a faculty member must, in addition to teaching, carry out research and publish. Otherwise he or she will be denied tenure and will be asked to leave. This leads to the syndrome of "publish or perish".

Prior to World War II, almost every faculty member had some practical experience and taught mostly practical design. Since World War II, the majority of faculty members have been fresh doctorates with limited practical experience. Thus, they tend to teach and stress theory. In recent years, there has been an outcry that the gap between what universities teach and what industries practice is widening. How to narrow the gap between theory and practice is a challenge in ECE education.

1.4 A course on signals and systems

The courses offered in EE and ECE programs evolve constantly. For example, *(passive RLC) Network Syntheses*, which was a popular course in the 1950s, no longer appears in present day ECE curricula. Courses on *automatic control* and *sampled or discrete-time control systems* first appeared in the 1950s; *linear systems* courses in the 1960s and *digital signal processing* courses in the 1970s. Because the mathematics used in the preceding courses is basically the same, it was natural and more efficient to develop a single course that provides a common background for follow-up courses in control, communication, filter design, electronics, and digital signal processing. This contributed to the advent of the course *signals and systems* in the 1980s. Currently, such a course is offered in every ECE program. The first book on the subject area was *Signals and Systems*, by A. V. Oppenheim and A. S. Willsky, published by Prentice-Hall in 1983 (see Reference [O1]). Since then, over thirty books on the subject area have been published in the U.S.. See the references at the end of this text. Most books follow the same outline as the aforementioned book: they introduce the Fourier series, Fourier transform, two- and one-sided Laplace and z-transforms, and their applications to signal analysis, system analysis, communication and feedback systems. Some of them also introduce state-space equations. Those books are developed mainly for those interested in communication, control, and digital signal processing.

This new text takes a fresh look on the subject area. Most signals are naturally generated; whereas systems are to be designed and built to process signals. Thus we discuss the role of signals in designing systems. The small class of systems studied can be described by four types of equations. Although they are mathematically equivalent, we show that only one type is suitable for design and only one type is suitable for implementation and real-time processing. We use operational amplifiers and simple RLC circuits as examples because they are the simplest possible physical systems available.

This text also discusses model reduction, thus it is also useful to those interested in microelectronics and sensor design.

1.5   Confession of the author

This section describes the author's personal evolution regarding teaching. I received my Ph.D. in 1966 and immediately joined the EE department at the State University of New York at Stony Brook. I have been teaching there ever since. My practical experience has been limited to a couple of summer jobs in industry.

As a consequence of my Ph.D. training, I became fascinated by mathematics for its absolute correctness and rigorous development — no ambiguity and no approximation. Thus in the first half of my teaching career, I basically taught mathematics. My research was in linear systems and focused on developing design methods using transfer functions and state-space equations. Because my design methods of pole placement and model matching are simpler both in concept and in computation, and more general than the existing design methods, I thought that these methods would be introduced in texts in control and be adopted in industry. When this expectation was not met, I was puzzled.

The steam engine developed in the late 18th century required three control systems: one to regulate the water level in the boiler, one to regulate the pressure and one to control the speed of the shaft using a centrifugal flyball. During the industrial revolution, textile manufacturing became machine based and required various control schemes. In mid 19th century, position control was used to steer big ships. The use of thermostats to achieved automatic temperature control began in the late 19th century. In developing the airplane in the early 20th century, the Wright brothers, after the mathematical calculation failed them, had to build a wind tunnel to test their design and finally led to a successful flight. Airplanes which required many control systems played an important role in World War I. In conclusion, control systems, albeit designed using empirical or trial-and-error methods, had been widely and successfully employed even before the advent of control and systems courses in engineering curricula in the 1930s.

The analysis and design methods discussed in most control texts, however, were all developed after 1932[2]. See Subsection 9.10.1. They are applicable only to systems that are describable by simple mathematical equations. Most, if not all, practical control systems, however, cannot be so described. Furthermore, most practical systems have some physical constraints and must be reliable, small in size, and competitive in price. These issues are not discussed in most control and systems texts. Thus textbook design methods may not be really used much in practice.[3] Neither, I suspect, will my design methods of pole placement and model matching. As a consequence, my research and publications gave me only personal satisfaction and, more importantly, job security.

In the latter half of my teaching career, I started to ponder what to teach in classes. First I stop teaching topics which are only of mathematical interests and focus on topics which seem to be useful in practice as evident from my cutting in half the first book

---

[2]The only exception is the Routh test which was developed in 1877.

[3]I hope that this perception is incorrect and that practicing engineers would write to correct me.

listed in page ii of this text from its second to third edition. I searched "applications" papers published in the literature. Such papers often started with mathematics but switched immediately to general discussion and concluded with measured data which are often erratic and defy mathematical descriptions. There was hardly any trace of using textbook design methods. On the other hand, so-called "practical" systems discussed in most textbooks, including my own, are so simplified that they don't resemble real-world systems. The discussion of such "practical" systems might provide some motivation for studying the subject, but it also gives a false impression of the reality of practical design. A textbook design problem can often be solved in an hour or less. A practical design may take months or years to complete; it involves a search for components, the construction of prototypes, trial-and-error, and repetitive testing. Such engineering practice is difficult to teach in a lecture setting. Computer simulations or, more generally, computer-aided design (CAD), help. Computer software has been developed to the point that most textbook designs can now be completed by typing few lines. Thus deciding what to teach in a course such as signals and systems is a challenge.

Mathematics has long been accepted as essential in engineering. It provides tools and skills to solve problems. Different subject areas clearly require different mathematics. For signals and systems, there is no argument about the type of mathematics needed. However, it is not clear how much of those mathematics should be taught. In view of the limited use of mathematics in practical system design, it is probably sufficient to discuss what is really used in practice. Moreover, as an engineering text, we should discuss more on issues involving design and implementation. With this realization, I started to question the standard topics discussed in most texts on signals and systems. Are they really used in practice or are they introduced only for academic reasons or for ease of discussion in class? Is there any reason to introduce the two-sided Laplace transform? Is the study of the Fourier series necessary? During the last few years, I have put a great deal of thought on these issues and will discuss them in this book.

1.6 A note to the reader

When I was an undergraduate student about fifty years ago, I did every assigned problem and was an "A" student. I believed that I understood most subjects well. This belief was reinforced by my passing a competitive entrance exam to a master's degree program in Taiwan. Again, I excelled in completing my degree and was confident for my next challenge.

My confidence was completely shattered when I started to do research under Professor Charles A. Desoer at the University of California, Berkeley. Under his critical and constant questioning, I realized that I did not understand my subject of study at all. More important, I also realized that my method of studying had been incorrect: I learned only the mechanics of solving problems without learning the underlying concepts. From that time on, whenever I studied a new topic, I pondered every statement carefully and tried to understand its implications. Are the implications still valid if some word in the statement is missing? Why? After some thought, I re-read the topic or article. It often took me several iterations of pondering and re-reading to fully grasp certain ideas and results. I also learned to construct simple examples to gain insight and, by keeping in mind the goals of a study, to differentiate between what is essential and what is secondary or not important. *It takes a great deal of time and thought to really understand a concept or subject.* Indeed, there is no simple concept. However every concept becomes very simple once it is fully understood.

Devotion is essential if one tries to accomplish some task. The task could be as small as studying a concept or taking a course; it could be as large as carrying out original research or developing a novel device. When devoted, one will put one's whole heart or, more precisely, one's full focus on the problem. One will engage the problem day in and day out, and try to think of every possible solution. Perseverance is important. One should not easily give up. It took Einstein five years to develop the theory of special relativity and another ten years to develop the theory of general relativity. No wonder Einstein once said, "I am no genius, I simply stay with a problem longer".

The purpose of education or, in particular, of studying this text is to gain some knowledge of a subject area. However, much more important is to learn how to carry out critical thinking, rigorous reasoning, and logical development. Because of the rapid change of technology, one can never foresee what knowledge will be needed in the future. Furthermore, engineers may be assigned to different projects many times during their professional life. Therefore, what you learn is not important. What is important is to learn how to learn. This is also true even if you intend to go into a profession other than engineering.

Students taking a course on signals and systems usually take three or four other courses at the same time. They may also have many distractions: part-time jobs, relationships, or the Internet. They simply do not have the time to really ponder a topic. Thus, I fully sympathize with their lack of understanding. When students come to my office to ask questions, I always insist that they try to solve the problems themselves by going back to the original definitions and then by developing the answers step by step. Most of the time, the students discover that the questions were not difficult at all. Thus, if the reader finds a topic difficult, he or she should go back and think about the definitions and then follow the steps logically. Do not get discouraged and give up. Once you give up, you stop thinking and your brain gets lazy. Forcing your brain to work is essential in understanding a subject.

Chapter 2

Signals

2.1 Introduction

This text studies signals that vary with time. Thus our discussion begins with time. Even though Einstein's relativistic time is used in the global positioning system (GPS), we show that time can be considered to be absolute and uniform in our study and be represented by a real number line. We show that a real number line is very rich and consists of infinitely many numbers in any finite segment. We then discuss where $t = 0$ is and show that ∞ and $-\infty$ are concepts, not numbers.

A signal is defined as a function of time. If a signal is defined over a continuous range of time, then it is a continuous-time (CT) signal. If a signal is defined only at discrete instants of time, then it is a discrete-time (DT) signal. We show that a CT signal can be approximated by a staircase function. The approximation is called the pulse-amplitude modulation (PAM) and leads naturally to a DT signal. We also discuss how to construct a CT signal from a DT signal.

We then introduce the concept of impulses. The concept is used to justify mathematically PAM. We next discuss digital procession of analog signals. Even though the first step in such a processing is to select a sampling period T, we argue that T can be suppressed in real-time and non-real-time procession. We finally introduce some simple CT and DT signals to conclude the chapter.

2.2 Time

We are all familiar with time. It was thought to be absolute and uniform. Let us carry out the following thought experiment to see whether it is true. Suppose a person, named Leo, is standing on a platform watching a train passing by with a constant speed v as shown in Figure 2.1. Inside the train, there is another person, named Bill. It is assumed that each person carries an identical watch. Now we emit a light beam from the floor of the train to the ceiling. To the person inside the train, the light beam will travel vertically as shown in Figure 2.1(a). If the height of the ceiling is h, then the elapsed time for the light beam to reach the ceiling is, according to Bill's watch,

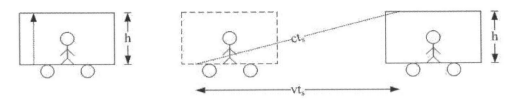

Figure 2.1: (a) A person observing a light beam inside a train that travels with a constant speed. (b) The same event observed by a person standing on the platform.

$t_v = h/c$, where $c = 3 \times 10^8$ meters per second is the speed of light. However, to the person standing on the platform, the time for the same light beam to reach the ceiling will be different as shown in Figure 2.1(b). Let us use t_s to denote the elapsed time according to Leo's watch for the light beam to reach the ceiling. Then we have, using the Pythagorean theorem,

$$(ct_s)^2 = h^2 + (vt_s)^2 \tag{2.1}$$

Here we have used the fundamental postulate of Einstein's special theory of relativity that the speed of light is the same to all observers no matter stationary or traveling at any speed even at the speed of light.[1] Equation (2.1) implies $(c^2 - v^2)t_s^2 = h^2$ and

$$t_s = \frac{h}{\sqrt{c^2 - v^2}} = \frac{h}{c\sqrt{1 - (v/c)^2}} = \frac{t_v}{\sqrt{1 - (v/c)^2}} \tag{2.2}$$

We see that if the train is stationary or $v = 0$, then $t_s = t_v$. If the train travels at 86.6% of the speed of light, then we have

$$t_s = \frac{t_v}{\sqrt{1 - 0.866^2}} = \frac{t_v}{\sqrt{1 - 0.75}} = \frac{t_v}{0.5} = 2t_v$$

It means that for the same event, the time observed or experienced by the person on the platform is twice of the time observed or experienced by the person inside the speeding train. Or the watch on the speeding train will tick at half the speed of a stationary watch. Consequently, a person on a speeding train will age slower than a person on the platform. Indeed, time is not absolute.

The location of an object such as an airplane, an automobile, or a person can now be readily determined using the global positioning system (GPS). The system consists of 24 satellites orbiting roughly 20,200 km (kilometer) above the ground. Each satellite carries atomic clocks and continuously transmits a radio signal that contains its identification, its time of emitting, its position, and others. The location of an object can then be determined from the signals emitted from four satellites or, more precisely, from the distances between the object and the four satellites. See Problems 1.1 and 1.2. The distances are the products of the speed of light and the elapsed times. Thus the synchronization of all clocks is essential. The atomic clocks are orbiting with a high speed and consequently run at a slower rate as compared to clocks on the ground.

---

[1]Under this postulate, a man holding a mirror in front of him can still see his own image when he is traveling with the speed of light. However he cannot see his image according to Newton's laws of motion.

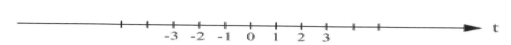

Figure 2.2: Time as represented by the real line.

They slows down roughly 38 microseconds per day. This amount must be corrected each day in order to increase the accuracy of the position computed from GPS signals. This is a practical application of the special theory of relativity.

Other than the preceding example, there is no need for us to be concerned with relativistic time. For example, the man-made vehicle that can carry passengers and has the highest speed is the space station orbiting around the Earth. Its average speed is about 7690 m/s. For this speed, the time experienced by the astronauts on the space station, comparing with the time on the Earth, is

$$t_s = \frac{t_v}{\sqrt{1 - (7690/300000000)^2}} = 1.00000000032853 t_v$$

To put this in prospective, the astronauts, after orbiting the Earth for one year (365 × 24 × 3600s), may feel 0.01 second younger than if they remain on the ground. Even after staying in the space station for ten years, they will look and feel only 0.1 second younger. No human can perceive this difference. Thus we should not be concerned with Einstein's relativistic time and will consider time to be absolute and uniform.

2.2.1 Time – Real number line

Time is generally represented by a real number line as shown in Figure 2.2. The real line or the set of all real numbers is very rich. For simplicity, we discuss only positive real line. It contains all positive integers such as 1, 2,.... There are infinitely many of them. A real number is called a *rational number* if it can be expressed as a ratio of two integers. We list all positive rational numbers in Table 2.1:

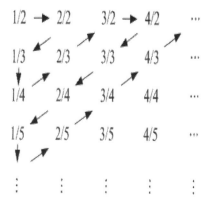

Table 2.1: Arrangement of all positive rational numbers.

There are infinitely many of them. We see that all rational numbers can be arranged in order and be counted as indicated by the arrows shown. If a rational number is expressed in decimal form, then it must terminate with zeros or continue on without ending but with a repetitive pattern. For example, consider the real number

$$x = 8.148900567156715671\cdots$$

with the pattern 5671 repeated without ending. We show that x is a rational number. We compute

$$
\begin{aligned}
10000x - x &= 81489.00567156715671\cdots - 8.14890056715671\cdots \\
&= 81480.8567710000000\cdots = \frac{81480856771}{1000000}
\end{aligned}
$$

which implies

$$x = \frac{81480856771}{9999 \times 10^6} = \frac{81480856771}{9999000000}$$

It is a ratio of two integers and is therefore a rational number.

In addition to integers and rational numbers, the real line still contains infinitely many *irrational numbers*. An irrational number is a real number with infinitely many digits after the decimal point and without exhibiting any repetitive pattern. Examples are $\sqrt{2}$, e, and π (see Section 3.6). The set of irrational numbers is even richer than the set of integers and the set of rational numbers. The set of integers is clearly countable. We can also count the set of rational numbers as shown in Table 2.1. This is not possible for the set of irrational numbers. We argue this by contradiction. Suppose it would be possible to list *all* irrational numbers between 0 and 1 in order as

$$
\begin{aligned}
x_p &= 0.p_1 p_2 p_3 p_4 \cdots \\
x_q &= 0.q_1 q_2 q_3 q_4 \cdots \\
x_r &= 0.r_1 r_2 r_3 r_4 \cdots \\
&\vdots
\end{aligned}
\tag{2.3}
$$

Even though the list contains all irrational numbers between 0 and 1, we still can create a new irrational number as

$$x_n = 0.n_1 n_2 n_3 n_4 \cdots$$

where n_1 be any digit between 0 and 9 but different from p_1, n_2 be any digit different from q_2, n_3 be any digit different from r_3, and so forth. This number is different from all the irrational numbers in the list and is an irrational number lying between 0 and 1. This contradicts the assumption that (2.3) contains all irrational numbers. Thus it is not possible to arrange all irrational numbers in order and then to count them. Thus the set of irrational numbers is *uncountably* infinitely many. In conclusion, the real line consists of three infinite sets: integers, rational numbers, and irrational numbers. We mention that every irrational number occupies a unique point on the real line, but we cannot pin point its exact location. For example, $\sqrt{2}$ is an irrational number lying between the two rational number 1.414 and 1.415, but we don't know where it is exactly located.

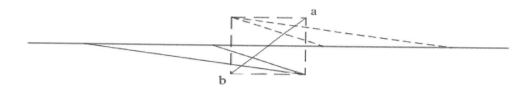

Figure 2.3: (a) Infinite real line and its finite segment. (b) Their one-to-one correspondence.

Because a real line has an infinite length, it is reasonable that it contains infinitely many real numbers. What is surprising is that any finite segment of a real line, no matter how small, also contains infinitely many real numbers. Let $[a, b]$, be a nonzero segment. We draw the segment across a real line as shown in Figure 2.3. From the plot we can see that for any point on the infinite real line, there is a unique point on the segment and vise versa. Thus the finite segment $[a, b]$ also has infinitely many real numbers on it. For example, in the interval $[0, 1]$, there are only two integers 0 and 1. But it contains the following rational numbers

$$\frac{1}{n}, \quad \frac{2}{n}, \quad \cdots, \quad \frac{n-1}{n}$$

for all integer $n \geq 2$. There are infinitely many of them. In addition, there are infinitely many irrational numbers between $[0, 1]$ as listed in (2.3). Thus the interval $[0, 1]$ contains infinitely many real numbers. The interval $[0.99, 0.991]$ also contains infinitely many real numbers such as

$$x = 0.990 n_1 n_2 n_3 \cdots n_N$$

where n_i can assume any digit between 0 and 9, and N be any positive integer. In conclusion, any nonzero segment, no matter how small, contains infinitely many real numbers.

A real number line consists of rational numbers (including integers) and irrational numbers. The set of irrational numbers is much larger than the set of rational numbers. It is said that if we throw a dart on the real line, the probability of hitting a rational number is zero. Even so *the set of rational numbers consists of infinitely many numbers and is much more than enough for our practical use*. For example, the number π is irrational. However it can be approximated by the rational number 3.14 or 3.1416 in practical application.

2.2.2   Where are time 0 and time $-\infty$?

By convention, we use a real number line to denote time. When we draw a real line, we automatically set 0 at the center of the line to denote time 0, and the line (time) is extended to both sides. At the very end of the right-hand side will be time $+\infty$ and at the very end of the left-hand side will be time $-\infty$. Because the line can be extended forever, we can never reach its ends or $\pm\infty$.

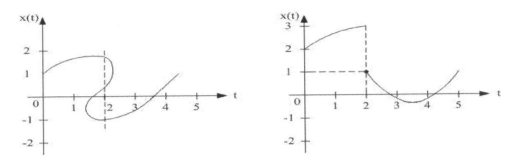

Figure 2.4: (a) Not a signal. (b) A CT signal that is discontinuous at $t = 2$.

Where is time 0? According to the consensus of most astronomers, our universe started with a Big Bang, occurred roughly 13.7 billion years ago. Thus $t = 0$ should be at that instant. Furthermore, the time before the Big Bang is completely unknown. Thus a real line does not denote the actual time. It is only a model and we can select any time instant as $t = 0$. If we select the current time as $t = 0$ and accept the Big Bang theory, then the universe started roughly at $t = -13.7 \times 10^9$ (in years). It is indeed a very large negative number. And yet it is still very far from $-\infty$; it is no closer to $-\infty$ than the current time. Thus $-\infty$ is *not a number* and has no physical meaning nor mathematical meaning. It is only an abstract concept. So is $t = +\infty$.

In engineering, where is time 0 depends on the question asked. For example, to track the maintenance record of an aircraft, the time $t = 0$ is the time when the aircraft first rolled off its manufacturing line. However, on each flight, $t = 0$ could be set as the instant the aircraft takes off from a runway. The time we burn a music CD is the actual $t = 0$ as far as the CD is concerned. However we may also consider $t = 0$ as the instant we start to play the CD. Thus where $t = 0$ is depends on the application and is not absolute. In conclusion, we can select the current time, a future time, or even a negative time (for example, two days ago) as $t = 0$. The selection is entirely for the convenience of study. If $t = 0$ is so chosen, then we will encounter time only for $t \geq 0$ in practice.

In signal analysis, mathematical equations are stated for the time interval $(-\infty, \infty)$. Most examples however are limited to $[0, \infty)$. In system analysis, we use exclusively $[0, \infty)$. Moreover, $+\infty$ will be used only symbolically. It may mean, as we will discussed in the text, only 50 seconds away.

2.3   Continuous-time (CT) signals

We study mainly signals that are functions of time. Such a signal will be denoted by $x(t)$, where t denotes time and x is called the *amplitude* or *value* of the signal and can assume only real numbers. Such a signal can be represented as shown in Figure 2.4 with the horizontal axis, a real line, to denote time and the vertical axis, also a real line, to denote amplitude. The only condition for $x(t)$ to denote a *signal* is that x must assume a unique real number at every t. The plot in Figure 2.4(a) is not a signal because the amplitude of x at $t = 2$ is not unique. The plot in Figure 2.4(b) denotes

Figure 2.5: Clock signal.

a signal in which $x(2)$ is defined as $x(2) = 1$. In mathematics, $x(t)$ is called a *function* of t if $x(t)$ assumes a unique value for every t. Thus we will use signals and functions interchangeably.

A signal $x(t)$ is called a *continuous-time (CT) signal* if $x(t)$ is defined at every t in a continuous range of time and its amplitude can assume any real number in a continuous range. Examples are the ones in Figures 1.1 through 1.4 and 2.4(b). Note that a continuous-time signal is not necessary a *continuous function* of t as shown in Figure 2.4(b). A CT signal is continuous if its amplitude does not jump from one value to another as t increases. All real-world signals such as speech, temperature, and the speed of an automobile are continuous functions of time.

We show in Figure 2.5 an important CT signal, called a *clock signal*. Such a signal, generated using a quartz-crystal oscillator, is needed from digital watches to supercomputers. A 2-GHz (2×10^9 cycles per second) clock signal will repeat its pattern every 0.5×10^{-9} second or half a nanosecond (ns). Our blink of eyes takes about half a second within which the clock signal already repeats itself one billion times. Such a signal is beyond our imagination and yet is a real-world signal.

2.3.1   Staircase approximation of CT signals – Sampling

We use an example to discuss the approximation of a CT signal using a staircase function. Consider the CT signal $x(t) = 4e^{-0.3t} \cos 4t$. It is plotted in Figure 2.6 with dotted lines for t in the range $0 \le t < 5$ or for t in $[0, 5)$. Note the use of left bracket and right parenthesis. A bracket includes the number ($t = 0$) and a parenthesis does not ($t \ne 5$). The signal is defined for infinitely many t in $[0, 5)$ and it is not possible to store such a signal in a digital computer.

Now let us discuss an approximation of $x(t)$. First we select a $T > 0$, for example $T = 0.3$. We then approximate the dotted line by the staircase function denoted by the solid line in Figure 2.6(a). The amplitude of $x(t)$ for all t in $[0, T)$ is approximated by $x(0)$; the amplitude of $x(t)$ for all t in $[T, 2T)$ is approximated by $x(T)$ and so forth. If the staircase function is denoted by $x_T(t)$, then we have

$$x_T(t) = x(nT) \quad \text{for } nT \le t < (n+1)T \tag{2.4}$$

for $n = 0, 1, 2, \ldots$, and for some $T > 0$. The approximation of $x(t)$ by $x_T(t)$ with $T = 0.3$ is poor as shown in Figure 2.6(a). If we select $T = 0.1$, then the approximation is better as shown in Figure 2.6(b). If $T = 0.001$, then we cannot tell the difference between $x(t)$ and $x_T(t)$ as shown in Figure 2.6(c). A staircase function $x_T(t)$ changes amplitude only at $t = nT$, for $n = 0, 1, 2, \ldots$. Thus it can be uniquely specified by $x(nT)$, for $n = 0, 1, , 2, \ldots$. We call T the *sampling period*; nT *sampling instants* or *sample times*; and $x(nT)$ the *sampled values* or *samples* of $x(t)$. The number of samples

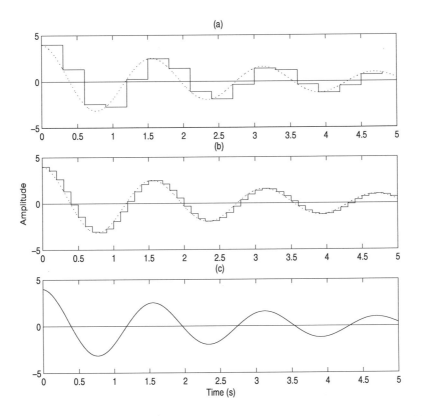

Figure 2.6: (a) CT signal approximated by a staircase function with $T = 0.3$ (b) With $T = 0.1$ (c) With $T = 0.001$.

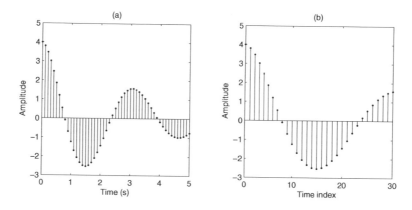

Figure 2.7: (a) DT signal plotted against time. (b) Against time index.

in one second is given by $1/T := f_s$ and should be called the *sampling rate*.[2] However $f_s = 1/T$ is more often called the *sampling frequency* with unit in Hz (cycles per second).

The approximation of $x(t)$ by $x_T(t)$ is called *pulse-amplitude modulation (PAM)*. If the amplitude of $x(nT)$ is coded using 1s and 0s, called the binary digits or bits, then the approximation is called *pulse-code modulation (PCM)*. If the amplitude is represented by a pulse of a fixed height but different width, then it is called *pulse-width modulation (PWM)*. If the amplitude is represented by a fixed pulse (fixed height and fixed narrow width) at different position, then it is called *pulse-position modulation (PPM)*. The signals in Figures 1.1 and 1.2 are obtained using PCM as we will discuss in Section 3.7.1.

2.4 Discrete-time (DT) signals

The staircase function shown in Figure 2.6(b) is a CT signal because it is defined for every t in $[0, 5)$. However it changes amplitudes only at discrete time instants at nT for $T = 0.1$ and $n = 0, 1, 2, \ldots$. Thus the staircase function can be uniquely specified by $x(nT)$ and be plotted as shown in Figure 2.7(a). The signal in Figure 2.7(a) is called a *discrete-time (DT) signal* because it is defined only at discrete instants of time. Although discrete time instants need not be equally spaced, we study only the case where they are equally spaced. In this case, the time instant can be expressed as $t = nT$, where T is the *sampling period* and n is called the *time index* and can assume only integers. In this case, a DT signal can be expressed as

$$x[n] := x(nT) = x(t)|_{t=nT} \qquad n = 0, 1, 2, \ldots \qquad (2.5)$$

and be plotted against time index as shown in Figure 2.7(b). Note the use of brackets and parentheses. In this book, we adopt the convention that variables inside a pair of

---

[2]We use $A := B$ to denote that A, by definition, equals B. We use $A =: B$ to denote that B, by definition, equals A.

brackets can assume only integers; whereas, variables inside a pair of parentheses can assume real numbers. A DT signal consists of a sequence of numbers, thus it is also called a DT sequence or a time sequence.

The DT signal in (2.5) consists of samples of $x(t)$ and is said to be obtained by *sampling* a CT signal. Some signals are inherently discrete time such as the one in Figure 1.4(b). The number of shares in Figure 1.4(b) is plotted against calendar days, thus the DT signal is not defined during weekends and holidays and the sampling instants are not equally spaced. However if it is plotted against *trading days*, then the sampling instants will be equally spaced. The plot in Figure 1.4(c) is not a signal because its amplitude is specified as a range, not a unique value. However if we consider only its closing price, then it is a DT signal. If we plot it against trading days, then its sampling instants will be equally spaced.

If a DT signal is of a finite length, then it can be stored in a computer using two sequences of numbers. The first sequence denotes time instants and the second sequence denotes the corresponding values. For example, a DT signal of length N starting from $t = 0$ and with sampling period T can be stored in a computer as

```
t=nT; n=0:N-1;
x=[x_1 x_2 x_3 ··· x_N]
```

where n=0:N-1 denotes the set of integers from 0 to $N - 1$. Note that the subscript of x starts from 1 and ends at N. Thus both t and x have length N.

A number α can be typed or entered into a computer in decimal form or in fraction. It is however coded using 1s and 0s before it is stored and processed in the computer. If the number of bits used in representing α is B, then the number of available values is 2^B. If α is not one of the available values, it must be approximated by its closest available value. This is called *quantization*. The error due to quantization, called *quantization error*, depends on the number of bits used. It is in the order of $10^{-7}|\alpha|$ using 32 bits or in the order of $10^{-16}|\alpha|$ using 64 bits. Generally there is no need to pay any attention to such small errors.

The amplitude of a DT signal can assume any value in a continuous range. There are infinitely many possible values in any continuous range. If the amplitude is binary coded using a finite number of bits, then it can assume a value only from a finite set of values. Such a signal is called a *digital signal*. Sampling a CT signal yields a DT signal. Binary coding a DT signal yields a digital signal. All signals processed in computers and microprocessors are digital signals. The study of digital signals however is complicated and no book carries out the study. Thus *in analysis and design, we study only DT signals. In implementation, all DT signals must be binary coded.* If quantization errors are large, they are studied using statistical methods. This topic is outside the scope of this text and will not be discussed. Note that quantization errors can be reduced by simply increasing the number of bits used.

2.4.1 Interpolation – Construction

A DT signal can be obtained from a CT signal by sampling. We now discuss how to construct a CT signal from a DT signal. This requires *interpolation* between sampling instants. The simplest interpolation scheme is to hold the current sample constant

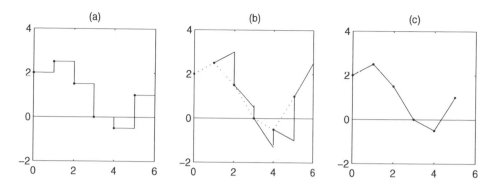

Figure 2.8: Construction of a CT signal from a DT signal specified by the six dots at time indices $n = 0 : 5$: (a) Zero-order hold. (b) First-order hold. (c) Linear interpolation.

until the arrival of the next sample as shown in Figure 2.8(a). This is called *zero-order hold*. If we connect the current and the immediate past samples by a straight line and then extend it until the arrival of the next sample, then the resulting CT signal will be as shown in Figure 2.8(b). This is called *first-order hold*. Note that the first segment is not plotted because it requires the value at $n = -1$ which is not specified. The interpolation shown in Figure 2.8(c) is obtained by connecting two immediate samples by a straight line and is called the *linear interpolation*. The CT signal in Figure 2.8(c) is continuous and appears to be the best among the three.

Consider a time sequence $x(nT)$. It is assumed that $x(nT)$ is running on real time. By this, we mean that $x(0)$ appears at time 0 (which we can select). $x(T)$ will appear T seconds later, $x(2T)$ $2T$ seconds later and so forth. Let $y(t)$ be the constructed CT signals shown in Figure 2.8. If the value of $y(t)$ can appear on real time, then it is called *real-time* processing.[3] For the linear interpolation, the value of $y(t)$ for t in $[0, T)$ is unknown before the arrival of $x(T)$ at time $t = T$, thus $y(t)$ cannot appear on real time. Thus the linear interpolation is not a real-time processing even though it yields the best result. For the zero-order hold, the value of $y(t)$ for t in $[0, T)$ is determined only by $x(0)$ and thus can appear on real time. So is the first-order hold. Because of its simplicity and real-time processing, the zero-order hold is most widely used in practice.

2.5 Impulses

In this section we introduce the concept of impulses. It will be used to justify the approximation of a CT signal by a staircase function. We will also use impulses to derive, very simply, many results to illustrate some properties of signals and systems. Thus it is a very useful concept.

---

[3]This issue will be discussed further in Section 2.6.1.

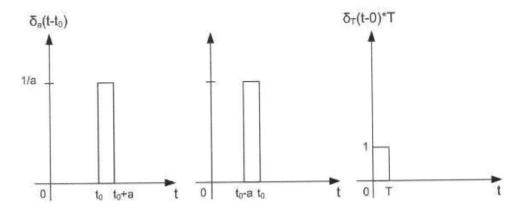

Figure 2.9: (a) $\delta_a(t - t_0)$. (b) Triangular pulse with area 1. (c) $\delta_T(t - 0) \times T = T\delta_T(t)$.

Consider the pulse $\delta_a(t - t_0)$ defined by

$$\delta_a(t - t_0) = \left\{ \begin{array}{ll} 1/a & t_0 \leq t < t_0 + a \\ 0 & t < t_0 \text{ and } t \geq t_0 + a \end{array} \right. \tag{2.6}$$

for any real t_0 and any real $a > 0$, and shown in Figure 2.9(a). It is located at t_0 and has width a and height $1/a$. Its area is 1 for every $a > 0$. As a approaches zero, the pulse has practically zero width but infinite height and yet its area remains to be 1. Let us define

$$\delta(t - t_0) := \lim_{a \to 0} \delta_a(t - t_0) \tag{2.7}$$

It is called the Dirac delta function, δ-function, or *impulse*. The impulse is located at $t = t_0$ and has the property

$$\delta(t - t_0) = \left\{ \begin{array}{ll} \infty & \text{if } t = t_0 \\ 0 & \text{if } t \neq t_0 \end{array} \right. \tag{2.8}$$

and

$$\int_{-\infty}^{\infty} \delta(t - t_0)dt = \int_{t_0}^{t_{0+}} \delta(t - t_0)dt = 1 \tag{2.9}$$

where t_{0+} is a time infinitesimally larger than t_0. Because $\delta(t - t_0)$ is zero everywhere except at $t = t_0$, the integration interval from $-\infty$ to ∞ in (2.9) can be reduced to the immediate neighborhood of $t = t_0$. The integration width from t_0 to t_{0+} is practically zero but still includes the whole impulse at $t = t_0$. If an integration interval does not cover any impulse, then its integration is zero such as

$$\int_{-\infty}^{2.5} \delta(t - 2.6)dt = 0 \quad \text{and} \quad \int_{3}^{10} \delta(t - 2.6)dt = 0$$

where the impulse $\delta(t - 2.6)$ is located at $t_0 = 2.6$. If an integration interval covers wholly an impulse, then its integration is 1 such as

$$\int_{-10}^{3} \delta(t - 2.6)dt = 1 \quad \text{and} \quad \int_{0}^{10} \delta(t - 2.6)dt = 1$$

If integration intervals touch an impulse such as

$$\int_{-\infty}^{2.6} \delta(t - 2.6)dt \quad \text{and} \quad \int_{2.6}^{10} \delta(t - 2.6)dt$$

then ambiguity may occur. If the impulse is defined as in Figure 2.9(a), then the former is 0 and the latter is 1. However, we may also define the impulse using the pulse shown in Figure 2.9(b). Then the former integration is 1 and the latter is 0. To avoid these situations, we assume that *whenever an integration interval touches an impulse, the integration covers the whole impulse and equals 1*. Using this convention, we have

$$\int_{t_0}^{t_0} \delta(t - t_0)dt = 1 \tag{2.10}$$

If $f(t)$ assumes a unique value at every t, it is called a function in mathematics. The impulse in (2.8) is not a function because its value at $t = t_0$ is infinity which is not defined. Thus it is called a *generalized* function. For any (ordinary) function $f(t)$, we have

$$\int_{t_0-}^{t_0+} f(t)dt = 0 \quad \text{or, simply,} \quad \int_{t_0}^{t_0} f(t)dt = 0 \tag{2.11}$$

where the integration interval is assumed to cover the entire value of $f(t_0)$. However no matter what value $f(t_0)$ assumes, such as $f(t_0) = 10, 10^{10}$, or 100^{100}, the integration is still zero. This is in contrast to (2.10). Equation (2.11) shows that the value of an ordinary function at any isolated time instant is immaterial in an integration.

Let A be a real number. Then the impulse $A\delta(t - t_0)$ is said to have *weight A*. This can be interpreted as the pulse in Figure 2.9(a) to have area A by changing the height or width or both. See Problem 2.8. Thus the impulse defined in (2.7) has weight 1. Note that there are many ways to define the impulse. In addition to the ones in Figures 2.9(a) and (b), we may also use the isosceles triangle shown in Figure 2.10(a) located at t_0 with base width a and height $2/a$ to define the impulse. We can also define the impulse as

$$\delta(t - t_0) = \lim_{a \to 0} \frac{\sin((t - t_0)/a)}{\pi(t - t_0)} \tag{2.12}$$

which is plotted in Figure 2.10(b) for $t_0 = 4$ and $a = 0.1$. Indeed there are many ways of defining the impulse.

The mathematics involving impulses is very complex. See Reference [Z1]. All we need to know in this text are the definition given in (2.7) and the next two properties:

$$f(t)\delta(t - t_0) = f(t_0)\delta(t - t_0) \tag{2.13}$$

and

$$\int_{-\infty}^{\infty} f(t)\delta(t - t_0)dt = \int_{-\infty}^{\infty} f(t_0)\delta(t - t_0)dt = f(t_0)\int_{-\infty}^{\infty} \delta(t - t_0)dt = f(t_0)$$

or

$$\int_{-\infty}^{\infty} f(t)\delta(t - t_0)dt = \int_{a}^{b} f(t)\delta(t - t_0)dt = f(t)|_{t-t_0=0} = f(t)|_{t=t_0} \tag{2.14}$$

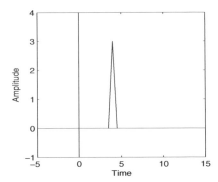 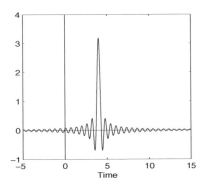

Figure 2.10: (a) Triangular pulse with area 1 located at $t_0 = 4$. (b) $\sin((t-4)/0.1)/\pi(t-4)$.

for any function $f(t)$ that is continuous at $t = t_0$, any $a \leq t_0$, and any $b \geq t_0$. Equation (2.13) follows from the property that $\delta(t - t_0)$ is zero everywhere except at t_0 and (2.14) follows directly from (2.13), (2.9) and (2.10). The integration in (2.14) is called the *sifting* or *sampling* property of the impulse. We see that whenever a function is multiplied by an impulse in an integration, we simply move the function outside the integration and then replace the integration variable by the variable obtained by equating the argument of the impulse to zero.[4] The sifting property holds for any integration interval which covers or touches the impulse. The property will be repeatedly used in this text and its understanding is essential. Check your understanding with Problem 2.9.

2.5.1   Pulse amplitude modulation (PAM)

Consider the staircase function shown in Figure 2.6(a) with $T = 0.3$. We show in this subsection that the staircase function approaches the original $x(t)$ as $T \to 0$. Thus staircase or PAM is a mathematically sound approximation of a CT signal.

We call the staircase function $x_T(t)$. It is decomposed as a sum of pulses as shown in Figure 2.11. The pulse in Figure 2.11(a) is located at $t = 0$ with height $x(0) = 4$ and width $T = 0.3$. The pulse can be expressed in terms of the pulse defined in (2.6). Note that $\delta_T(t - 0)$ is by definition located at $t = 0$ with width T and height $1/T$. Thus $T\delta_T(t)$ has height 1 as shown in Figure 2.9(c). Consequently the pulse in Figure 2.11(a) can be expressed as $x(0)\delta_T(t - 0)T$. The pulse in Figure 2.11(b) is located at $t = T$ with height $x(T) = 1.325$ and width T. Thus it can be expressed as $x(T)\delta_T(t - T)T$. The pulse in Figure 2.11(c) is located at $t = 2T = 0.6$ with height $x(2T) = -2.464$ and width T. Thus it can be expressed as $x(2T)\delta_T(t - 2T)T$. Proceeding forward, we can express $x_T(t)$ as

$$x_T(t) \quad = \quad x(0)\delta_T(t - 0)T + x(T)\delta_T(t - T)T + x(2T)\delta_T(t - 2T)T + \cdots$$

---

[4]An impulse can also be used to *shift* a function. See Problem 2.10. We use in this text only its sifting property.

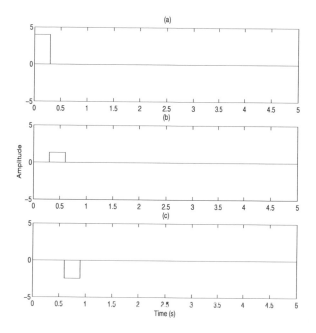

Figure 2.11: Decomposition of the staircase function in Figure 2.6(a).

$$= \sum_{n=0}^{\infty} x(nT)\delta_T(t - nT)T \tag{2.15}$$

This equation expresses the staircase function in terms of the samples of $x(t)$ at sampling instants nT.

Let us define $\tau := nT$. As $T \to 0$, τ becomes a continuous variable and we can write T as $d\tau$. Furthermore, the summation in (2.15) becomes an integration and $\delta_T(t - \tau)$ becomes an impulse. Thus (2.15) becomes, as $T \to 0$,

$$x_T(t) = \int_0^{\infty} x(\tau)\delta(t - \tau)d\tau$$

This equation reduces to, using the sifting property in (2.14),

$$x_T(t) = \int_0^{\infty} x(\tau)\delta(t - \tau)d\tau = x(\tau)|_{\tau=t} = x(t) \tag{2.16}$$

This shows that the staircase function equals $x(t)$ as $T \to 0$. Thus PAM is a mathematically sound approximation of a CT signal.

2.5.2   Bounded variation

A CT signal can be approximated by a staircase function as shown in Figure 2.6. The approximation becomes exact if the sampling period T approaches zero as shown in the preceding subsection. However not every CT signal can be so approximated. For example, consider the signal defined by

$$\alpha(t) = \begin{cases} 1 & \text{if } t \text{ is a rational number} \\ 0 & \text{if } t \text{ is an irrational number} \end{cases} \tag{2.17}$$

It is defined at every t and is a CT signal. However it has infinitely many discontinuities in every finite interval. This signal simply cannot be approximated by a staircase function no matter how small T is chosen. To exclude this type of signals, we define a CT signal to be of *bounded variation* if it has a finite number of discontinuities and a finite number of maxima and minima in every finite time interval. The signal in (2.17) is not of bounded variation. So is the CT signal defined by

$$\beta(t) = \sin(1/t) \quad \text{for } t \geq 0 \tag{2.18}$$

This signal will oscillate with increasing frequency as $t \to 0$ and has infinitely many maxima and minima in $[0,1]$. Note that $\beta(t)$ is a continuous-function of t for all $t > 0$. The signals $\alpha(t)$ and $\beta(t)$ are mathematically contrived and will not arise in the real world. Thus we assume from now on that every CT signal encountered in this text is of bounded variation.

2.6 Digital procession of analog signals

A CT signal is also called an *analog signal* because its waveform is often analogous to that of the physical variable. A DT signal with its amplitudes coded in binary form is called a *digital signal*. Even though our world is an analog one, because of many advantages of digital technique over analog technique, many analog signals are now processed digitally. For example, music and pictures are analog signals but they are now stored, transmitted, and processed in digital form. We discuss briefly how this is achieved.

Consider the arrangement shown in Figure 2.12. The first block, counting from left, denotes a transducer. The second box is an analog lowpass filter. The filter is called an *anti-aliasing* filter that can reduce the effect of frequency aliasing as we will discuss in a later chapter; it can also eliminate noises which often arise in using a transducer. The third box is called a *sample-and-hold* circuit that will convert a CT signal $x(t)$ into a staircase signal $x_T(t)$. The fourth block is called an *encoder* that will code the amplitude of $x(nT)$ into binary form to yield a digital signal. The combination of the third and fourth blocks is called an analog-to-digital converter (ADC). The binary coding can be in series format or parallel format. It can also be coded to detect and correct errors. The sampling period T or sampling frequency ($f_s := 1/T$) and the number of bits used to represent amplitudes depends on application. The telephone transmission uses eight bits and 8 kHz sampling frequency. An audio CD uses 16 bits plus some additional bits for error correcting, control, and synchronization, and 44.1 kHz sampling frequency.

The digital signal at the output of ADC can then be used for transmission (via copper wires, fiber-optic cable, or wireless), storage (in the hard drive of a computer or in a compact disc), or procession (amplification or filtering). A personal computer

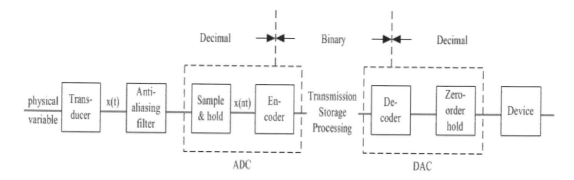

Figure 2.12: Digital processing of analog signals

usually has a built-in ADC which may contain several sampling frequencies and several coding schemes with 8 or 16 bits. The user has the freedom in selecting each of them.

After transmission, storage, or processing, a digital signal can be converted back to an analog signal as shown in the right-hand-side blocks in Figure 2.12. The *decoder* transforms a binary-coded digital signal into a DT signal. This DT signal can be displayed on a monitor or plotted on a recorder. A DT signal is a sequence of numbers and cannot drive a physical device such as a speaker, an analog filter, or a motor. To do so, it must be transformed into a CT signal using, for example, a zero-order hold. The combination of the decoder and the zero-order hold is called a digital-to-analog converter (DAC). A personal computer usually has a built-in DAC as well.

2.6.1 Real-time and non-real-time processing

The digital processing of an analog signal always involves a sampling period. For example, the sampling period is $T = 1/8000 = 0.000125$ in second or 125 μs for telephone transmission and $T = 1/44100 = 0.0000226$ in second or 22.6 μs for audio CD. Tracking such T is inconvenient. Fortunately, in actual processing and in the study of DT signals and systems, we can ignore T or simply assume T to be 1 as we explain next.

Let $u(nT)$, for $n = 0 : N$, be a signal to be processed and $y(nT)$, for $n = 0 : N$ be the processed signal. There are two types of processing: non-real-time and real-time. In non-real-time processing, we start to compute $y(nT)$ only after the entire $u(nT)$ has become available as shown in Figure 2.13(a). In this case, the nT in $u(nT)$ and the nT in $y(nT)$ do not denote the same real time as shown. Most computer computations are non-real-time processing. In this processing, time and the sampling period do not play any role. What is important is the order of the entries in $u(nT)$ and $y(nT)$. Note that because of the very high speed of present-day computers, simple processing or simulation can often be completed in a very short time. For example, to process the signal in Figure 1.3 will take, as will be discussed in Chapter 11, less than 0.02 second. Thus the two $t = 0$ in Figure 2.13(a) may be very close in real time. Nevertheless it is still a non-real-time processing.

Telephone conversation is carried out in real time. In real-time processing, the time

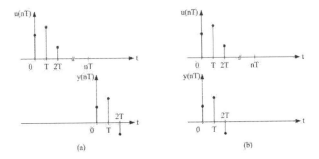

Figure 2.13: (a) Non-real-time processing. (b) Real-time processing.

axes of $u(nT)$ and $y(nT)$ are the same real time as shown in Figure 2.13(b). To simplify the discussion, we assume that $y(nT)$ can be expressed as

$$y(nT) = 3u([n-1]T) + 4u([n-2]T) - u([n-3]T) \qquad (2.19)$$

Because $y(nT)$ does not depend on $u(nT)$, we can start to compute $y(nT)$ right after receiving $u([n-1]T)$ or right after time instant $[n-1]T$.[5] Depending on the speed of the processor or hardware used and on how it is programmed, the amount of time needed to compute $y(nT)$ will be different. For convenience of discussion, suppose one multiplication requires 30 ns (20×10^{-9} s) and one addition requires 20 ns. Because (2.19) involves two additions and two multiplications and if they are carried out sequentially, then computing $y(nT)$ requires 100 ns or 0.1 μs. If this amount of computing time is less than T, we store $y(nT)$ in memory and then deliver it to the output terminal at time instant $t = nT$. If the amount of time is larger than T, then $y(nT)$ is not ready for delivery at time instant $t = nT$. Furthermore when $u(nT)$ arrives, we cannot start computing $y([n+1]T)$ because the processor is still in the middle of computing $y(nT)$. Thus if T is less than the amount of computing time, then the processor cannot function properly. On the other hand, if T is larger than the computing time, then $y(nT)$ can be computed and then stored in memory. Its delivery at nT is controlled separately. In other words, even in real-time procession, there is no need to pay any attention to the sampling period T so long as it is large enough to carry out the necessary computation.

In conclusion, in processing DT signals $x(nT)$, the sampling period T does not play any role. Thus we can suppress T or assume $T = 1$ and use $x[n]$, a function of time index n, in the study of of DT signals and systems.

2.7 CT step and real-exponential functions - time constant

A CT signal $x(t)$ can be defined for all t in $(-\infty, \infty)$. It is called a *positive-time* signal if $x(t)$ is identically zero for all $t < 0$. It is a *negative-time* signal if $x(t)$ is identically zero for all $t > 0$. In other words, a positive-time signal can assume nonzero value only

---

[5]The case where $y(nT)$ depends on $u(nT)$ will be discussed in Section 6.8.1.

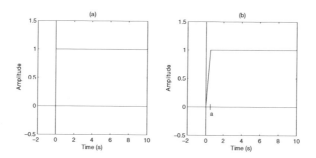

Figure 2.14: (a) Step function. (b) Function $q_a(t)$.

for positive time $(t \geq 0)$ and a negative-time signal can assume nonzero value only for negative time $(t \leq 0)$. A signal is called a *two-sided* signal if it is not positive time nor negative time. Thus a two-sided signal must assume nonzero value for some $t > 0$ and some $t < 0$. In this section we introduce two simple positive-time signals.

Step function: The (unit) step function is defined as

$$q(t) := \begin{cases} 1 & \text{for } t \geq 0 \\ 0 & \text{for } t < 0 \end{cases} \tag{2.20}$$

and shown in Figure 2.14(a). The value of $q(t)$ at $t = 0$ is defined as 1. In fact, it can be defined as 0, 0.5, -10, 10^{10}, or any other value and will not affect our discussion. The information or energy of CT signals depends on not only amplitude but also on *time duration*. An isolated t has zero width (zero time duration), and its value will not contribute any energy. Thus the value of $q(t)$ at $t = 0$ is immaterial. See (2.11).

The step function defined in (2.20) has amplitude 1 for $t \geq 0$. If it has amplitude -2, we write $-2q(t)$. We make distinction between *amplitude* and *magnitude*. Both are required to be real-valued. An amplitude can be negative, zero, or positive. A magnitude can be zero or positive but cannot be negative.

Consider the function $q_a(t)$ shown in Figure 2.14(b). It is zero for $t < 0$ and 1 for $t \geq a > 0$, where a is a small positive number. The function increases from 0 at $t = 0$ to 1 at $t = a$ with a straight line. Thus the slope of the straight line is $1/a$. If we take the derivative of $q_a(t)$, then $dq_a(t)/dt$ equals 0 for $t < 0$ and $t > a$ and $1/a$ for t in $[0, a]$. This is the pulse defined in Figure 2.9(a) with $t_0 = 0$ or $\delta_a(t)$. As $a \to 0$, $q_a(t)$ becomes $q(t)$ and $\delta_a(t)$ becomes the impulse $\delta(t)$. Thus we have

$$\delta(t) = \frac{dq(t)}{dt} \tag{2.21}$$

This is another way of defining impulse.

Real exponential function — Time constant: Consider the exponential function

$$x(t) = \begin{cases} e^{-at} & \text{for } t \geq 0 \\ 0 & \text{for } t < 0 \end{cases} \tag{2.22}$$

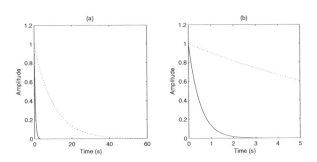

Figure 2.15: (a) e^{-at} with $a = 0.1$ (dotted line) and $a = 2$ (solid line) for t in $[0,\ 60]$. (b) Segment of (a) for t in $[0,\ 5]$.

where a is real and nonnegative (zero or positive). If a is negative, the function grows exponentially to infinity. No device can generate such a signal. If $a = 0$, the function reduces to the step function in (2.20). If $a > 0$, the function decreases exponentially to zero as t approaches ∞. We plot in Figure 2.15(a) two real exponential functions with $a = 2$ (solid line) and 0.1 (dotted line). The larger a is, the faster e^{-at} vanishes. In order to see e^{-2t} better, we replot them in Figure 2.15(b) using a different time scale.

A question we often ask in engineering is: at what time will e^{-at} become zero? Mathematically speaking, it becomes zero only at $t = \infty$. However, in engineering, a function can be considered to become zero when its magnitude remains less than 1% of its peak value. For example, in a scale with its reading from 0 to 100, a reading of 1 or less can be considered to be 0. We give an estimate for e^{-at} to reach roughly zero. Let us define $t_c := 1/a$. It is called the *time constant*. Because

$$\frac{x(t + t_c)}{x(t)} = \frac{e^{-a(t+t_c)}}{e^{-at}} = \frac{e^{-at}e^{-at_c}}{e^{-at}} = e^{-at_c} = e^{-1} = 0.37 \qquad (2.23)$$

for all t, the amplitude of e^{-at} decreases to 37% of its original amplitude whenever the time increases by one time constant $t_c = 1/a$. Because $(0.37)^5 = 0.007 = 0.7\%$, the amplitude of e^{-at} decreases to less than 1% of its original amplitude in five time constants. Thus we often consider e^{-at} to have reached zero in five time constants. For example, the time constant of $e^{-0.1t}$ is $1/0.1 = 10$ and $e^{-0.1t}$ reaches zero in $5 \times 10 = 50$ seconds as shown in Figure 2.15(a). The signal e^{-2t} has time constant $1/2 = 0.5$ and takes $5 \times 0.5 = 2.5$ seconds to reach zero as shown in Figure 2.15(b).

2.7.1   Time shifting

This section discusses the shifting of a CT signal. The procedure is very simple. We select a number of t and then do direct substitution as we illustrate with an example.

Given the signal $x(t)$ shown in Figure 2.16(a). From the plot, we can see that $x(0) = 0, x(1) = 1.5, x(2) = 1, x(3) = 0.5$, and $x(4) = 0$. We now plot the signal defined by $x(t - 1)$. To plot $x(t - 1)$, we select a number of t and then find its values from the given $x(t)$. For example, if $t = -1$, then $x(t - 1) = x(-2)$, which equals 0 as we can read from Figure 2.16(a). If $t = 0$, then $x(t - 1) = x(-1)$, which equals 0. If

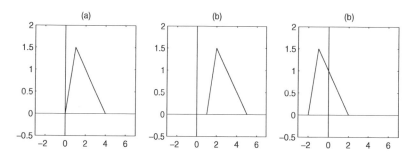

Figure 2.16: (a) A given function $x(t)$. (b) $x(t-1)$. (c) $x(t+2)$.

$t = 1$, then $x(t-1) = x(0)$, which equals 0. Proceeding forward, we have the following

t	-1	0	1	2	3	4	5	6
$x(t-1)$	0	0	0	1.5	1	0.5	0	0

From these, we can plot in Figure 2.16(b) $x(t-1)$. Using the same procedure we can plot $x(t+2) = x(t-(-2))$ in Figure 2.16(c). We see that $x(t-t_0)$ simply shifts $x(t)$ from $t = 0$ to t_0. If $t_0 > 0$, $x(t-t_0)$ shifts $x(t)$ to the right and is positive time if $x(t)$ is positive time. If $t_0 < 0$, $x(t-t_0)$ shifts $x(t)$ to the left and may not be positive time even if $x(t)$ is positive time.

2.8 DT impulse sequences

This section discusses DT impulse sequences. As discussed in Section 2.6.1, in processing and manipulation of DT signals, the sampling period does not play any role. Thus the sampling period T will be suppressed unless stated otherwise.

The *impulse sequence* at $n = n_0$ is defined as

$$\delta_d[n - n_0] = \begin{cases} 1 & \text{for} \quad n = n_0 \\ 0 & \text{for} \quad n \neq n_0 \end{cases} \tag{2.24}$$

where n_0 is any integer. In other words, we have $\delta_d[0] = 1$ and $\delta_d[n] = 0$ for all nonzero integer n. The sequence is also called the *Kronecker delta sequence* and is the DT counterpart of the CT impulse defined in (2.8). Unlike the CT impulse which cannot be generated in practice, the impulse sequence can be easily generated. Note that $-2\delta_d[n] = -2\delta_d[n - 0]$ is an impulse sequence at $n = 0$; it is zero for all integer n except at $n = 0$ where the amplitude is -2.

A DT signal is defined to be *positive time* if $x[n] = 0$ for all $n < 0$. Thus a DT positive-time signal can assume nonzero value only for $n \geq 0$. Consider the DT positive-time signal shown in Figure 2.17(a) with $x[0] = 1.8, x[1] = 2.5, x[2] = -0.8, x[3] = 0.5$, and so forth. Such a signal can be represented graphically as shown. It can also be expressed in tabular form as

n	0	1	2	3	$\cdots$
$x[n]$	1.8	2.2	-0.8	0.5	$\cdots$

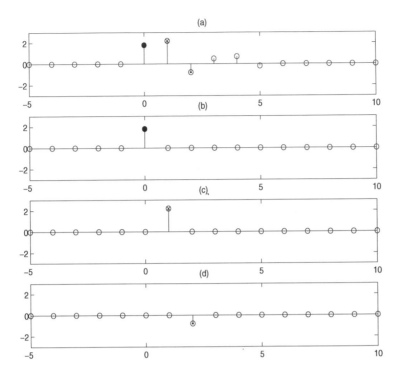

Figure 2.17: (a) DT signal $x[n]$. (b) $x[0]\delta_d[n]$. (c) $x[1]\delta_d[n-1]$. (d) $x[2]\delta_d[n-2]$.

Can we express it as
$$x[n] = 1.8 + 2.2 - 0.8 + 0.5 + \cdots$$
for $n \geq 0$? The answer is negative. (Why?) Then how do we express it as an equation?

Let us decompose the DT signal shown in Figure 2.17(a) as in Figures 2.17(b), (c), (d), and so forth. That is, the sum of those in Figures 2.17(b), (c), (d), ... yields Figure 2.17(a). Note that the summation is carried out at every n.

The sequence in Figure 2.17(b) is the impulse sequence $\delta_d[n]$ with height $x[0]$ or $1.8\delta_d[n-0]$. The sequence in Figure 2.17(c) is the sequence $\delta_d[n]$ shifted to $n=1$ with height $x[1]$ or $2.2\delta_d[n-1]$. The sequence in Figure 2.17(d) is $-0.8\delta_d[n-2]$. Thus the DT signal in Figure 2.17(a) can be expressed as

$$x[n] = 1.8\delta_d[n-0] + 2.2\delta_d[n-1] - 0.8\delta_d[n-2] + 0.5\delta_d[n-3] + \cdots$$

or

$$x[n] = \sum_{k=0}^{\infty} x[k]\delta_d[n-k] \qquad (2.25)$$

This holds for every integer n. Note that in the summation, n is fixed and k ranges from 0 to ∞. For example, if $n = 10$, then (2.25) becomes

$$x[10] = \sum_{k=0}^{\infty} x[k]\delta_d[10-k]$$

As k ranges from 0 to ∞, every $\delta_d[10 - k]$ is zero except $k = 10$. Thus the infinite summation reduces to $x[10]$ and the equality holds. Note that (2.25) is the DT counterpart of (2.16).

To conclude this section, we mention that (2.24) must be modified as

$$\delta_d([n - n_0]T) = \delta_d(nT - n_0T) = \begin{cases} 1 & \text{for } n = n_0 \\ 0 & \text{for } n \neq n_0 \end{cases}$$

if the sampling period $T > 0$ is to be expressed explicitly. Moreover, (2.25) becomes

$$x(nT) = \sum_{k=-\infty}^{\infty} x(kT)\delta_d(nT - kT) \qquad (2.26)$$

if the signal starts from $-\infty$. This equation will be used in Chapter 4.

2.8.1 Step and real-exponential sequences - time constant

In this section we discuss two DT positive-time signals or sequences.

Step sequence: The (unit) step sequence or the DT (unit) step function is defined as

$$q_d[n] := \begin{cases} 1 & \text{for } n \geq 0 \\ 0 & \text{for } n < 0 \end{cases}$$

It has amplitude 1 for $n \geq 0$. If it has amplitude -2, we write $-2q_d[n]$. The shifting discussed in Section 2.7.1 is directly applicable to the DT case. Thus $-2q_d[n - 3]$ shifts $-2q_d[n]$ to the right by 3 samples. That is, $-2q_d[n - 3] = 0$ for all $n < 3$ and $-2q_d[n - 3] = -2$ for all $n \geq 3$.

Real exponential sequence — Time constant Consider the sequence

$$x[n] = \begin{cases} b^n & \text{for } n \geq 0 \\ 0 & \text{for } n < 0 \end{cases} \qquad (2.27)$$

where b is a real positive constant. We call it a real exponential sequence or DT exponential signal because it is the sample of a CT real exponential function e^{-at} with some sampling period T or

$$x[n] = e^{-at}\big|_{t=nT} = e^{-anT} = \left(e^{-aT}\right)^n =: b^n$$

with $b = e^{-aT}$. If $b = 1$, the exponential sequence reduces to a step sequence. If $0 \leq b < 1$, b^n decreases exponentially to zero as $n \to \infty$. If $1 < b$, b^n increases exponentially to ∞ as $n \to \infty$. Note that if b is negative, then b^n will grow or vanish oscillatorily with the highest possible frequency as we will discuss in Chapter 11.

We consider in the following only $0 \leq b < 1$. We plot in Figure 2.18(a) b^n with $b = 0.5$ (solid-line stems) and 0.9 (dotted-line stems) for $n = 0 : 50$. The smaller b is, the faster b^n vanishes. In order to see 0.5^n better, we replot them in Figure 2.18(b) for $n = 0 : 10$.

As in the CT case, we may ask at what time will b^n become zero? Mathematically speaking, it becomes zero only at $n = \infty$. However, in engineering, a sequence can

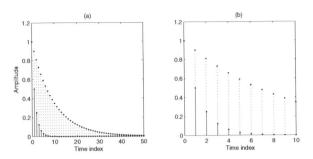

Figure 2.18: (a) b^n with $b = 0.5$ (solid-line stems) and $b = 0.9$ (dotted-line stems) for $n = 0 : 50$. (b) Segment of (a) for $n = 0 : 10$.

be considered to have become zero when its magnitude remains less than 1% of its peak value. As discussed in Section 2.7, for the CT case, we have defined the time constant $t_c := 1/a$ for e^{-at}, and it takes five time constants for e^{-at} to become zero. Because of $b = e^{-aT}$ or $\ln b = -aT$, we may define $\bar{t}_c = -T/\ln b$ (in seconds) or $\bar{t}_c = -1/\ln b$ (in samples) to be the *time constant* of b^n, for $0 \le b < 1$ and it takes five time constants for b^n to become zero. Note that, because $b < 1$, the time constant $\bar{t}_c$ is positive. For examples, the time constant of 0.5^n is $\bar{t}_c = -T/\ln 0.5 = 1.44T$ and it takes $5 \times 1.44T = 7.2T$ in seconds or 7.2 in samples[6] for 0.5^n to become zero. This is indeed the case as shown in Figure 2.18(b). The time constant of 0.9^n is $\bar{t}_c = -T/\ln 0.9 = 9.49T$ and it takes $5 \times 9.49T = 47.45T$ in seconds or 47.45 in samples for 0.9^n to become zero. This is indeed the case as shown in Figure 2.18(a).

Problems

2.1 Consider a point A on a straight line. If we know the distance of A from a known point, can we determine the position of A?

2.2 Consider a point A on a plane. Can you determine the position of A from three distances from three known points? How?

2.3 Is an integer a rational number? Why?

2.4 Verify that the midpoint of any two rational numbers is a rational number. Using the procedure, can you list an infinite number of rational numbers between $(3/5)$ and $(2/3)$? List only three of them.

2.5 Consider $x(t)$ defined by $x(t) = 0.5t$, for $0 \le t \le 3$ and $x(t) = 0$, for $t < 0$ and $t \ge 3$. Plot $x(t)$ for t in $[0, 5]$. Is $x(t)$ a signal? If not, modify the definition to make it a signal.

---

[6]Strictly speaking, the number of samples must be an integer. This fact will be ignored because we give only an estimate.

2.6 Find the samples of the modified $x(t)$ in Problem 2.5 if the sampling period is $T = 0.5$.

2.7 Construct, using zero-order, first-order holds, and linear interpolation, CT signals for t in $[0, 3]$, from the DT signal $x[-1] = 0, x[0] = 1, x[1] = 0, x[2] = -1, x[3] = 2, x[4] = 1, x[5] = 1, x[6] = 0$ with sampling period $T = 0.5$.

2.8 What are the mathematical expressions of the pulses shown in Figure 2.19 as $a \to 0$?

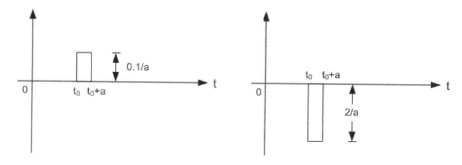

Figure 2.19: Pulses

2.9 What are the values of the following integrals involving impulses:

1. $\int_{t=-\infty}^{\infty} (\cos t)\delta(t)dt$

2. $\int_{t=0}^{\pi/2} (\sin t)\delta(t - \pi/2)dt$

3. $\int_{t=\pi}^{\infty} (\sin t)\delta(t - \pi/2)dt$

4. $\int_{t=0}^{\infty} \delta(t + \pi/2) \left[\sin(t - \pi)\right] dt$

5. $\int_{t=-\infty}^{\infty} \delta(t)(t^3 - 2t^2 + 10t + 1)dt$

6. $\int_{t=0}^{0} \delta(t)e^{2t}dt$

7. $\int_{t=0}^{0} 10^{10}e^{2t}dt$

[*Answers:* 1, 1, 0, 0, 1, 1, 0.]

2.10 Verify

$$\int_{-\infty}^{\infty} x(\tau)\delta(t - \tau - 3)d\tau = x(t - 3)$$

and

$$\int_{-\infty}^{\infty} x(t - \tau)\delta(\tau - t_0)d\tau = x(t - t_0)$$

Thus an impulse can be used to shift a function.

2.11 Consider the signal shown in Figure 2.20. Is it continuous? How many discontinuities has it in $(-\infty, \infty)$? Is it of bounded variation?

Figure 2.20: A sequence of square pulses.

2.12 Consider the signal $x(t)$ in Figure 2.16(a). Plot $x(-t)$ for t in $[-5, 5]$. Can you conclude that $x(-t)$ flips $x(t)$, with respect to $t = 0$, to the negative time?

2.13 What is the time constant of the positive-time signal $x(t) = e^{-0.4t}$, for $t \geq 0$? Use it to sketch roughly the function for t in $[0, 20]$.

2.14 What is the time constant of $x(t) = -2e^{-1.2t}$? Use it to sketch roughly the function for t in $[0, 10]$.

2.15 Consider $x(t)$ shown in Figure 2.21. Plot $x(t-1)$, $x(t+1)$, $x(t-1) + x(t+1)$, and $x(t-1)x(t+1)$. Note that the addition of two straight lines is a straight line, but their product is not.

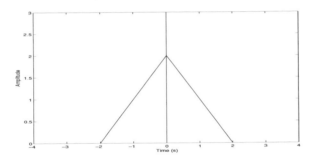

Figure 2.21: A given function.

2.16 Given the DT signal $x[n]$ defined by $x[0] = -2, x[1] = 2, x[3] = 1$ and $x[n] = 0$ for the rest of n. Plot $x[n]$ and express it as an equation using impulse sequences.

2.17 Given the $x[n]$ in Problem 2.16. Plot $x[n-3]$ and $x[-n]$.

2.18 What is the sampled sequence of $e^{-0.4t}$, for $t \geq 0$, with sampling period $T = 0.5$? What are its time constants in seconds and in samples?

2.19 What is the sampled sequence of $-2e^{-1.2t}$, for $t \geq 0$, with sampling period $T = 0.2$? What are its time constants in seconds and in samples?

2.20 Plot $\delta_2(t)$, $\delta_{0.1}(t+6)$, and $\delta_{0.2}(t-3)$. Can you plot $\delta_{(-0.2)}(t-3)$? Why not?

Chapter 3

Some mathematics and MATLAB

3.1 Introduction

In this chapter we introduce some mathematics and MATLAB. Even though we encounter only real-valued signals, we need the concept of complex numbers. Complex numbers will be used to define frequency and will arise in developing frequency contents of signals. They also arise in analysis and design of systems. Thus the concept is essential in engineering.

We next introduce matrices and some simple manipulations. We discuss them only to the extent needed in this text. We then discuss the use of mathematical notations and point out a confusion which often arises in using Hz and rad/s. A mathematical notation can be defined in any way we want. However once it is defined, all uses involving the notation must be consistent.

We discuss a number of mathematical conditions. These conditions are widely quoted in most texts on signals and systems. Rather than simply stating them as in most texts, we discuss their implications and differences. It is hoped that the discussion will provide better appreciation of these concepts. Even if you do not fully understand them, you should proceed to the next topic. After all, you rarely encounter these conditions in practice, because all real-world signals automatically meet these conditions.

We then give a proof of an irrational number. The proof will give a flavor of logical development.

In the remainder of this chapter, we introduce MATLAB. It is an important software package in engineering and is widely available. For example, every PC at Stony Brook for students' use contains MATLAB. The reader is assumed to have access to MATLAB. One way to learn MATLAB is to pick up a book and to study it throughout. Our approach is to study only what is needed as we proceed. This chapter introduces MATLAB to the extent of generating Figures 1.1-1.3, and all figures in Chapter 2. The discussion is self-contained. A prior knowledge of MATLAB is helpful but not

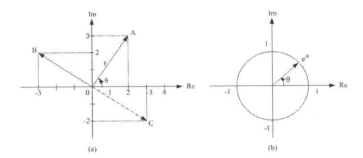

Figure 3.1: (a) Complex plane. (b) The complex number $e^{j\theta}$.

essential.

3.2   Complex numbers

Consider the polynomial equation $x^2 - 4 = 0$ of degree 2. Clearly it has solutions $x = 2$ and $x = -2$. Consider next the equation $x^2 + 4 = 0$. If x is required to be a real number, then the equation has no solution. Let us define $j := \sqrt{-1}$. It has the properties $j^2 = -1$, $j^3 = j^2 j = -j$ and $j^4 = 1$. Let α and β be any real numbers. Then we call $j\beta$ an *imaginary number* and $\alpha + j\beta$, a *complex number*. The complex number consists of a real part α and an imaginary part β. Imaginary numbers or, more generally, complex numbers do not arise in the real world. They are introduced for mathematical necessity in solving polynomial equations. For example, the polynomial equation $x^2 + 4 = 0$ has no real-valued solution but has the complex numbers $\pm j2$ as its solutions. By introducing complex numbers, then *every polynomial equation of degree n has n solutions*.

A real number can be represented by a point on a real number line. To represent a complex number, we need two real number lines, perpendicular to each other, as shown in Figure 3.1. It is called a *complex plane*. The horizontal axis, called the *real axis*, denotes the real part and the vertical axis, called the *imaginary axis*, denotes the imaginary part. For example, the complex number $2 + 3j$ can be represented by the point denoted by A shown in Figure 3.1(a) and $-3 + 2j$ by the point denoted by B.

Let us draw a line from the origin $(0 + j0)$ toward A with an arrow as shown in Figure 3.1(a). We call the line with an arrow a *vector*. Thus a complex number can be consider a vector in a complex plane. The vector can be specified by its *length r* and its *angle* or *phase* θ as shown. Because the length cannot be negative, r is called *magnitude*, not *amplitude*. The angle is measured from the positive real axis. By convention, an angle is positive if measured counterclockwise and negative if measured clockwise. Thus a complex number A can be represented as

$$A = \alpha + j\beta = re^{j\theta} \tag{3.1}$$

The former is said to be in *rectangular form* and the latter in *polar form*. Either form can be obtained from the other. Substituting Euler's formula

$$e^{j\theta} = \cos\theta + j\sin\theta$$

into (3.1) yields

$$\alpha = r\cos\theta \quad \text{and} \quad \beta = r\sin\theta \tag{3.2}$$

Thus the real part α is the *projection* of the vector A on the real axis and the imaginary part β is the projection of A on the imaginary axis. Using (3.2), $(\cos\theta)^2 + (\sin\theta)^2 = 1$, and $\tan\theta = \sin\theta/\cos\theta$, we can obtain

$$\alpha^2 + \beta^2 = r^2 \quad \text{and} \quad \beta/\alpha = \tan\theta$$

which implies

$$r = \sqrt{\alpha^2 + \beta^2} \geq 0 \quad \text{and} \quad \theta = \tan^{-1}[\beta/\alpha] \tag{3.3}$$

For example, for $A = 2 + 3j$, we have

$$r = \sqrt{2^2 + 3^2} = \sqrt{4 + 9} = \sqrt{13} = 3.6$$

and

$$\theta = \tan^{-1}(3/2) = 56^o$$

Thus we also have $A = 3.6e^{j56^o}$. Note that even though $\sqrt{13} = \pm 3.6$, we cannot take -3.6 because we require r to be positive. Note that r and θ can also be obtained by measurement using a ruler and a protractor.

If we use a calculator to compute the angle of a complex number, we may obtain an incorrect answer. For example, consider $-3 + j2 = re^{j\theta}$. Using (3.3), we can obtain $r = 3.6$, $\theta = \tan^{-1}(2/(-3)) = -34^o$, and write $-3 + j2 = 3.6e^{-j34^o}$. Is this correct? The best way of checking this is to plot $3.6e^{-j34^o}$ as shown in Figure 3.1(a). The complex number $-3 + j2$ is the vector B and the complex number $3.6e^{-j34^o}$ is the vector C. They are clearly different vectors. Note that we have $\tan(-34^o) = -2/3$ and $\tan 146^o = -2/3$. Thus the correct polar form of $-3 + j2$ is $3.6e^{j146^o}$. This can be easily verified from the plot in Figure 3.1(a). Thus in transforming a complex number from rectangular form into polar form, it is always a good practice to physically or, at least, mentally plot the complex number on a complex plane.

Angles can be expressed in terms of degrees or radians. An angle θ in degrees equals $\theta \times \pi/180$ in radians. All mathematical expressions involving angles must use radians. Note that the equation $2 + j3 = 3.6e^{j56^o}$ is only marginally acceptable, and the equation $2 + j3 = 3.6e^{j56}$ is incorrect. Because 56 in degrees equals $56 \times \pi/180 = 0.98$ in radians, the correct expression should be $2 + j3 = 3.6e^{j0.98}$. We use degrees in the preceding discussion because we are more familiar with degrees. Furthermore protractors are calibrated in degrees, not in radians. We mention that $-3 = -3e^{j0}$ is *not* in polar form because we require r in (3.1) to be positive. The polar form of -3 is $3e^{j\pi}$. If a is a positive real number, then the polar form of $-a$ is $a^{j\pi}$. Note that $-a$ is a vector emitting from the origin along the negative real axis. Its angle or phase is π.

Before proceeding, we mention that the complex number

$$e^{j\theta} = \cos\theta + j\sin\theta$$

has magnitude 1 and phase θ. It is a vector of unit length as shown in Figure 3.1(b) or simply a point on the unit circle. From the plot we can readily obtain

$$e^{j0} = 1 + 0 \times j = 1$$

$$\begin{aligned}
e^{j(\pi/2)} &= 0 + 1 \times j = j \\
e^{-j(\pi/2)} &= 0 + (-1) \times j = -j \\
e^{j\pi} &= -1 + 0 \times j = -1
\end{aligned}$$

Of course, we can also obtain them using (3.2). But it is much simpler to obtain them graphically.

To conclude this section, we introduce the concept of complex conjugation. Consider the complex number $A = \alpha + j\beta$, where α and β are real. The complex conjugate of A, denoted by A^*, is defined as $A^* = \alpha - j\beta$. Following this definition, we have that A is real if and only if $A^* = A$. We also have $AA^* = |A|^2$ and $[AB]^* = A^*B^*$ for any complex numbers A and B.

We now compute the complex conjugate of $re^{j\theta}$, where r and θ are real or $r^* = r$ and $\theta^* = \theta$. We have

$$[re^{j\theta}]^* = r^*[e^{j\theta}]^* = re^{j^*\theta^*} = re^{-j\theta}$$

This can also be verified as

$$[re^{j\theta}]^* = [r\cos\theta + jr\sin\theta]^* = r\cos\theta - jr\sin\theta = r(\cos\theta - j\sin\theta) = re^{-j\theta}$$

Thus complex conjugation simply changes j to $-j$ or $-j$ to j if all variables involved are real valued. As a final remark, if $A = \alpha + j\beta = 0$, we require $\alpha = 0$ and $\beta = 0$. However, if $A = re^{j\theta} = 0$, we require only $r = 0$; whereas θ can be any real number.

3.2.1 Angles - Modulo

In expressing a complex number in polar form $re^{j\theta}$, the angle is measured from the positive real axis. By convention, an angle is positive if measured counterclockwise and negative if measured clockwise. For example, the vector A in Figure 3.1(a) has angle 56° or -304°. In other words, 56 and -304 denote the same angle. Now we introduce the concept of *modulo* to make the discussion precise.

Two angles θ_1 and θ_2 are considered the same angle if they differ by 360 or its integer multiple or, equivalently, their difference can be divided by 360 wholly. This is expressed mathematically as

$$\theta_1 = \theta_2 \quad \text{(modulo 360) or (mod 360)} \tag{3.4}$$

if

$$\theta_1 - \theta_2 = k \times 360 \tag{3.5}$$

for some integer k (negative, 0, or positive). Clearly, we have $56 \neq -304$. But we can write $56 = -304 \pmod{360}$ because $56 - (-304) = 360$. More generally we have

$$56 = 416 = 776 = \cdots = -304 = -664 \quad \text{(mod 360)}$$

The equation is easy to comprehend. After measuring 56 degrees, if we continue to measure 360 degrees (one complete cycle), we will obtain 416 degrees and come back to the same vector. If we measure one more cycle, we will obtain 776 degrees. After measuring 56 degrees, if we measure one complete cycle in the opposite direction,

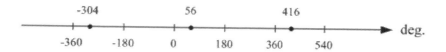

Figure 3.2: Representation of an angle on a real number line.

we will obtain $(56 - 360 = -304)$ degrees. Thus the use of modulo is convenient in expressing repetitive nature of degrees.

Let us use a real number line to denote degrees as shown in Figure 3.2. The three dots all represent the same angle, thus the representation of an angle on the real line is not unique. Because a degree will repeat itself every 360 degrees, we may say that the angle is periodic with period 360. In order to have a unique representation of an angle, we must restrict the representation to a segment of 360. Any segment of 360 will do. But the one from -180 to 180 will be most convenient. To be precise, we require a degree θ to lie in $(-180, 180]$ $(-180 < \theta \le 180)$ or $[-180, 180)$ $(-180 \le \theta < 180)$. If we use $[-180, 180]$, then the representation of 180 is not unique, because it can also be expressed as -180. If we use $(-180, 180)$, then the range is not enough because it does not include 180. Although either $(-180, 180]$ or $[-180, 180)$ can be used, we select the former and call $(-180, 180]$ the *principal range* of angles. From now on, all angles are required to lie inside the principal range.

3.3 Matrices

In this section we discuss some matrix notations and operations. An $m \times n$ matrix has m rows and n columns of entries enclosed by a pair of brackets and is said to have *dimension* or *order* m by n. Its entries can be numbers or variables. The entry at the kth-row and lth-column is called the (k, l)th entry. Every matrix will be denoted by a boldface letter. A 1×1 matrix is called a *scalar* and will be denoted by a regular-face letter. An $m \times 1$ matrix is also called a *column vector* and a $1 \times n$ matrix is also called a *row vector*.

Two matrices equal each other if and only if they have the same dimension and all corresponding entries equals each other. For matrices, we may define additions, subtractions, multiplications and divisions. Divisions are defined through inverses which will not be used in this text and will not be discussed. For additions and subtractions, the two matrices must be of the same order and the operation is carried out at the corresponding entries in the usual way.

The multiplication of a scalar and an $m \times n$ matrix is an $m \times n$ matrix with every of its entries multiplied by the scalar. For example, if $d = 3$ and

$$\mathbf{M} = \left[\begin{array}{ccc} 2 & 3 & -1 \\ 4 & -1.5 & 0 \end{array} \right]$$

then we have

$$d\mathbf{M} = \mathbf{M}d = \left[\begin{array}{ccc} 6 & 9 & -3 \\ 12 & -4.5 & 0 \end{array} \right]$$

Note that the positional order of d and $\mathbf{M}$ can be interchanged.

In order for the multiplication of an $m \times n$ matrix $\mathbf{M}$ and a $p \times q$ matrix $\mathbf{P}$, denoted by $\mathbf{MP}$, to be defined, we require $n = p$ (their inner dimensions must be the same) and the resulting $\mathbf{MP}$ is an $m \times q$ matrix ($\mathbf{MP}$ has their outer dimensions). For example, let $\mathbf{M}$ be a 2×3 matrix and $\mathbf{P}$ be a 3×2 matrix as

$$\mathbf{M} = \begin{bmatrix} 2 & 3 & -1 \\ 4 & -1.5 & 0 \end{bmatrix} \quad \text{and} \quad \mathbf{P} = \begin{bmatrix} 2.5 & 0 \\ 1 & 2 \\ -3 & 0.8 \end{bmatrix} \tag{3.6}$$

Then $\mathbf{MP}$ is a 2×2 matrix defined by

$$\begin{aligned} \mathbf{MP} &= \begin{bmatrix} 2 \times 2.5 + 3 \times 1 + (-1) \times (-3) & 2 \times 0 + 3 \times 2 + (-1) \times 0.8 \\ 4 \times 2.5 + (-1.5) \times 1 + 0 \times (-3) & 4 \times 0 + (-1.5) \times 2 + 0 \times 0.8 \end{bmatrix} \\ &= \begin{bmatrix} 11 & 5.2 \\ 8.5 & -3 \end{bmatrix} \end{aligned}$$

Note that the (k,l)th entry of $\mathbf{MP}$ is the sum of the products of the corresponding entries of the kth row of $\mathbf{M}$ and the lth column of $\mathbf{P}$. Using the same rule, we can obtain

$$\mathbf{PM} = \begin{bmatrix} 5 & 7.5 & -2.5 \\ 10 & 0 & -1 \\ -2.8 & -10.2 & 3 \end{bmatrix}$$

It is a 3×3 matrix. In general, we have $\mathbf{MP} \neq \mathbf{PM}$. Thus the positional order of matrices is important and should not be altered. The only exception is the multiplication by a scalar such as $d = 3$. Thus we can write $d\mathbf{MP} = \mathbf{M}d\mathbf{P} = \mathbf{MP}d$.

Example 3.4.1 Consider

$$\begin{bmatrix} x_1 \\ x_2 \\ x_3 \end{bmatrix} = \begin{bmatrix} 2 & 3 & -1 \\ 4 & -1.5 & 0 \\ 0 & 5 & 2 \end{bmatrix} \begin{bmatrix} 2.5 \\ 1 \\ -3 \end{bmatrix} + \begin{bmatrix} 0 \\ 2 \\ 0.8 \end{bmatrix} (-3) \tag{3.7}$$

We have

$$\begin{bmatrix} x_1 \\ x_2 \\ x_3 \end{bmatrix} = \begin{bmatrix} 2 \times 2.5 + 3 \times 1 + (-1)(-3) \\ 4 \times 2.5 + (-1.5) \times 1 + 0 \times (-3) \\ 0 \times 2.5 + 5 \times 1 + 2 \times (-3) \end{bmatrix} + \begin{bmatrix} 0 \times (-3) \\ 2 \times (-3) \\ 0.8 \times (-3) \end{bmatrix}$$

or

$$\begin{bmatrix} x_1 \\ x_2 \\ x_3 \end{bmatrix} = \begin{bmatrix} 11 \\ 8.5 \\ -1 \end{bmatrix} + \begin{bmatrix} 0 \\ -6 \\ -2.4 \end{bmatrix} = \begin{bmatrix} 11 \\ 2.5 \\ -3.4 \end{bmatrix}$$

Thus we have $x_1 = 11, x_2 = 2.5$, and $x_3 = -3.4$. $\square$

The *transpose* of $\mathbf{M}$, denoted by a prime as $\mathbf{M}'$, changes the first row into the first column, the second row into the second column, and so forth. Thus if $\mathbf{M}$ is of order $n \times m$, then $\mathbf{M}'$ has order $m \times n$. For example, for the $\mathbf{M}$ in (3.6), we have

$$\mathbf{M}' = \begin{bmatrix} 2 & 4 \\ 3 & -1.5 \\ -1 & 0 \end{bmatrix}$$

The use of transpose can save space. For example, the 3×1 column vector in (3.7) can be written as $[x_1 \quad x_2 \quad x_3]'$.

A great deal more can be said about matrices. The preceding discussion is all we need in this text.

3.4   Mathematical notations

We can define a notation any way we want. However once it is defined, then all equations involving the notation must be consistent. For example, if we define

$$X(\omega) := \frac{3}{2 + j5\omega} \tag{3.8}$$

then we can write

$$X(2\pi f) = \frac{3}{2 + j5 \cdot 2\pi f} \tag{3.9}$$

and

$$X(f) = \frac{3}{2 + j5f} \tag{3.10}$$

But we *cannot* write

$$X(f) = \frac{3}{2 + j5 \cdot 2\pi f} = \frac{3}{2 + j10\pi f} \tag{3.11}$$

because it is inconsistent with (3.10). However, (3.10) and (3.11) are used simultaneously in some engineering texts. To avoid the confusion, we may define a new variable $\bar{X}$ as

$$\bar{X}(f) := \frac{3}{2 + j10\pi f} \tag{3.12}$$

then we have $\bar{X}(f) = X(2\pi f)$ and no confusion will arise. Under the definition of (3.8), we have

$$\frac{3}{2 - j5\omega} = X(-\omega) \quad \text{and} \quad \frac{3}{2 + j\omega} = X(\omega/5)$$

We give one more example. Let us define

$$X(\omega) := \int_{t=-\infty}^{\infty} x(t)e^{-j\omega t}dt = \int_{\tau=-\infty}^{\infty} x(\tau)e^{-j\omega\tau}d\tau$$

There are two variables ω and t, where t is the integration or dummy variable. The dummy variable can be denoted by any notation such as τ as shown. From the definition, we can write

$$X(-\omega) = \int_{t=-\infty}^{\infty} x(t)e^{j\omega t}dt$$

and

$$X(\omega - \omega_c) := \int_{t=-\infty}^{\infty} x(t)e^{-j(\omega-\omega_c)t}dt$$

but we *cannot* write

$$X(f) = \int_{t=-\infty}^{\infty} x(t)e^{-j2\pi ft}dt$$

3.5 Mathematical conditions used in signals and systems

In this section we introduce some mathematical conditions. These conditions are quoted in most texts on signals and systems without much discussion. Our discussion is more thorough; we discuss their relationships and implications. Glance through this section. It will show the intricacy of mathematical conditions and the need of precise statements; missing a word may mean a different condition. You should proceed to the next section even if you do not grasp fully all the discussion.

We study in this text both CT and DT signals. The mathematics needed to analyze them are very different. Because the conditions for the latter are simpler than those for the former, we study first the DT case. For convenience, all mathematical conditions are stated for complex-valued signals defined for all n or t in $(-\infty, \infty)$. They are however directly applicable to real-valued signals defined for n or t in $[0, \infty)$.

Consider a DT signal or sequence $x[n]$. The sequence $x[n]$ is defined to be *summable* in $(-\infty, \infty)$ if

$$\left| \sum_{n=-\infty}^{\infty} x[n] \right| < \bar{M} < \infty \tag{3.13}$$

for some finite $\bar{M}$. In other words, if $x[n]$ is not summable, then their sum will approach either ∞ or $-\infty$. The sequence is defined to be *absolutely summable* if

$$\sum_{n=-\infty}^{\infty} |x[n]| < M < \infty \tag{3.14}$$

for some finite M. We first use an example to discuss the difference of the two definitions. Consider $x[n] = 0$, for $n < 0$, and

$$x[n] = \begin{cases} 1 & \text{for } n \text{ even} \\ -1 & \text{for } n \text{ odd} \end{cases} \tag{3.15}$$

for $n \geq 0$. The sequence assumes 1 and -1 alternatively as n increases from 0 to infinity. Let us consider

$$S := \sum_{n=0}^{N} x[n] = 1 - 1 + 1 - 1 + 1 - 1 + \cdots + (-1)^N \tag{3.16}$$

The sum is *unique* no matter how large N is and is either 1 or 0. However if $N = \infty$, the sum is no longer unique and can assume any value depending on how it is added up. It could be 0 if we group it as

$$S = (1 - 1) + (1 - 1) + \cdots$$

or 1 if we group it as

$$S = 1 - (1 - 1) - (1 - 1) - \cdots$$

If we write $1 = 9 - 8$ and carry out the following grouping

$$S = (9 - 8) - (9 - 8) + (9 - 8) - \cdots = 9 - (8 - 8) - (9 - 9) - \cdots$$

then the sum in (3.16) with $N = \infty$ appears to be 9. This is a paradox involving ∞. Even so, the summation is always finite and the sequence is summable.

Next we consider

$$S_a = \sum_{n=0}^{\infty} |x[n]| = 1 + |-1| + 1 + |-1| + \cdots = 1 + 1 + 1 + 1 + \cdots$$

in which we take the absolute value of $x[n]$ and then sum them up. Because there is no more cancellation, the sum is infinity. Thus the sequence in (3.15) is summable but not absolutely summable.

If a sequence $x[n]$ is absolutely summable, then it must be *bounded* in the sense

$$|x[n]| < M_1 < \infty$$

for some finite M_1 and for all n. Roughly speaking, if $|x[n_1]| = \infty$, for some n_1, then the sum including the term will also be infinity and the sequence cannot be absolutely summable.

If a sequence $x[n]$ is absolutely summable, then the magnitude of the sequence must approach zero as $n \to +\infty$. Its mathematical meaning is that for any arbitrarily small $\epsilon > 0$, there exists an N such that

$$|x[n]| < \epsilon \quad \text{for all } |n| \geq N \tag{3.17}$$

We prove this by contradiction. If $x[n]$ does not approach zero as $|n| \to \infty$, then there are infinitely many[1] n with $|x[n]| \geq \epsilon$. Thus their sum will be infinity. In conclusion, *if $x[n]$ is absolutely summable, then it is bounded for all n and approaches zero as $|n| \to \infty$.*

We now discuss the CT counterpart. A CT signal $x(t)$ is defined to be *integrable* in $(-\infty, \infty)$ if

$$\left| \int_{-\infty}^{\infty} x(t) dt \right| < \bar{M} < \infty$$

for some finite $\bar{M}$. It is *absolutely integrable* in $(-\infty, \infty)$ if

$$\int_{-\infty}^{\infty} |x(t)| dt < M < \infty$$

for some finite M. Consider $x(t) = \sin 3t$, for $t \geq 0$ and $x(t) = 0$ for $t < 0$ as plotted in Figure 3.3(a). Its integration is the algebraic sum of all positive and negative areas. Because of the cancellation of positive and negative areas, the integration cannot be infinity. Thus $\sin 3t$ is integrable. We plot the magnitude of $\sin 3t$ or $|\sin 3t|$ in Figure 3.3(b). All areas are positive and there is no more cancellation. Thus its total area is infinity and $\sin 3t$ is not absolutely integrable.

An absolutely summable DT signal is bounded and approaches 0 as $n \to \pm\infty$. The corresponding statement for the CT case however is not necessarily true. For example, consider the CT signal $x(t)$ shown in Figure 3.4. The triangle located at $t = n$, where n is a positive integer, is defined to have base width $2/n^3$ and height n. The triangle

---

[1] If there are only finitely many, then we can find an N to meet (3.17).

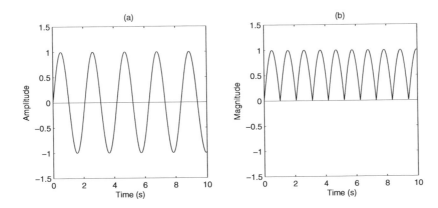

Figure 3.3: (a) $\sin 3t$. (b) $|\sin 3t|$.

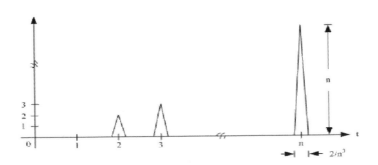

Figure 3.4: CT signal that is absolutely integrable is not bounded nor approaches zero as $t \to \infty$.

is defined for all $n \geq 2$. The CT signal so defined is not bounded, nor does it approach zero as $t \to \infty$. Because the area of the triangle located at $t = n$ is $1/n^2$, we have

$$\int_{-\infty}^{\infty} |x(t)| dt = \frac{1}{2^2} + \frac{1}{3^2} + \cdots + \frac{1}{n^2} + \cdots$$

The infinite summation can be shown to be finite (Problem 3.18). Thus the CT signal is absolutely integrable. In conclusion, unlike the DT case, an absolutely integrable CT signal may not be bounded and may not approach zero as $|t| \to \infty$.

3.5.1 Signals with finite total energy – Real-world signals

Let $x(t)$ be a real-valued voltage signal. Its application across a 1-Ω resistor will yield the current $i(t) = x(t)/1$. Thus the power delivered by $x(t)$ is $x(t)i(t) = x^2(t)$ and the total energy delivered by $x(t)$ in $(-\infty, \infty)$ is

$$E = \int_{t=-\infty}^{\infty} [x(t)]^2 dt \qquad (3.18)$$

Now if $x(t)$ is complex-valued, then the total energy delivered by $x(t)$ is defined as

$$E = \int_{t=-\infty}^{\infty} x(t) x^*(t) dt = \int_{t=-\infty}^{\infty} |x(t)|^2 dt \qquad (3.19)$$

where $x^*(t)$ is the complex conjugate of $x(t)$. Note that (3.19) reduces to (3.18) if $x(t)$ is real. If E in (3.19) is finite, $x(t)$ is defined to be *squared absolutely integrable* or to have finite total energy.

Consider a DT real-valued signal $x(nT)$. It is a sequence of numbers. Such a sequence cannot drive any physical device and its total energy is zero. Now if we use $x(nT)$ to construct a CT signal using a zero-order hold, then the resulting CT signal is a staircase function as shown in Figure 2.6 with solid lines. If we use $x_T(t)$ to denote the staircase function, then $x_T(t)$ has the total energy

$$E = \int_{t=-\infty}^{\infty} [x_T(t)]^2 dt = T \sum_{n=-\infty}^{\infty} [x(nT)]^2 \qquad (3.20)$$

Thus *mathematically* we may define the total energy of complex-valued $x[n]$ defined for all n in $(-\infty, \infty)$ as

$$E_d = \sum_{n=-\infty}^{\infty} x[n] x^*[n] = \sum_{n=-\infty}^{\infty} |x[n]|^2 \qquad (3.21)$$

If E_d is finite, $x[n]$ is defined to be *squared absolutely summable* or to have finite total energy.

Now we show that if $x[n]$ is absolutely summable, then it has finite total energy. As shown in the preceding section, if $x[n]$ is absolutely summable, then it is bounded, that is, there exists a finite M_1 such that $|x[n]| < M_1$ for all n. We substitute this into one of $|x[n]| \times |x[n]|$ in (3.21) to yield

$$E_d := \sum_{n=-\infty}^{\infty} |x[n]||x[n]| < M_1 \sum_{n=-\infty}^{\infty} |x(nT)| \qquad (3.22)$$

If $x[n]$ is absolutely summable, the last infinite summation is less than some finite M. Thus we have $E_d < MM_1 < \infty$. In conclusion, an absolutely summable $x[n]$ has finite total energy.

Do we have a similar situation in the CT case? That is, if $x(t)$ is absolutely integrable, will it have finite total energy? The answer is negative. For example, consider $x(t) = t^{-2/3}$ for t in $[0, 1]$ and $x(t) = 0$ for t outside the range. Using the formula

$$\int t^a dt = \frac{1}{a+1} t^{a+1}$$

where a is a real number other than -1,[2] we can compute

$$\int_{-\infty}^{\infty} |x(t)| dt = \int_{t=0}^{1} t^{-2/3} dt = \frac{1}{-2/3+1} \left. t^{1/3} \right|_{t=0}^{1} = 3(1-0) = 3 \qquad (3.23)$$

and

$$\int_{-\infty}^{\infty} |x(t)|^2 dt = \int_{t=0}^{1} t^{-4/3} dt = \frac{1}{-4/3+1} \left. t^{-1/3} \right|_{t=0}^{1} = -3(1-\infty) = \infty \qquad (3.24)$$

Thus the signal is absolutely integrable but has infinite energy. Note that the impulse defined in (2.7) and the signal in Figure 3.4 are absolutely integrable but have infinite energy. See Problems 3.22 and 3.23. In conclusion, an absolutely integrable CT signal may not have finite total energy or absolute integrability does not imply squared absolute integrability.

Why is the CT case different from the DT case? An absolutely summable DT signal is always bounded. This is however not the case in the CT case. The signal $x(t) = t^{-2/3}$ for t in $[0, 1]$ is not bounded. Nor are the impulse and the signal in Figure 3.4. Thus boundedness is an important condition. Now we show that if $x(t)$ is absolutely integrable and bounded for all t in $(-\infty, \infty)$, then it has finite total energy. Indeed, if $x(t)$ is bounded, there exists a finite M_1 such that $|x(t)| < M_1$, for all t. We substitute this into one of $|x(t)| \times |x(t)|$ in (3.19) to yield

$$E = \int_{t=-\infty}^{\infty} |x(t)||x(t)| dt < M_1 \int_{t=-\infty}^{\infty} |x(t)| dt$$

If $x(t)$ is absolutely integrable, the last integral is less than some finite M. Thus we have $E < MM_1 < \infty$ and $x(t)$ has finite total energy. Note that the proof must use both conditions: boundedness and absolute integrability. If we use only one of them, then the proof cannot be completed.

Before proceeding, we discuss the sinc function defined by

$$\text{sinc}(t) := \frac{\sin t}{t} \qquad (3.25)$$

for all t in $(-\infty, \infty)$. It is plotted in Figure 3.5. Because $\sin k\pi = 0$ for all integer k, we have sinc $(k\pi) = 0$ for all integers k except $k = 0$. For $k = 0$ or at $t = 0$, we have, using l'Hopital's rule,

$$\text{sinc}(0) = \left. \frac{\sin t}{t} \right|_{t=0} = \left. \frac{\cos t}{1} \right|_{t=0} = 1$$

---

[2]If $a = -1$, then the integration equals $\ln t$.

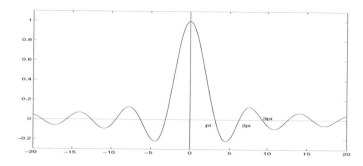

Figure 3.5: The sinc function sinc $(t) = \sin t/t$.

as shown. This function was used in (2.12) (see Problem 3.32) and will appear again later. The function has the following properties

$$\int_{-\infty}^{\infty} \frac{\sin t}{t} dt = \pi \tag{3.26}$$

$$\int_{-\infty}^{\infty} \left| \frac{\sin t}{t} \right| dt = \infty \tag{3.27}$$

and

$$\int_{-\infty}^{\infty} \left| \frac{\sin t}{t} \right|^2 dt = 1 \tag{3.28}$$

See Reference [W2]. Thus the sinc function is integrable and squared absolutely integrable but is not absolutely integrable. It also shows that squared absolute integrability does not imply absolute integrability.

In signal analysis, a signal is required to be absolutely integrable and/or squared absolutely integrable in order to ensure the existence of its spectrum and to have finite energy (Chapters 4 and 5). In practice, a signal must start somewhere and cannot last forever. If we select the starting time instant as $t = 0$, then the signal $x(t)$ is defined only for t in $[0, L]^3$, for some finite L. The signal is implicitly assumed to be zero outside the range and is said to have a *finite duration*. Clearly a practical signal cannot grow to ∞ or $-\infty$, thus the signal is bounded for all t in $[0, L]$. If $x(t)$ is defined only for t in $[0, L]$ and is bounded by M, then

$$\int_{-\infty}^{\infty} |x(t)| dt = \int_{0}^{L} |x(t)| dt \leq M \int_{0}^{L} dt = ML$$

and

$$\int_{-\infty}^{\infty} |x(t)|^2 dt = \int_{0}^{L} |x(t)|^2 dt \leq M^2 \int_{0}^{L} dt = M^2 L$$

Thus every real-world signal is automatically absolutely integrable and has finite total energy. In system analysis, we will use the concept of absolute integrability to establish a stability condition (Chapter 9) and then use it to develop a simpler condition.

---

[3] If a signal does not contain impulses, there is no difference in using the time interval $[0, L]$, $[0, L)$, $(0, L]$ or $(0, L)$.

3.5.2   Can we skip the study of CT signals and systems?

Because the mathematics used in the study of DT signals and systems is considerably simpler than the one for the CT case and because most CT signals are now processed digitally, one may raise the question: Can we study only DT signals and systems without studying the CT case? The answer is negative for the following reasons:

1. Most real-world signals are CT or analog. To process them digitally, we must select a sampling period. To do so, we need to know frequency contents of CT signals. Moreover, the frequency of DT sinusoids must be defined, as we will discuss in Subsection 4.6.1, through the frequency of CT sinusoids. Thus we must study CT sinusoids before studying DT sinusoids.

2. All signals in a computer are actually analog signals. However they can be *modeled* as 1s or 0s and there is no need to pay any attention to their analog behavior. However, as we can see from Figure 2.12, the interface between analog and digital world requires anti-aliasing filter, sample-and-hold circuit, ADC and DAC. They are all analog systems.

3. Both CT and DT systems can be classified as FIR and IIR. In the CT case, we study exclusively IIR. See Subsection 8.8.1. In the DT case, we study and design both FIR and IIR systems. Because there is no CT counterpart, DT FIR systems must be designed directly. However, direct design of DT IIR systems is complicated. It is simpler to design first CT systems and then transform them into the DT case. See References [C5, C7]. This is the standard approach in designing DT IIR systems.

In view of the preceding discussion, we cannot study only DT signals and systems. Thus we study both cases in this text. Because of our familiarity with CT physical phenomena and examples, we study first the CT case. The only exception is to use DT systems with finite memory to introduce some systems' concepts because there are no simple CT counterparts.

3.6   A mathematical proof

In this section, we show that $\sqrt{2}$ is an irrational number.[4] The purpose is to show that we can give any reasonable definitions and then derive logically and rigorously results. This is a beauty of mathematics.

We assume that all integers are self evident. Next we define an integer to be *even* if it can be divided by 2 without remainder. It is *odd* if its division by 2 yields 1 as its remainder. Thus every even integer can be expressed as $2k_1$ for some integer k_1 and every odd integer can be expressed as $2k_2 + 1$ for some integer k_2.

Lemma Let p be an integer. Then p is even if and only if p^2 is even.

First we show that if p is even, then p^2 is even. Indeed if p is even, there exists an integer k such that $p = 2k$. Clearly we have $p^2 = 4k^2 = 2(2k^2) =: 2k_1$, where $k_1 = 2k^2$ is an integer. Thus p^2 is even.

---

[4]This section may be skipped without loss of continuity.

Next we show that if p^2 is even, then p is even. We show this by contradiction. Suppose p^2 is even but p is odd. If p is odd, then it can be expressed as $p = 2k + 1$ for some integer k and we have $p^2 = (2k+1)^2 = 4k^2 + 4k + 1 = 2(2k^2 + 2k) + 1 =: 2k_2 + 1$. Because $k_2 = 2k^2 + 2k$ is an integer, p^2 is odd. This contradicts our assumption that p^2 is even. Thus we conclude that p cannot be odd. This completes the proof of the Lemma.

Let us *define* a number to be a *rational number* if it can be expressed as a ratio of two integers. Otherwise, it is an *irrational number*. We now show that $\sqrt{2}$ is an irrational number. We again prove it by contradiction.

Suppose $\sqrt{2}$ is a rational number. Then there exists two integers p and q such that $\sqrt{2} = p/q$. Furthermore, p and q cannot be both even. Indeed, if p and q are even, then they have common factor 2 and can be canceled. Thus at least one of p and q is odd.

If $\sqrt{2} = p/q$, then we have

$$2 = \frac{p^2}{q^2}$$

or $p^2 = 2q^2$ which implies p^2 to be even. It follows from the lemma that p is even or there exists an integer k such that $p = 2k$. Substituting this into $p^2 = 2q^2$ yields $4k^2 = 2q^2$ or $q^2 = 2k^2$. Thus q^2 and, consequently, q are even. In other words, if $\sqrt{2}$ is a rational number, then both p and q are even. This contradicts our assumption that at least one of p and q is odd. Thus $\sqrt{2}$ cannot be a rational number.

The preceding argument was essentially contained in Euclid's *Elements of Geometry* (circa 350 BC). A formal proof uses the fact that any integer can be expressed as a unique product of prime numbers, and that p/q can be reduced to have no common factor. Even though we do not introduce those concepts and results, our proof is complete.

The proofs of e, the base of natural logarithm, and π, the ratio of the circumference of a circle and its diameter, to be irrational however are much more difficult. The former was first proved in 1737, and the latter in 1768.

3.7 MATLAB

MATLAB runs on a number of windows. When we start a MATLAB program in MS Windows, a window together with a prompt ">>" will appear. It is called the *command window*. It is the primary window where we interact with MATLAB.

A sequence of numbers, such as -0.6, -0.3, 0, 0.3, 0.6, 0.9, 6/5, can be expressed in MATLAB as a *row vector* by typing after the prompt >> as

```
>> t=[-0.6,-0.3,0,0.3,0.6,0.9,6/5]
```

or

```
>> t=[-0.6 -0.3 0 0.3 0.6 0.9 6/5]
```

Note that the sequence of entries must be bounded by a pair of brackets and entries can be separated by commas or spaces. If we type "enter" at the end of the line, MATLAB will execute and store it in memory, and then display it on the monitor as t= -0.6000 -0.3000 0   0.3000 0.6000 0.9000 1.2000

From now on, it is understood that every line will be followed by "enter". If we type

>> t=[-0.6 -0.3 0 0.3 0.6 0.9 1.2];

then MATLAB will execute and store it in memory but will not display it on the monitor. In other words, the semicolon at the end of a statement suppresses the display. Note that we have named the sequence t for later use.

The MATLAB function a:b:c generates a sequence of numbers starting from a, adding b repeatedly untile c but not beyond. For example, 0:1:5 generates {0, 1, 2, 3, 4, 5}. The functions 0:2:5 and x=0:3:5 generate, respectively, {0, 2, 4} and {x=0, 3}. If b=1, a:b;c can be reduced to a:c. Using n=-2:4, the sequence t can be generated as t=0.3*n or t=n*0.3. It can also be generated as t=-0.6:0.3:1.2.

Suppose we want to compute the values of $x(t) = 4e^{-0.3t} \cos 4t$ at $t = 0 : 0.3 : 1.5$. Typing in the command window

>> t=0:0.3:1.5;
>> x=4*exp(-0.3*t).*cos(4*t)

will yield

x = 4.0000 1.3247 − 2.4637 − 2.7383 0.2442 2.4489

These are the values of $x(t)$ at those t, denoted by x. There are six entries in x. MATLAB automatically assigns the entries by x(k), with k=1:6. We call the integer k the *internal index*. Internal indices start from 1 and cannot be zero or negative. If we continue to type x(0), then an error message will occur because the internal index cannot be 0. If we type

>> x(2),x(7)

then it will yield 1.3247 and the error message "Index exceeds matrix dimensions". Typing

>> x([1:2:6])

will yield

4.0000 − 2.4637 0.2442

They are the values of $x(1)$, $x(3)$, and $x(5)$. Note that two or more commands or statements can be typed in one line separated by commas or semicolons. Those followed by commas will be executed and displayed; those followed by semicolons will be executed but not displayed.

We have used * and .* in expressing x=4*exp(-0.3*t).*cos(4*t). The multiplication '*', and (right) division '/' in MATLAB (short for MATrix LABoratory) are defined for matrices. See Section 3.4. Adding a dot such as .* or ./, an operation becomes element by element. For example, exp(-0.3*t) and cos(4*t) with t=0:0.3:1.5 are both 1×6 row vectors, and they cannot be multiplied in matrix format. Typing .* will change the multiplication into element by element. All operations in this section involve element by element and all * and / can be replaced by .* and ./. For example, typing in the command window

```
>> t=0:0.3:1.5;x=4*exp(-0.3*t)*cos(4*t)
```

will yield an error message. Typing

```
>> t=0:0.3:1.5;x=4.*exp(-0.3.*t).*cos(4.*t)
```

will yield the six values of x. Note that the addition '+' and subtraction '-' of two matrices are defined element by element. Thus there is no need to introduce '.' into '+' and '-', that is, '.+' and '.-' are not defined.[5]

Next we use t and x to generate some plots. MATLAB can generate several plots in a figure. The function subplot(m,n,k) generates m rows and n columns of plots with k=1:mn indicating its position, counting from left to right and from top to bottom. For example, subplot(2,3,5) generates two rows and three columns of plots (for a total of $2 \times 3 = 6$ plots) and indicates the 5th or the middle plot of the second row. If there is only one plot, there is no need to use subplot(1,1,1).

If a program consists of several statements and functions, it is more convenient to develop a program as a file. Clicking the new file icon on the command window toolbar will open an *edit window*. We type in the edit window the following

```
%Program 3.1 (f36.m)
n=0:20;T=0.3;t=n*T;
x=4.*exp(-0.3.*t).*cos(4.*t);
subplot(1,2,1)
stem(n,x,'.'),title('(a)'),axis square
xlabel('Time index'),ylabel('Amplitude')
subplot(1,2,2)
stem(t,x),title('(b)'),axis square
xlabel('Time (s)'),ylabel('Amplitude')
```

and then save it with the file name f36 (for Figure 3.6). This is called an m-file because an extension .m is automatically attached to the file name. All m-files are resided in the subdirectory "work" of MATLAB. For easy reference, we call the file Program 3.1. Now we explain the program. All text after a percent sign (%) is taken as a comment statement and is ignored by MATLAB. Thus it can be omitted. The next two lines generate $x(t)$ at $t = n * T$. Note the use of .* for all multiplications in x. The function subplot(1,2,k) generates one row and two columns of boxes as shown in Figure 3.6. The third index k denotes the k-th box, counting from left to right. The first plot or subplot(1,2,1) is titled (a) which is generated by the function title('(a)'). The second plot or subplot(1,2,2) is titled (b). The function stem(p,q,'linetype'), where p and q are two sequences of numbers of the same length, plots stems with height specified by q at the horizontal positions specified by p with ends of stems specified by the linetype. If 'linetype' is missing, the default is small hollow circle. For example, Figure 3.6(a) is generate by the function stem(n,x,'.') which uses stems with dots at their ends to plot x against its *time index* n as shown. Figure 3.6(b) is generated by stem(t,x) which uses hollow circles (default) to plot x against time t. We label

---

[5]Try in MATLAB >> 2.+3 and >> 2.0.+3. The former yields 5; the latter yields an error message. Why?

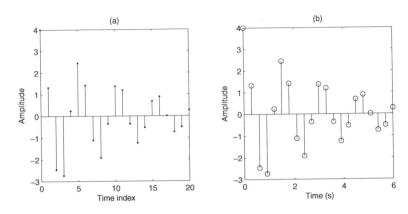

Figure 3.6: (a) Plot of a DT signal against time index. (b) Against time.

their vertical or y-axes as 'Amplitude' by typing `ylabel('Amplitude')` and label their horizontal or x-axes by typing `xlabel('text')` where 'text' is 'Time index' or 'Time (s)'. We also type `axis square`. Without it, the plots will be of rectangular shape.

Now if we type in the command window the following

>> f36

then the plots as shown in Figure 3.6 will appear in a *figure window*. If the program contains errors, then error messages will appear in the command window and no figure window will appear. Note that the program can also be run from the edit window by clicking 'Debug'. If there are errors, we must correct them and repeat the process until a figure appears. If the figure is not satisfactory, we may modify the program as we will discuss shortly. Thus the use of an edit window is convenient in generating a figure.

Once a satisfactory figure is generated in a figure window, it can be printed as a hard copy by clicking the printer icon on the figure window. It can also be saved as a file, for example, an eps file by typing on the command window as

>> print -deps f36.eps

where we have named the file f36.eps.

We next discuss how to generate CT signals from DT signals. This requires interpolation between sampling instants. The MATLAB function `stairs(t,x)` carries out zero-order hold. The function `plot(t,x,'linetype')` carries out linear interpolation[6] using the specified line type. Note that the 'linetype' is limited to solid line (default, no linetype), dotted line (':'), dashed line ('- -'), and dash-and-dotted line ('-.'). If the 'linetype' is 'o' or '.', then no interpolation will be carried out. We now modify Program 3.1 as

```
%Program 3.2 (f37.m)
n=0:20;T=0.3;t=n*T;
x=4.*exp(-0.3.*t).*cos(4.*t);
```

---

[6]It connects neighboring points by straight lines. See Figure 2.8(c)

```
subplot(2,2,1)
stairs(t,x),title('(a)')
ylabel('Amplitude'),xlabel('Time (s)')
subplot(2,2,2)
plot(t,x,':'),title('(b)')
ylabel('Amplitude'),xlabel('Time (s)')
subplot(2,2,3)
stairs(t,x),title('(c)')
ylabel('Amplitude'),xlabel('Time (s)')
axis([0 6 -4 5])
hold on
plot([0 6],[0 0])
subplot(2,2,4)
plot(t,x,':',[-1 6],[0 0],[0 0],[-5 5]),title('(d)')
ylabel('Amplitude'),xlabel('Time (s)')
axis([-1 6 -5 5])
```

This program will generate two rows and two columns of plots using `subplot (2,2,k)` with k=1:4. The first plot, titled (a), is generated by `stairs(t,x)` which carries out zero-order hold as shown in Figure 3.7(a). The second plot, titled (b), is generated by `plot(t,x,':')` which carries out linear interpolation using dotted lines as shown in Figure 3.7(b). Note that `plot(t,x)` without the 'linetype' carries out linear interpolation using solid lines. See Problems 3.28 and 3.29.

The ranges of the horizontal (x-) and vertical (y-) axes in Figures 3.7(a) and (b) are automatically selected by MATLAB. We can specify the ranges by typing `axis([x_(min) x_(max) y_(min) y_(max)])`. For example, we added `axis([0 6 -4 5])` for `subplot(2,2,3)` or Figure 3.7(c). Thus its horizontal axis ranges from 0 to 6 and its vertical axis ranges from −4 to 5 as shown. In Figure 3.7(c) we also draw the horizontal axis by typing `plot([0 6],[0 0])`. Because the new plot (horizontal axis) is to be superposed on the original plot, we must type "hold on". Without it, the new plot will replace the original plot.

The ranges of the horizontal (x-) and vertical (y-) axes in Figure 3.7(d) are specified by `axis([-1 6 -5 5])`. The dotted line is generated by `plot(t,x,':')`; the horizontal axis by `plot([-1 6],[0 0])`; the vertical axis by `plot([0 0],[-5 5])`. The three plot functions can be combined into one as in the program. In other words, the plot function may contain one or more *pairs* such as `plot(x1, y1, x2, y2, ...)` with or without 'linetype' and each pair must be of the same length.

Next we list the program that generate Figure 2.6:

```
%Program 3.3(f26.m)
L=5;t=0:0.0001:L;
x=4*exp(-0.3*t).*cos(4*t);
T=0.3;n=0:L/T;
xT=4*exp(-0.3*n*T).*cos(4*n*T);
subplot(3,1,1)
```

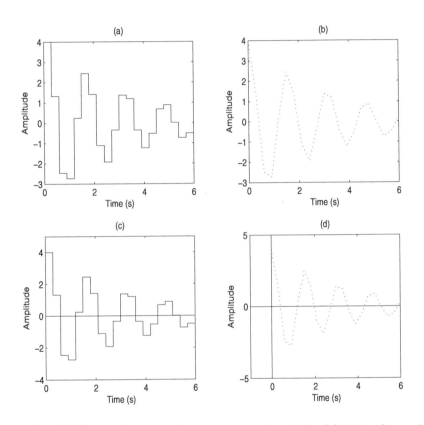

Figure 3.7: (a) A CT signal generated using zero-order hold. (b) Using linear interpolation. (c) Improvement of (a). (d) Improvement of (b).

```
plot(t,x,':',[0 5],[0 0]),title('(a)')
hold on
stairs(n*T,xT)

T=0.1;n=0:L/T;
xT=4*exp(-0.3*n*T).*cos(4*n*T);
subplot(3,1,2)
plot(t,x,':',[0 5],[0 0]),title('(b)')
hold on
stairs(n*T,xT)
ylabel('Amplitude')

T=0.001;n=0:L/T;
xT=4*exp(-0.3*n*T).*cos(4*n*T);
subplot(3,1,3)
plot(t,x,':',[0 5],[0 0]),title('(c)')
hold on
stairs(n*T,xT)
xlabel('Time (s)')
```

The first two lines compute the amplitudes of $x(t)$ from $t = 0$ to $t = 5$ with increment 0.0001. The next two lines compute the samples of $x(t)$ or $x(nT)$ with the sampling period $T = 0.3$ and the time index from $n = 0$ up to the integer equal to or less than L/T, which can be generated in MATLAB as n=0:L/T. The program generates three rows and one column of plots as shown in Figure 2.6. The first line after subplot(3,1,1) plots in Figure 2.6(a) $x(t)$ for t in [0, 5] with a dotted line and with linear interpolation. It also plots the horizontal axis specified by the pair [0 5], [0 0] with a solid line (default). In order to superpose the staircase function specified by stairs(n*T, xT) on Figure 2.6(a), we type "hold on". Without it, the latter plot will replace, instead of superpose on, the previous plot. The programs to generate Figures 2.6(b) and (c) are identical to the one for Figure 2.6(a) except that $T = 0.3$ is replaced by $T = 0.1$ and then by $T = 0.001$. In addition, we add "Amplitude" to the vertical axis in Figure 2.6(b) by typing ylabel('Amplitude') and add "Time (s)" to the horizontal axis in Figure 2.6(c) by typing xlabel('Time (s)'). In conclusion, MATLAB is a simple and yet powerful tool in generating figures.

3.7.1 How Figures 1.1–1.3 are generated

We now discuss how the plot in Figure 1.2 is obtained. We first connect a microphone to a PC. In MS-Windows, there is a *Sound Recorder* inside *Entertainments* which is a part of *Accessories*. Click 'File' on the tool bar of the sound recorder and then click 'Properties'. It will open a dialog box named 'Properties for sound'. Clicking 'Convert Now ...' will open 'Sound Selection' dialog box.[7] In 'Format', we select PCM. In 'Attributes', we may select 8 Bit or 16 Bit, Stereo or Mono, nine possible sampling frequencies from 8k to 48 kHz. To record the sound generated by a 128-Hz tuning

---

[7]Without converting, the default is PCM, 20.050 kHz, 18 Bit, and Stereo.

fork, we select PCM, 8 kHz, 8 Bit and Mono. The sound was recorded and saved in the subdirectory "work" in MATLAB as `f12.wav`. Note that the extension .wav is automatically added by MS-Windows.

In an edit window of MATLAB, we then type the following

```
%Program 3.4 (f12.m)
x=wavread('f12.wav');
N=length(x),n=0:N-1;T=1/8000;
subplot(3,1,1)
plot(n*T,x),title('(a) 128-Hz tuning fork')
ylabel('Amplitude')
subplot(3,1,2)
plot(n*T,x),title('(b) Segment of (a)')
ylabel('Amplitude')
axis([0.6 1.01 -0.2 0.2])
subplot(3,1,3)
plot(n*T,x),title('(c) Segment of (b)')
xlabel('Time (s)'),ylabel('Amplitude')
axis([0.9 1.0097 -0.2 0.2])
```

and save it as `f12.m`. The first line uses the function `wavread` to load the MS file into MATLAB. Its output is called x which consists of a vertical string of entries. The number of entries, N, can be obtained by using the function `length` as shown in the second line. These N entries are the samples of the sound at `n*T` with $n = 0 : N-1$. The function `plot` connects these samples by linear interpolation as shown in Figure 1.2(a). We plot its small segment from 0.6 to 1.01 seconds in Figure 1.2(b). It is achieved by typing the function `axis([x_(min) x_(max) y_(min) y_(max)])` =axis([0.6 1.01 -0.2 0.2]). We see that the function `axis` can be used to zoom in x-axis. Figure 1.2(c) zooms in an even smaller segment of Figure 1.2(a) using the function `axis([0.9 1.0097 -0.2 0.2])`.

In the control window, if we type

```
>> f12
```

then `N=10600` will appear in the control window and Figure 1.2 will appear in a figure window. Note that in the second line of the program, we use a comma after `N=length(x)`, thus N is not suppressed and appears in the control window. This is how Figure 1.2 is generated.

The sound "signals and systems" in Figure 1.1 was recorded using PCM, 24 kHz, 8 Bit, and mono. We list in the following the program that generates Figure 1.1:

```
%Program 3.5 (f11.m)
xb=wavread('f11.wav');Nb=length(xb)
x=xb([2001:53500]);
N=length(x),n=0:N-1;T=1/24000;
subplot(2,1,1)
plot(n*T,x),title('(a) Signals and systems')
```

```
ylabel('Amplitude'),axis([0 2.2 -0.2 0.2])
subplot(2,1,2)
plot(n*T,x),title('(b) Segment of (a)')
axis([1.375 1.38 -0.2 0.2])
xlabel('Time (s)'),ylabel('Amplitude')
```

The first line loads the MS file, called f11.wav into MATLAB. The output is called xb. The number of data in xb is 66000. Because the sampling frequency is 24000 Hz, xb contains roughly 3 seconds of data. However the sound "signals and systems" lasts only about 2.3 seconds. Thus xb contains some unneeded data. We use x=xb([2001:53500]) to pick out the part of xb whose internal indices starting from 2001 up to 53500, with increment 1. This range is obtained by trial and error. The rest of the program is similar to Program 3.4.

The piano middle C in Figure 1.3 was recorded using PCM, 22.05 kHz, 8 Bit, and mono. We list in the following the program that generates Figure 1.3:

```
%Program 3.6 (f13.m)
xb=wavread('f13.wav');Nb=length(xb)
x=xb([3001:52000]);
N=length(x),n=0:N-1;T=1/22050;
subplot(2,1,1)
plot(n*T,x),title('(a)')
ylabel('Amplitude'),axis([0 1 -0.6 0.6])
subplot(2,1,2)
plot(n*T,x),title('(b)')
axis([0.1 0.15 -0.6 0.6])
xlabel('Time (s)'),ylabel('Amplitude')
```

It is similar to Program 3.5 and its discussion is omitted.

Problems

3.1 Plot $A = -3 - j4$ on a complex plane and then use a ruler and a protractor to express A in polar form. It is important when you draw a plan, you must select or assign a scale.

3.2 Express the following complex numbers in polar form. Use $\sqrt{2} = 1.4$ and draw mentally each vector on a complex plane.

$$-3, \quad -1005, \quad 1 - j1, \quad 1 + j1, \quad -10j, \quad 20j, \quad -1 - j1, \quad -1 + j1, \quad 51$$

3.3 Let a be a real number. If $a < 0$, what is its polar form?

3.4 The equation $e^{j\pi} + 1 = 0$ is discussed in Reference [C11, *The Great Equations*]. It relates two irrational numbers e and π, the imaginary number $j = \sqrt{-1}$, and the real number 1. Can you mentally justify its validity?

3.5 What are the values of $e^{jn\pi}$, for all integer n? [*Answer:* 1 if n is even and -1 if n is odd.]

3.6 What is the value of $e^{jnk2\pi}$, where n and k are any integers? [*Answer:* 1.]

3.7 If A is real, show $A \cdot A = |A|^2$. If A is complex, do we still have $A \cdot A = |A|^2$?

3.8 Let x and y be any two complex numbers. Verify $(xy)^* = x^*y^*$, $(e^{xy})^* = e^{x^*y^*}$, and

$$\left(\int x(t)e^{j\omega t}dt \right)^* = \int x^*(t)e^{-j\omega t}dt$$

where ω and t are two real-valued variables.

3.9 The unit matrix $\mathbf{I}_n$ is defined as an $n \times n$ square matrix with all its diagonal elements equal to 1 and the rest zero such as

$$\mathbf{I}_3 = \begin{bmatrix} 1 & 0 & 0 \\ 0 & 1 & 0 \\ 0 & 0 & 1 \end{bmatrix}$$

Consider the 2×3 matrix $\mathbf{M}$ defined in (3.6). Is $\mathbf{I}_3\mathbf{M}$ defined? Verify $\mathbf{I}_2\mathbf{M} = \mathbf{M}$ and $\mathbf{M}\mathbf{I}_3 = \mathbf{M}$.

3.10 Given $\mathbf{x} = [1\ 2\ 3]$ and $\mathbf{y} = [2\ 5\ -3]$. What are $\mathbf{xy}$, $\mathbf{xy}'$, and $\mathbf{x}'\mathbf{y}$?

3.11 Verify

$$\sum_{n=0}^{N} a_n b_n = [a_0\ a_1\ \cdots\ a_N][b_0\ b_1\ \cdots\ b_N]'$$

where the prime denotes the transpose.

3.12 Verify that the following two equations

$$\begin{aligned} x_1 &= 3y_1 - 5y_2 - 2u \\ x_2 &= 0.5y_1 + 0.2y_2 \end{aligned}$$

can be expressed in matrix format as

$$\begin{bmatrix} x_1 \\ x_2 \end{bmatrix} = \begin{bmatrix} 3 & -5 \\ 0.5 & 0.2 \end{bmatrix} \begin{bmatrix} y_1 \\ y_2 \end{bmatrix} + \begin{bmatrix} -2 \\ 0 \end{bmatrix} u$$

3.13 Let us define

$$X(\omega) := \int_0^\infty x(t)e^{-j\omega t}dt$$

Express the following in terms of X:

$$\int_0^\infty x(t)e^{j\omega t}dt \quad \text{and} \quad \int_0^\infty x(t)e^{-st}dt$$

[*Answer:* $X(-\omega)$, $X(s/j)$]

3.14 Can we define $x[n]$ to be summable as

$$\sum_{n=-\infty}^{\infty} x[n] < \bar{M} < \infty$$

for some finite $\bar{M}$? [*Answer:* No.]

3.15 If $x(t) \geq 0$, for all t, is there any difference between whether it is integrable or absolute integrable?

3.16 Show that the summation in (3.16) with $N = \infty$ can equal 5.

3.17 If a sequence is absolutely summable, it is argued in Section 3.5 that the sequence approaches zero as $n \to \infty$. Is it true that if a sequence is summable, then the sequence approaches zero as $n \to \infty$? If not, give a counter example. Note the missing of 'absolutely' in the second statement.

3.18* [8] Use the integration formula

$$\int_1^{\infty} \frac{1}{t^2} dt = \frac{-1}{t}\bigg|_{t=1}^{\infty} = 1$$

to show

$$\frac{1}{2^2} + \frac{1}{3^2} + \cdots < 1$$

Thus the CT signal in Figure 3.4 is absolutely integrable.

3.19* For $n \geq 2$, we have $1/n < 1$. Thus we have

$$\sum_{n=1}^{\infty} \frac{1}{n} < \sum_{n=1}^{\infty} 1 = \infty$$

Can we conclude that $\sum_{n=1}^{\infty}(1/n) < M$, for some finite M? Verify

$$\sum_{n=1}^{\infty} \frac{1}{n} = \infty \tag{3.29}$$

Thus we have $\infty < \infty$. This is a paradox involving ∞. It also show that the use of $|A| < \infty$ to denote that $|A|$ is summable is not very precise. A more precise notation is that there exists a finite M such that $|A| < M$ or $|A| < M < \infty$.

3.20 We showed in Subsection 3.5.1 that absolute summable imply squared absolute summable. Does squared absolute summable imply absolute summable? If not, give a counter example.

---

[8]Study the conclusion of a problem with an asterisk. Do the problem only if you are interested in.

3.21 If $|x[n]| < M_1$, for all n, then we have

$$E_d = \sum_{n=-\infty}^{\infty} |x[n]|^2 < (M_1)^2 \sum_{n=-\infty}^{\infty} 1 = \infty$$

Can we conclude that E_d is finite? Can you give the reason that we substitute $|x[n]| < M_1$ only once in (3.22)?

3.22 Is the pulse in Figure 2.9(a) absolutely integrable for all $a > 0$? Verify that its total energy approaches infinity as $a \to 0$. Thus it is not possible to generate an impulse in practice.

3.23* Let us define $x(t) = n^4 t$, for t in $[0, 1/n^3]$, for any integer $n \geq 2$. It is a right triangle with base width $1/n^3$ and height n. Verify

$$\int_{t=0}^{1/n^3} [x(t)]^2 dt = \frac{1}{3n}$$

Use this and the result in Problem 3.19 to verify that the signal in Figure 3.4 has infinite energy.

3.24 Verify that $x(t) = 1/t$, for t in $[1, \infty)$ is squared absolutely integrable but not absolutely integrable in $[1, \infty)$. Thus squared absolutely integrable does not imply absolutely integrable.

3.25 Is the lemma in Section 3.6 still valid if we delete the preamble "Let p be an integer."?

3.26 In MATLAB, if $N = 5$, what are $n = 0 : 0.5 : N/2$? What are $n = 0 : N/2$?

3.27 Given $n = 0 : 5$ and $x = [2 \ -1 \ 1.5 \ 3 \ 4 \ 0]$, type in MATLAB `stem(n,x)`, `stem(x,n)`, `stem(x,x)`, `stem(n,x,'.')`, and `stem(n,x,'fill')`. Study their results.

3.28 Given $n = 0 : 5$ and $x = [2 \ -1 \ 1.5 \ 3 \ 4 \ 0]$, type in MATLAB `plot(n,x)`, `plot(n,x,':')`, `plot(n,x,'--')`, and `plot(n,x,'-.')`. Study their results. Do they carry out linear interpolation?

3.29 Given $n = 0 : 5$ and $x = [2 \ -1 \ 1.5 \ 3 \ 4 \ 0]$, type in MATLAB `plot(n,x,'o')`, `plot(n,x,'.')` and study their results. Do they carry out linear interpolation? What is the result of `plot(n,x,n,x,'o')`?

3.30 Given $m = 0 : 2$ and $x = [2 \ -1 \ 1.5 \ 3 \ 4 \ 0]$. In MATLAB what are the results of typing `plot(m,x)`, `plot(m,x(m))`, and `plot(m,x(m+1))`?

3.31 Program 3.1 plots one row of two plots. Modify it so that the two plots are in one column.

3.32 Verify that the function in (2.12) can be expressed as

$$\frac{\sin((t - t_0)/a)}{\pi(t - t_0)} = \frac{1}{a\pi} \text{sinc} \ ((t - t_0)/a)$$

where sinc is defined in (3.25).

Chapter 4

Frequency spectra of CT and DT signals

4.1   Introduction

In this chapter, we introduce the concept of frequency spectra or, simply, spectra for CT signals. This concept is needed in discussing bandwidths of signals in communication, and in determining a sampling period if a CT signal is to be processed digitally. It is also needed in specifying a system to be designed. Thus the concept is important in engineering.

This chapter begins with the definition of frequency using a spinning wheel. Using a spinning wheel, we can define easily positive and negative frequencies. We then introduce the Fourier series for periodic signals and then the Fourier transform for periodic and aperiodic signals. We give reasons for downplaying the former and for stressing the latter. The Fourier transform of a CT signal, if it exists, is defined as the frequency spectrum.

Frequency spectra of most, if not all, real-world signals, such as those in Figure 1.1 through 1.4, cannot be expressed in closed form and cannot be computed analytically. Thus we downplay their analytical computation. Instead we stress their general properties and some concepts which are useful in engineering.

The last part of this chapter discusses the DT counterpart of what has been discussed for CT signals. We first show that the frequency of DT sinusoids cannot be defined directly and is defined through CT sinusoids. We then discuss sampling of pure sinusoids which leads naturally to aliased frequencies and a simplified version of the sampling theorem. Finally we apply the (CT) Fourier transform to modified DT signals to yield frequency spectra of DT signals.

4.1.1   Frequency of CT pure sinusoids

Consider a wheel with mark A as shown in Figure 4.1(a). It is assumed that the wheel rotates with a constant speed. Then the *period*, denoted by P, is defined as the time (in seconds) for the mark A to complete one cycle. The number of cycles the mark A

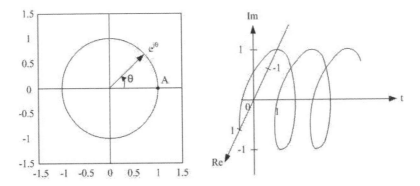

Figure 4.1: (a) Mark A spins on a wheel. (b) Plot of A against time.

rotates in a second is defined as the *frequency* with unit in Hz (cycles per second). Note that the preceding definition holds only if the wheel rotates with a constant speed.

Every point on the wheel, as discussed in Section 3.3, can be denoted by $re^{j\theta}$, where r is the radius of the wheel. Our definition of frequency is independent of r, thus we assume $r = 1$. If the wheel starts to rotate from $t = 0$ onward, then the angle or phase θ will increase with time. If the wheel rotates with a constant angular speed ω_0 (rad/s) in the counterclockwise direction, then we have

$$\theta = \omega_0 t$$

and the rotation of mark A can be expressed as $x(t) = e^{j\omega_0 t}$. We call $e^{j\omega_0 t}$ a *complex exponential* function. Because one cycle has 2π radians, we have $P\omega_0 = 2\pi$ or

$$P = \frac{2\pi}{\omega_0}$$

If we use f_0 to denote the frequency, then we have $f_0 = 1/P$ with unit in Hz (cycles per second). One cycle has 2π radians, thus the frequency can also be defined as

$$\omega_0 := 2\pi f_0 = \frac{2\pi}{P} \tag{4.1}$$

with unit in rad/s. In mathematical equations, all frequencies must be expressed in the unit of rad/s.

When A rotates on a wheel, the independent variable t appears as a parameter on the wheel. Now we plot the rotation of A against time as shown in Figure 4.1(b). Because A is represented by a complex number, its plotting against time requires three dimensions as shown. Such a plot is difficult to visualize. Using Euler's formula

$$e^{j\omega_0 t} = \cos\omega_0 t + j\sin\omega_0 t \tag{4.2}$$

we can plot the imaginary part $\sin\omega_0 t$ against time and the real part $\cos\omega_0 t$ against time as shown in Figure 4.2 with $\omega_0 = 3$ (rad/s). They are respectively the projections of $e^{j\omega_0 t}$ on the imaginary and real axes. They are much easier to visualize. They have period $2\pi/3 = 2.09$ or they repeat every 2.09 seconds.

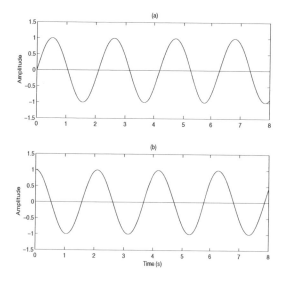

Figure 4.2: (a) $\sin 3t$. (b) $\cos 3t$.

If $\omega_0 > 0$, then $e^{j\omega_0 t}$ or A in Figure 4.1(a) rotates in the counterclockwise direction and $e^{-j\omega_0 t}$ or A rotates in the clockwise direction. Thus we will encounter both positive and negative frequencies. Physically there is no such thing as a negative frequency; the negative sign merely indicates its direction of rotation. In theory, the wheel can spin as fast as desired. Thus we have

$$\text{Frequency range of } e^{j\omega_0 t} = (-\infty, \infty) \tag{4.3}$$

To conclude this section, we mention that if we use $\sin \omega_0 t$ or $\cos \omega_0 t$ to define the frequency, then the meaning of negative frequency is not clear because of $\sin(-\omega_0 t) = -\sin \omega_0 t = \sin(\omega_0 t + \pi)$ and $\cos(-\omega_0 t) = \cos \omega_0 t$. Thus it is best to use a spinning wheel or $e^{j\omega_0 t}$ to define the frequency. However most discussion concerning $e^{j\omega t}$ is directly applicable to $\sin \omega_0 t$ and $\cos \omega_0 t$. For convenience, we call $e^{j\omega t}$, $\sin \omega t$, and $\cos \omega t$ *pure sinusoids*.

4.2 CT periodic signals

A CT signal $x(t)$ is said to be *periodic* with *period* P if $x(t) = x(t + P)$ for all t in $(-\infty, \infty)$. Note that if $x(t) = x(t + P)$, for all t, then we have $x(t) = x(t + 2P) = x(t+3P) = \cdots$, for all t. The smallest such P, denoted by P_0, is called the *fundamental period* and $\omega_0 = 2\pi/P_0$ is called the *fundamental frequency* of $x(t)$. If $x(t)$ is periodic with fundamental period P_0, then it will repeat itself every P_0 seconds. If a signal never repeat itself or $P_0 = \infty$, then it is said to be *aperiodic*. For example, the signals in Figures 1.1 through 1.3 are aperiodic. The clock signal shown in Figure 2.5 is periodic with fundamental period P as shown. The signals $\sin 3t$ and $\cos 3t$ shown in Figure 4.2 are periodic with fundamental period $2\pi/3 = 2.09$ seconds and fundamental frequency $\omega_0 = 3$ rad/s. For pure sinusoids, there is no difference between frequency and fundamental frequency.

Let $x_i(t)$, for $i = 1, 2$, be periodic with fundamental period P_{0i} and fundamental frequency $\omega_{0i} = 2\pi/P_{0i}$. Is their linear combination $x(t) = a_1 x_1(t) + a_2 x_2(t)$, for any constants a_1 and a_2 periodic? The answer is affirmative if $x_1(t)$ and $x_2(t)$ have a common period. That is, if there exist integers k_1 and k_1 such that $k_1 P_{01} = k_2 P_{02}$ or

$$\frac{P_{01}}{P_{02}} = \frac{\omega_{02}}{\omega_{01}} = \frac{k_2}{k_1} \tag{4.4}$$

then $x(t)$ is periodic. In words, if the ratio of the two fundamental periods or fundamental frequencies can be expressed as a ratio of two integers, then $x(t)$ is periodic. We give examples.

Example 4.2.1 Is $x(t) = \sin 3t - \cos \pi t$ periodic? The periodic signal $\sin 3t$ has fundamental frequency 3 and $\cos \pi t$ has fundamental frequency π. Because π is an irrational number, $3/\pi$ cannot be expressed as a ratio of two integers. Thus $x(t)$ is not periodic.□

Example 4.2.2 Consider

$$x(t) = 1.6 \sin 1.5t + 3 \cos 2.4t \tag{4.5}$$

Clearly we have

$$\frac{\omega_{02}}{\omega_{01}} = \frac{2.4}{1.5} = \frac{24}{15} = \frac{8}{5} =: \frac{k_1}{k_2} \tag{4.6}$$

Thus the ratio of the two fundamental frequencies can be expressed as a ratio of two integers 24 and 15, and $x(t)$ is periodic. □

If $x(t)$ in (4.5) is periodic, what is its fundamental period or fundamental frequency? To answer this question, we introduce the concept of *coprimeness*. Two integers are said to be coprime, if they have no common integer factor other than 1. For example, 24 and 15 are not coprime, because they have integer 3 as their common factor. After canceling the common factor, the pair 8 and 5 becomes coprime. Now if we require k_2 and k_1 in (4.4) to be coprime, then the fundamental period of $x(t)$ is $P_0 = k_1 P_{01} = k_2 P_{02}$ and the fundamental frequency is

$$\omega_0 = \frac{2\pi}{P_0} = \frac{2\pi}{k_1 P_{01}} = \frac{2\pi}{k_2 P_{02}} = \frac{\omega_{01}}{k_1} = \frac{\omega_{02}}{k_2} \tag{4.7}$$

We discuss how to use (4.7) to compute the fundamental frequency.

A real number α is called a *divisor* of β if β/α is an integer. Thus β has divisors β/n, for all positive integer n. For example, 2.4 has the following divisors

$$2.4, \ 1.2, \ 0.8, \ 0.6, \ 0.48, \ 0.4, \ 0.3, \ 0.2, \ 0.1, \ 0.01, \ \ldots \tag{4.8}$$

and 1.5 has the following divisors

$$1.5, \ 0.75, \ 0.5, \ 0.375, \ 0.3, \ 0.25, \ 0.1, \ 0.01, \ 0.001, \ \ldots \tag{4.9}$$

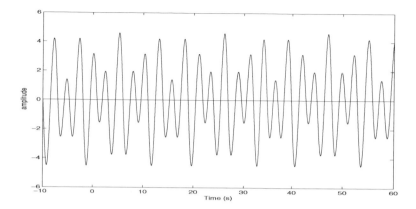

Figure 4.3: $x(t) = 1.6 \sin 1.5t + 3 \cos 2.4t$.

From (4.8) and (4.9), we see that 2.4 and 1.5 have many common divisors. The largest one is called the *greatest common divisor (gcd)*. From (4.8) and (4.9), we see that the gcd of 2.4 and 1.5 is 0.3.[1]

Because k_1 and k_2 in (4.7) are integers, ω_0 is a common divisor of ω_{01} and ω_{02}. Now if k_1 and k_2 are coprime, then ω_0 is the gcd of ω_{01} and ω_{02}. For the signal in (4.5), its fundamental frequency is the gcd of 1.5 and 2.4 or $\omega_0 = 0.3$ and its fundamental period is $P_0 = 2\pi/0.3 = 20.9$. We plot in Figure 4.3 the signal in (4.5) for t in $[-10, 60]$. From the plot, we can see that its fundamental period is indeed 20.9 seconds.

4.2.1 Fourier series of CT periodic signals

We introduced in the preceding section the fundamental frequency. It turns out that a periodic signal with fundamental frequency ω_0 can be expressed as a linear combination of pure sinusoids with frequencies $m\omega_0$, for some or all integers m. Such an expression is called a Fourier series. We mention that a pure sinusoids with frequency $m\omega_0$ is called an mth *harmonic*. It is called *fundamental* if $m = 1$ and *dc* if $m = 0$. Note that a dc is just a constant, associated with frequency zero. Before giving a formal definition, we give an example.

Consider the periodic signal in (4.5). Its fundamental frequency, as discussed earlier, is $\omega_0 = 0.3$. Using $\omega_0 = 0.3$, we can write (4.5) as

$$x(t) = 1.6 \sin 5\omega_0 t + 3 \cos 8\omega_0 t \tag{4.10}$$

Indeed, the periodic signal in Figure 4.3 is a linear combination of its fifth and eighth harmonics. It contains no fundamental nor dc.

Equation (4.10) is one possible form of Fourier series. Using $\sin(\theta + \pi/2) = \sin(\theta + 1.57) = \cos\theta$, we can write (4.10) as

$$x(t) = 1.6 \sin(5\omega_0 t + 0) + 3 \sin(8\omega_0 t + 1.57) \tag{4.11}$$

---

[1]The gcd of two real numbers can also be obtained using the Euclidean algorithm with integer quotients.

This is another form of Fourier series. Using $\cos(\theta - \pi/2) = \sin\theta$, we can write (4.10) also as

$$x(t) = 1.6\cos(5\omega_0 t - 1.57) + 3\cos(8\omega_0 t + 0) \tag{4.12}$$

This is yet another form of Fourier series.

Using Euler's formulas

$$\sin\theta = \frac{e^{j\theta} - e^{-j\theta}}{2j} \qquad \cos\theta = \frac{e^{j\theta} + e^{-j\theta}}{2}$$

we can write (4.10) as

$$
\begin{aligned}
x(t) &= 1.6\frac{e^{j5\omega_0 t} - e^{-j5\omega_0 t}}{2j} + 3\frac{e^{j\omega_0 t} + e^{-j8\omega_0 t}}{2} \\
&= -0.8je^{j5\omega_0 t} + 0.8je^{-j5\omega_0 t} + 1.5e^{j8\omega_0 t} + 1.5e^{-j8\omega_0 t} \tag{4.13}
\end{aligned}
$$

This is also a Fourier series.

In conclusion, a Fourier series may assume many forms; it may use sine and cosine functions, exclusively sine functions, exclusively cosine functions, or complex exponentials. The first three forms involve only positive frequencies and real numbers and are called *real Fourier series*. The last form in (4.13) however involves positive and negative frequencies and complex numbers and is called a *complex Fourier series*. Even though the complex Fourier series is most complex, it can be easily modified to yield the Fourier transform. Thus we discuss only the latter. The discussion will be brief because we use mainly the Fourier transform in this text.

Consider a periodic signal $x(t)$ with fundamental period P_0 and fundamental frequency $\omega_0 = 2\pi/P_0$. Then it can be be expressed in *complex Fourier series* as

$$x(t) = \sum_{m=-\infty}^{\infty} c_m e^{jm\omega_0 t} \tag{4.14}$$

where m is an integer ranging from $-\infty$ to ∞ and is called the *frequency index*. The number c_m, called *Fourier coefficients*, can be computed from

$$c_m = \frac{1}{P_0}\int_{t=-P_0/2}^{P_0/2} x(t)e^{-jm\omega_0 t}dt = \frac{1}{P_0}\int_{t=0}^{P_0} x(t)e^{-jm\omega_0 t}dt \tag{4.15}$$

See Problem 4.38. Note that the preceding integration can be carried out over any one period. For example, the periodic signal in Figure 4.3 with fundamental frequency $\omega_0 = 0.3$ has its complex Fourier series in (4.13) with $c_5 = -0.8j$, $c_8 = 1.5$, $c_{(-5)} = 0.8j$, $c_{(-8)} = 1.5$ and $c_m = 0$ for all integers m other than ± 5 and ± 8. Note that these coefficients can be computed using (4.15). Because $x(t)$ is already expressed in terms of pure sinusoids, the c_m can be obtained more easily using Euler's equation as we did in (4.13).

Example 4.2.3 (Sampling function) Consider the periodic signal shown in Figure 4.4. It consists of a sequence of impulses with weight 1 as shown. The sequence extends

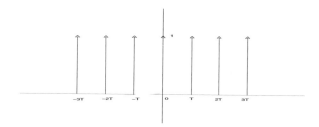

Figure 4.4: A sequence of impulses.

all the way to $\pm\infty$. The impulse at $t = 0$ can be denoted by $\delta(t)$, the impulses at $t = \pm T$ can be denoted by $\delta(t \mp T)$ and so forth. Thus the signal can be expressed as

$$r(t) := \sum_{n=-\infty}^{\infty} \delta(t - nT) \tag{4.16}$$

The signal is periodic with fundamental period $P_0 - T$ and fundamental frequency $\omega_0 = 2\pi/T =: \omega_s$. We call the signal $r(t)$ the *sampling function* for reasons to be given in the next chapter.

Using (4.14) we can express $r(t)$ as

$$r(t) = \sum_{m=-\infty}^{\infty} c_m e^{jm\omega_s t}$$

where, using (4.15) and (2.14),

$$c_m = \frac{1}{T} \int_{t=-T/2}^{T/2} \delta(t)e^{-jm\omega_s t} dt = \frac{1}{T} \left. e^{-jm\omega_s t}\right|_{t=0} = \frac{1}{T}$$

for all m. Thus the complex Fourier series of the periodic signal $r(t)$ is, with $\omega_s = 2\pi/T$,

$$r(t) = \frac{1}{T} \sum_{m=-\infty}^{\infty} e^{jm\omega_s t} \tag{4.17}$$

Thus the periodic signal in Figure 4.4 can be expressed as in (4.16) and as in (4.17) or

$$r(t) = \sum_{n=-\infty}^{\infty} \delta(t - nT) = \frac{1}{T} \sum_{m=-\infty}^{\infty} e^{jm\omega_s t} \tag{4.18}$$

This equation will be used in the next chapter. $\square$

A great deal more can be said about the Fourier series. However we will skip the discussion for reasons to be given shortly.

4.3   Frequency spectra of CT aperiodic signals

The CT Fourier series is developed for CT periodic signals. Now we modify it so that the resulting equation is applicable to aperiodic signals as well. Note that a periodic signal becomes aperiodic if its fundamental period P_0 becomes ∞.

Let us multiply P_0 to (4.15) to yield

$$X(m\omega_0) := P_0 c_m = \int_{t=-P_0/2}^{P_0/2} x(t)e^{-jm\omega_0 t}dt \tag{4.19}$$

where we have defined $X(m\omega_0) := P_0 c_m$. This is justified by the fact that c_m is associated with frequency $m\omega_0$. We then use $X(m\omega_0)$ and $\omega_0 = 2\pi/P_0$ or $1/P_0 = \omega_0/2\pi$ to write (4.14) as

$$x(t) = \frac{1}{P_0}\sum_{m=-\infty}^{\infty} P_0 c_m e^{jm\omega_0 t} = \frac{1}{2\pi}\sum_{m=-\infty}^{\infty} X(m\omega_0)e^{jm\omega_0 t}\omega_0 \tag{4.20}$$

A periodic signal with period P_0 becomes aperiodic if P_0 approaches infinity. Define $\omega := m\omega_0$. If $P_0 \to \infty$, then we have $\omega_0 = 2\pi/P_0 \to 0$. Thus, as $P_0 \to \infty$, $\omega = m\omega_0$ becomes a continuous variable and ω_0 can be written as $d\omega$. Furthermore, the summation in (4.20) becomes an integration. Thus the modified Fourier series pair in (4.19) and (4.20) becomes, as $P_0 \to \infty$,

$$X(\omega) = \mathcal{F}[x(t)] := \int_{t=-\infty}^{\infty} x(t)e^{-j\omega t}dt \tag{4.21}$$

for all ω in $(-\infty, \infty)$, and

$$x(t) = \mathcal{F}^{-1}[X(\omega)] := \frac{1}{2\pi}\int_{\omega=-\infty}^{\infty} X(\omega)e^{j\omega t}d\omega \tag{4.22}$$

for all t in $(-\infty, \infty)$. The set of two equations is the *CT Fourier transform* pair. $X(\omega)$ is the Fourier transform of $x(t)$ and $x(t)$ is the inverse Fourier transform of $X(\omega)$. Before proceeding, we mention that the CT Fourier transform is a linear operator in the sense that

$$\mathcal{F}[a_1 x_1(t) + a_2 x_2(t)] = a_1\mathcal{F}[x_1(t)] + a_2\mathcal{F}[x_2(t)]$$

for any constants a_1 and a_2. This can be directly verified from the definition in (4.21).

The Fourier transform is applicable to real- or complex-valued $x(t)$ defined for all t in $(-\infty, \infty)$. However all examples will be real valued and defined only for positive time unless stated otherwise.

Example 4.3.1 Consider the positive-time function

$$x(t) = e^{2t}$$

for $t \geq 0$. Its Fourier transform is

$$\begin{aligned} X(\omega) &= \int_{t=-\infty}^{\infty} x(t)e^{-j\omega t}dt = \int_{t=0}^{\infty} e^{2t}e^{-j\omega t}dt \\ &= \frac{1}{2-j\omega}e^{(2-j\omega)t}\Big|_{t=0}^{\infty} = \frac{1}{2-j\omega}\left[e^{(2-j\omega)t}\Big|_{t=\infty} - 1\right] \end{aligned}$$

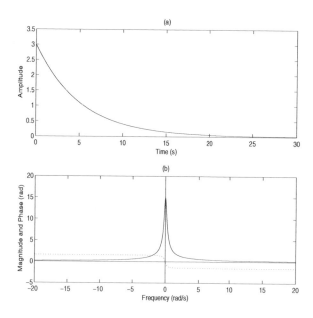

Figure 4.5: (a) Positive-time function $3e^{-0.2t}$. (b) Its magnitude spectrum (solid line) and phase spectrum (dotted line).

where we have used $e^{(2-j\omega)\times 0} = e^0 = 1$. Because the function

$$e^{(2-j\omega)t} = e^{2t}e^{-j\omega t} = e^{2t}[\cos \omega t - j\sin \omega t]$$

grows unbounded as $t \to \infty$, the value of $e^{(2-j\omega)t}$ at $t = \infty$ is not defined. Thus the Fourier transform of e^{2t} is not defined or does not exist.□

Example 4.3.2 Consider $x(t) = 1.5\delta(t - 0.8)$. It is an impulse at $t_0 = 0.8$ with weight 1.5. Using the sifting property of impulses, we have

$$X(\omega) = \int_{t=-\infty}^{\infty} 1.5\delta(t - 0.8)e^{-j\omega t}dt = 1.5 \left. e^{-j\omega t}\right|_{t=0.8} = 1.5e^{-j0.8\omega} \qquad (4.23)$$

□

Example 4.3.3 Consider the positive-time signal

$$x(t) = \begin{cases} 3e^{-0.2t} & \text{for } t \geq 0 \\ 0 & \text{for } t < 0 \end{cases} \qquad (4.24)$$

plotted in Figure 4.5(a). It vanishes exponentially to zero as shown. Its Fourier transform can be computed as

$$\begin{aligned} X(\omega) &= \int_{t=-\infty}^{\infty} x(t)e^{-j\omega t}dt = \int_{t=0}^{\infty} 3e^{-0.2t}e^{-j\omega t}dt \\ &= \frac{3}{-0.2 - j\omega} \left. e^{(-0.2-j\omega)t}\right|_{t=0}^{\infty} = \frac{3}{-0.2 - j\omega} \left[\left. e^{(-0.2-j\omega)t}\right|_{t=\infty} - 1 \right] \end{aligned}$$

Because $e^{-0.2t}e^{-j\omega t}$, for any ω, equals zero as $t \to \infty$, we have

$$X(\omega) = \frac{3}{-0.2 - j\omega}[0 - 1] = \frac{3}{0.2 + j\omega} \qquad (4.25)$$

for all ω in $(-\infty, \infty)$. This is the Fourier transform of the positive-time signal $3e^{-0.2t}$.
□

Example 4.3.1 shows that not every signal has a Fourier transform. In general, if a signal grows to ∞ or $-\infty$ as $t \to \pm\infty$, then its Fourier transform does not exist. In addition to be of bounded variation (see Section 2.5.2), if a signal is (1) absolutely integrable (see Section 3.5), or (2) squared absolutely integrable (see Section 3.5.1 and Problem 4.37), or (3) periodic (see Section 4.5), then its Fourier transform exists. These three sufficient conditions are mutually exclusive. We discuss here only the first condition. Bounded variation and absolute integrability are called the *Dirichlet* conditions and are quoted in most texts. If a signal $x(t)$ meets the Dirichlet conditions, then its Fourier transform $X(\omega)$ exists. Moreover $X(\omega)$ is bounded and is a continuous function of ω. Indeed, (4.21) implies

$$\begin{aligned} |X(\omega)| &= \left| \int_{t=-\infty}^{\infty} x(t)e^{-j\omega t}\,dt \right| \le \int_{t=-\infty}^{\infty} |x(t)||e^{-j\omega t}|\,dt \\ &= \int_{t=-\infty}^{\infty} |x(t)|\,dt < M < \infty \end{aligned}$$

for some constant M. The proof of continuity of $X(\omega)$ is more complex. See Reference [C7, pp. 92-93]. If $X(\omega)$ for a $x(t)$ exists, it is called the *frequency spectrum* or, simply, the *spectrum* of $x(t)$.

The impulse in Example 4.3.2 and the signal in Example 4.3.3 meet the Dirichlet conditions. Their Fourier transforms or spectra exist as derived in (4.23) and (4.25). They are indeed bounded and continuous. All real-world signals are absolutely integrable, as discussed in Subsection 3.5.1, and of bounded variation. Thus every real-world signal has a bounded and continuous Fourier transform or spectrum.

The signal shown in Figure 4.5(a) can be described by $x(t)$ in (4.24) or by $X(\omega)$ in (4.25). Because $x(t)$ is a function of time t, it is called the *time-domain* description of the signal. Because $X(\omega)$ is a function of frequency ω, it is called the *frequency-domain* description. Mathematically speaking, $x(t)$ and $X(\omega)$ contain the same amount of information and either one can be obtained from the other. However $X(\omega)$ will reveal explicitly the distribution of energy of a signal in frequencies as we will discuss in a later section.

Even though a signal is real-valued and positive-time ($x(t) = 0$, for all $t < 0$), its spectrum $X(\omega)$ is generally complex-valued and defined for all ω in $(-\infty, \infty)$. For a complex $X(\omega)$, we can express it as

$$X(\omega) = A(\omega)e^{j\theta(\omega)} \qquad (4.26)$$

where $A(\omega) := |X(\omega)| \ge 0$ is the magnitude of $X(\omega)$ and is called the *magnitude spectrum*. The function $\theta(\omega) := \angle X(\omega)$ is the phase of $X(\omega)$ and is called the *phase spectrum.* Note that both $A(\omega)$ and $\theta(\omega)$ are real-valued functions of real ω. For example, for the spectrum in (4.25), using $3 = 3^{j0}$ and

$$0.2 + j\omega = \sqrt{(0.2)^2 + \omega^2}e^{j\tan^{-1}(\omega/0.2)}$$

we can write it as

$$X(\omega) = \frac{3}{0.2 + j\omega} = \frac{3}{\sqrt{0.04 + \omega^2}}e^{j(0 - \tan^{-1}(\omega/0.2))}$$

Thus the magnitude spectrum of the signal in (4.24) is

$$A(\omega) := |X(\omega)| = \frac{3}{\sqrt{0.04 + \omega^2}} \tag{4.27}$$

and its phase spectrum is

$$\theta(\omega) := -\tan^{-1}(\omega/0.2) \tag{4.28}$$

We next discuss a general property of spectra of real-valued signals. A signal $x(t)$ is real-valued if its complex-conjugate equals itself, that is, $x(t) = x^*(t)$. Substituting (4.26) into (4.21), taking their complex conjugate, using $x^*(t) = x(t)$, $A^*(\omega) = A(\omega)$, $\theta^*(\omega) = \theta(\omega)$, and $(e^{j\theta})^* = e^{-j\theta}$, we have

$$\left[A(\omega)e^{j\theta(\omega)}\right]^* = \left[\int_{t=-\infty}^{\infty} x(t)e^{-j\omega t}dt\right]^*$$

or

$$A(\omega)e^{-j\theta(\omega)} = \int_{t=-\infty}^{\infty} x(t)e^{j\omega t}dt$$

See Section 3.2 and Problem 3.8. The right-hand side of the preceding equation equals $X(-\omega)$. See Section 3.4. Thus we have

$$A(\omega)e^{-j\theta(\omega)} = X(-\omega) = A(-\omega)e^{j\theta(-\omega)}$$

In conclusion, if $x(t)$ is real, we have $A(\omega) = A(-\omega)$ and $\theta(\omega) = -\theta(-\omega)$ or

$$\begin{align} |X(-\omega)| &= |X(\omega)| \quad \text{(even)} \\ \measuredangle X(-\omega) &= -\measuredangle X(\omega) \quad \text{(odd)} \end{align} \tag{4.29}$$

Indeed, $A(\omega)$ in (4.27) is even because

$$A(-\omega) = \frac{3}{\sqrt{0.04 + (-\omega)^2}} = \frac{3}{\sqrt{0.04 + \omega^2}} = A(\omega)$$

and $\theta(\omega)$ in (4.28) is odd because

$$\theta(-\omega) := -\tan^{-1}(-\omega/0.2) = +\tan^{-1}(\omega/0.2) = -\theta(\omega)$$

If $X(\omega)$ is complex, its plot against frequencies requires three dimensions and is difficult to visualize. We may plot its real part and imaginary part. However, these two plots have no physical meaning. Thus we plot its magnitude and phase against frequency. For the spectrum in (4.25) or $X(\omega) = 3/(0.2 + j\omega)$, we compute

$$\omega = 0: \quad X(0) = \frac{3}{0.2} = 15e^{j0}$$

$$\omega = 0.2: \quad X(0.2) = \frac{3}{0.2 + j0.2} = \frac{3}{0.2 \cdot \sqrt{2}e^{j\pi/4}} = 10.6e^{-j\pi/4}$$

$$\omega = 10: \quad X(10) = \frac{3}{0.2 + j10} \approx \frac{3}{j10} = -0.3j = 0.3e^{-j\pi/2}$$

From these, we can plot in Figure 4.5(b) its magnitude spectrum (solid line) and phase spectrum (dotted line). The more ω we compute, the more accurate the plot. Note that the phase approaches $\mp\pi/2 = \mp1.57$ rad as $\omega \to \pm\infty$. Figure 4.5(b) is actually obtained using MATLAB by typing

```
% Subprogram 4.1
w=-10:0.001:10;
X=3.0./(j*w+0.2);
plot(w,abs(X),w,angle(X),':')
```

The first line generates ω from -10 to 10 with increment 0.001. Note that every statement in MATLAB is executed and stored in memory. A semicolon (;) at its end suppresses its display on the monitor. The second line is (4.25). Note the use of dot division (./) for element by element division in MATLAB. If we use division without dot (/), then an error message will appear. The MATLAB function **abs** computes absolute value or magnitude and **angle** computes angle or phase. We see that using MATLAB to plot frequency spectra is very simple. As shown in Figure 4.5(b), the magnitude spectrum is an even function of ω, and the phase spectrum is an odd function of ω.

4.3.1   Why we stress Fourier transform and downplay Fourier series

We compare the Fourier series and the Fourier transform.

1. Consider a periodic signal $x(t) = x(t + P_0)$, where P_0 is the fundamental period. Then it can be expressed in the Fourier series as in (4.14) and (4.15). The expression is applicable for all t in $(-\infty, \infty)$. However if we specify the applicable time interval, then the Fourier series is applicable to any segment of the periodic signal.

   Consider a signal $x(t)$ that is identically zero outside the time interval $[-L, L]$, for some finite L. The signal inside $[-L, L]$ is entirely arbitrary. For example, it could be zero for all t in $[-14, 0)$ and assumes the waveform in Figure 1.2 for t in $[0, 14]$. If we mentally extend $x(t)$ periodically to $(-\infty, \infty)$ with period $P = 2L$, then it can be described by the following Fourier series

$$x(t) = \sum_{m=-\infty}^{\infty} c_m e^{jm\omega_0 t} \tag{4.30}$$

with $\omega_0 = 2\pi/P = \pi/L$ and

$$c_m = \frac{1}{2L} \int_{t=-L}^{L} x(t)e^{-jm\omega_0 t}dt \qquad (4.31)$$

Note that (4.30) holds only for t in $[-L, L]$. For the same signal $x(t)$, its Fourier transform is

$$X(\omega) = \int_{t=-L}^{L} x(t)e^{-j\omega t}dt \qquad (4.32)$$

Comparing (4.31) and (4.32) yields

$$c_m = X(m\omega_0)/2L \qquad (4.33)$$

for all m. In other words, the Fourier coefficients are simply the samples of the Fourier transform divided by $2L$ or $X(\omega)/2L$. On the other hand, the Fourier transform $X(\omega)$ can be computed from the Fourier coefficients as

$$X(\omega) = \sum_{m=-\infty}^{\infty} c_m \frac{2\sin[(\omega - m\omega_0)L]}{\omega - m\omega_0} \qquad (4.34)$$

for all ω. See Problem 4.18. Thus for a signal of a finite duration, there is a one-to-one correspondence between its Fourier series and Fourier transform.

For a signal of infinite duration, the Fourier series is applicable only if the signal is periodic. However the Fourier transform is applicable whether the signal is periodic or not. Thus the Fourier transform is more general than the Fourier series.

2. There is a unique relationship between a time function $x(t)$ for all t in $(-\infty, \infty)$ and its Fourier transform $X(\omega)$ for all ω in $(-\infty, \infty)$. If $x(t)$ is identically zero for some time interval, the information will be imbedded in $X(\omega)$ and the inverse Fourier transform of $X(\omega)$ will yield zero in the same time interval. For example, the Fourier transforms of a periodic signal of one period, two periods and ten periods are all different and their inverse Fourier transforms will yield zero outside one period, two periods, or ten periods. The Fourier series of one period, two periods, or ten periods are all the same. Its applicable time interval is specified separately.

3. The Fourier transform is a function of ω alone (does not contain explicitly t) and describes fully a time signal. Thus it is the frequency-domain description of the time signal. The Fourier coefficients cannot be used alone, they make sense only in the expression in (4.14) which is still a function of t. Thus the Fourier series is, strictly speaking, not a frequency-domain description. However, many texts consider the Fourier series to be a frequency-domain description and call its coefficients frequency spectrum.

4. The basic result of the Fourier series is that a periodic signal consists of fundamental and harmonics. Unfortunately, the expression remains the same no

matter how long the periodic signal is. The Fourier transform of a periodic signal will also exhibit fundamental and harmonics. Moreover, the longer the periodic signal, the more prominent the harmonics. See Problem 5.14 and Figure 5.19 of the next chapter.

In conclusion, the Fourier transform is a better description than the Fourier series in describing both periodic and aperiodic signals. Thus we downplay the Fourier series. The only result in the Fourier series that we will use in this text is the Fourier series of the sampling function discussed in Example 4.2.3.

4.4   Distribution of energy in frequencies

Every real-world signal satisfies, as discussed in Subsections 3.5.1 and 2.5.2 the Dirichlet conditions, thus its frequency spectrum is well defined. Furthermore the spectrum is bounded and continuous. Even so, its analytical computation is generally not possible. For example, there is no way to compute analytically the spectra of the signals in Figures 1.1 and 1.2. However they can be easily computed numerically. See the next chapter. Thus we will not discuss further analytical computation of spectra. Instead we discuss their physical meaning and some properties which are useful in engineering.

As discussed in Section 3.5.1, the total energy of $x(t)$ defined for t in $(-\infty, \infty)$ is

$$E = \int_{t=-\infty}^{\infty} x(t)x^*(t)dt = \int_{t=-\infty}^{\infty} |x(t)|^2 dt \qquad (4.35)$$

Not every signal has finite total energy. For example, if $x(t) = 1$, for all t, then its total energy is infinite. Sufficient conditions for $x(t)$ to have finite total energy are, as discussed in Section 3.5.1, $x(t)$ is bounded and absolutely integrable. All real-world signals have finite total energy.

It turns out that the total energy of $x(t)$ can also be expressed in terms of the spectrum of $x(t)$. Substituting (4.22) into (4.35) yields

$$E = \int_{t=-\infty}^{\infty} x(t)[x(t)]^* dt = \int_{t=-\infty}^{\infty} x(t) \left[\frac{1}{2\pi} \int_{\omega=-\infty}^{\infty} X^*(\omega)e^{-j\omega t}d\omega \right] dt$$

which becomes, after interchanging the order of integrations,

$$E = \frac{1}{2\pi} \int_{\omega=-\infty}^{\infty} X^*(\omega) \left[\int_{t=-\infty}^{\infty} x(t)e^{-j\omega t}dt \right] d\omega$$

The term inside the brackets is $X(\omega)$. Thus we have, using $X^*(\omega)X(\omega) = |X(\omega)|^2$,

$$\int_{t=-\infty}^{\infty} |x(t)|^2 dt = \frac{1}{2\pi} \int_{\omega=-\infty}^{\infty} |X(\omega)|^2 d\omega = \frac{1}{\pi} \int_{\omega=0}^{\infty} |X(\omega)|^2 d\omega \qquad (4.36)$$

where we have used the evenness of $|X(\omega)|$. Equation (4.36) is called a *Parseval's formula*. Note that the energy of $x(t)$ is independent of its phase spectrum. It depends only on its magnitude spectrum.

Equation (4.36) shows that the total energy of $x(t)$ can also be computed from its magnitude spectrum. More important, the magnitude spectrum reveals the distribution of the energy of $x(t)$ in frequencies. For example, the energy contained in the frequency range $[\omega_1, \omega_2]$ can be computed from

$$\frac{1}{2\pi} \int_{\omega=\omega_1}^{\omega_2} |X(\omega)|^2 d\omega \tag{4.37}$$

We mention that if the width of $[\omega_1, \omega_2]$ approaches zero and if $X(\omega)$ contains no impulses, then (4.37) is always zero. See (2.11). Thus for a real-world signal, it is meaningless to talk about its energy at an isolated frequency. Its energy is nonzero only over a nonzero frequency interval. For the signal in Figure 4.5(a), even though its energy is distributed over frequencies in $(-\infty, \infty)$, most of its energy is contained in the frequency range $[-5, 5]$ in rad/s as we can see from Figure 4.5(b).

4.4.1 Frequency Shifting and modulation

Let $x(t)$ be a CT signal with spectrum $X(\omega)$. Then we have

$$\begin{aligned} \mathcal{F}[e^{j\omega_0 t} x(t)] &= \int_{-\infty}^{\infty} e^{j\omega_0 t} x(t) e^{-j\omega t} dt - \int_{-\infty}^{\infty} x(t) e^{-j(\omega-\omega_0)t} dt \\ &= X(\omega - \omega_0) \end{aligned} \tag{4.38}$$

This is called *frequency shifting* because the frequency spectrum of $x(t)$ is shifted to ω_0 if $x(t)$ is multiplied by $e^{j\omega_0 t}$.

Consider a CT signal $x(t)$. Its multiplication by $\cos \omega_c t$ is called *modulation*. To be more specific, we call

$$x_m(t) := x(t) \cos \omega_c t$$

the *modulated signal*; $\cos \omega_c t$, the *carrier signal*; ω_c, the *carrier frequency*; and $x(t)$, the *modulating signal*. For example, if $x(t) = 3e^{-0.2t}$, for $t \geq 0$ as in Figure 4.5(a), and if $\omega_c = 1$, then $x_m(t) = x(t) \cos t$ is as shown in Figure 4.6(a). Note that because, using $|\cos \omega_c t| \leq 1$,

$$|x_m(t)| = |x(t)||\cos \omega_c t| \leq |x(t)|$$

the plot of $x_m(t)$ is bounded by $\pm x(t)$ as shown in Figure 4.6(a) with dashed lines. We plot in Figure 4.6(b) $x_m(t)$ for $\omega_c = 10$.

We now compute the spectrum of $x_m(t)$. Using Euler's formula, the linearity of the Fourier transform, and (4.38), we have

$$\begin{aligned} X_m(\omega) &:= \mathcal{F}[x_m(t)] = \mathcal{F}[x(t) \cos \omega_c t] = \mathcal{F}\left[x(t) \frac{e^{j\omega_c t} + e^{-j\omega_c t}}{2}\right] \\ &= 0.5[X(\omega - \omega_c) + X(\omega + \omega_c)] \end{aligned} \tag{4.39}$$

for all ω in $(-\infty, \infty)$. This is an important equation. We first give an example.

Example 4.4.1 The spectrum of $x(t) = 3e^{-0.2t}$, for $t \geq 0$, as computed in Example 4.3.3, is $X(\omega) = 3/(0.2 + j\omega)$. Thus the spectrum of $x_m(t) = x(t) \cos \omega_c t$ is

$$X_m(\omega) = 0.5 \left[\frac{3}{0.2 + j(\omega - \omega_c)} + \frac{3}{0.2 + j(\omega + \omega_c)} \right]$$

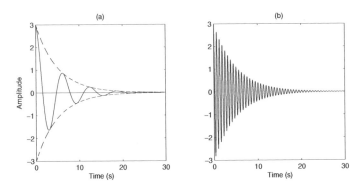

Figure 4.6: (a) $3e^{-0.2t}\cos 2t$. (b) $3e^{-0.2t}\cos 10t$.

$$= \quad 1.5\left[\frac{1}{(0.2+j\omega)-j\omega_c}+\frac{1}{(0.2+j\omega)+j\omega_c}\right]$$

$$= \quad 1.5\left[\frac{0.2+j\omega+j\omega_c+0.2+j\omega-j\omega_c}{(0.2+j\omega)^2-(j\omega_c)^2}\right]$$

or

$$X_m(\omega) \quad = \quad \frac{3(0.2+j\omega)}{(0.2+j\omega)^2+\omega_c^2} \tag{4.40}$$

We plot in Figure 4.7(a) $|X_m(\omega)|$ (solid line), $0.5|X(\omega-\omega_c)|$ (dotted line), and $0.5|X(\omega-\omega_c)|$ (dot-and-dashed line) for $\omega_c=1$. We plot in Figure 4.7(b) the corresponding phase spectra. We repeat the plots in Figures 4.8 for $\omega_c=10$.□

Recovering $x(t)$ from $x_m(t)$ is called *demodulation*. In order to recover $x(t)$ from $x_m(t)$, the waveform of $X_m(\omega)$ must contain the waveform of $X(\omega)$. Note that $X_m(\omega)$ is the sum of $0.5X(\omega-\omega_c)$ and $0.5X(\omega+\omega_c)$. The two shifted $0.5X(\omega\pm\omega_c)$ with $\omega_c=1$ overlap significantly as shown in Figures 4.7, thus $X_m(\omega)$ do not contain the waveform of $X(\omega)$ shown in Figure 4.5(b). On the other hand, for $\omega_c=10$, the overlapping of $0.5X(\omega\pm\omega_c)$ is negligible as in Figure 4.8, and we have $2X_m(\omega)\approx X(\omega-10)$ for ω in $[0,20]$.

We now discuss a condition for the overlapping to be negligible. Consider a signal with spectrum $X(\omega)$. We select ω_{max} such that $|X(\omega)|\approx 0$ for all $|\omega|>\omega_{max}$. In other words, most nonzero magnitude spectrum or most energy of the signal lies inside the frequency range $[-\omega_{max},\omega_{max}]$. Now if we select the carrier frequency ω_c to meet

$$\omega_c > \omega_{max} \tag{4.41}$$

then the overlapping of $X(\omega\pm\omega_c)$ will be negligible. Under the condition, it is possible to recover $X(\omega)$ from $X_m(\omega)$ and, consequently, to recover $x(t)$ from $x_m(t)$.

Modulation and demodulation can be easily explained and understood using the concept of spectra. It is not possible to do so directly in the time domain. Frequency spectra are also needed in selecting ω_c as in (4.41). Thus the Fourier analysis and its associated concept are essential in the study. However, It is important to mention that

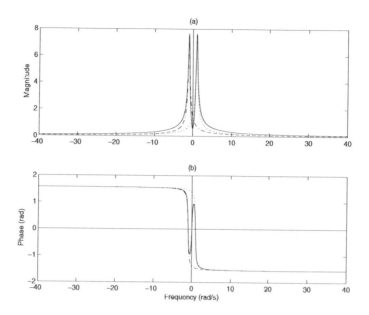

Figure 4.7: (a) Plots of $|X_m(\omega)|$ (solid line), $0.5|X(\omega - \omega_c)|$ (dotted line), and $0.5|X(\omega + \omega_c)|$ (dot and-dashed line) for $\omega_c = 1$. (b) Corresponding phase plots.

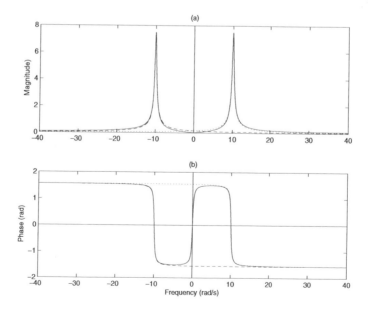

Figure 4.8: Plots of $|X_m(\omega)|$ (solid line), $0.5|X(\omega - \omega_c)|$ (dotted line), and $0.5|X(\omega + \omega_c)|$ (dot-and-dashed line) for $\omega_c = 10$. (b) Corresponding phase plots.

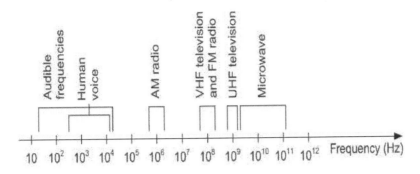

Figure 4.9: Various frequency bands

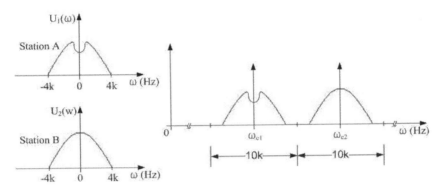

Figure 4.10: Frequency-division multiplexing.

once ω_c is selected, modulation and demodulation are carried out entirely in the time domain.

Modulation and demodulation are basic in communication. In communication, different applications are assigned to different frequency bands as shown in Figure 4.9. For example, the frequency band is limited from 540 to 1600 kHz for AM (amplitude modulation) radio transmission, and from 87.5 to 108 MHz for FM (frequency modulation) transmission. In AM transmission, radio stations are assigned carrier frequencies from 540 to 1600 kHz with 10 kHz increments as shown in Figure 4.10. This is called *frequency-division multiplexing*.

4.4.2   Time-Limited Bandlimited Theorem

In this subsection, we discuss a fundamental theorem between a time signal and its frequency spectrum.[2] A CT signal $x(t)$ is *bandlimited* to W if its frequency spectrum $X(\omega)$ is identically zero for all $|\omega| > W$. It is *time-limited* to b if $x(t)$ is identically zero for all $|t| > b$. It turns out that if a CT signal is time-limited, then it cannot be bandlimited and vice versa. The only exception is the trivial case $x(t) = 0$ for all t. To

---

[2]The derivation may be skipped. But it is important to understand the meaning and implication of the theorem.

establish the theorem, we show that if $x(t)$ is bandlimited to W and time-limited to b, then it must be identically zero for all t. Indeed, if $x(t)$ is bandlimited to W, then (4.22) becomes

$$x(t) = \frac{1}{2\pi} \int_{-W}^{W} X(\omega) e^{j\omega t} d\omega \qquad (4.42)$$

Its differentiation repeatedly with respect to t yields

$$x^{(k)}(t) = \frac{1}{2\pi} \int_{-W}^{W} X(\omega)(j\omega)^k e^{j\omega t} d\omega$$

for $k = 0, 1, 2, \ldots$, where $x^{(k)}(t) := d^k x(t)/dt^k$. Because $x(t)$ is time-limited to b, its derivatives are identically zero for all $|t| > b$. Thus we have

$$\int_{-W}^{W} X(\omega)(\omega)^k e^{j\omega a} d\omega = 0 \qquad (4.43)$$

for any $t = a$ with $a > b$. Next we use

$$e^c = 1 + \frac{c}{1!} + \frac{c^2}{2!} + \cdots = \sum_{k=0}^{\infty} \frac{c^k}{k!}$$

to rewrite (4.42) as

$$\begin{aligned} x(t) &= \frac{1}{2\pi} \int_{-W}^{W} X(\omega) e^{j\omega(t-a)} e^{j\omega a} d\omega \\ &= \frac{1}{2\pi} \int_{-W}^{W} X(\omega) \left[\sum_{k=0}^{\infty} \frac{(j\omega(t-a))^k}{k!} \right] e^{j\omega a} d\omega \\ &= \sum_{k=0}^{\infty} \frac{(j(t-a))^k}{2\pi k!} \int_{-W}^{W} X(\omega)(\omega)^k e^{j\omega a} d\omega \end{aligned}$$

which, following (4.43), is zero for all t. Thus a bandlimited and time-limited CT signal must be identically zero for all t. In conclusion, no nontrivial CT signal can be both time-limited and bandlimited. In other words, if a CT signal is time limited, then its spectrum cannot be bandlimited or its nonzero magnitude spectrum will extend to $\pm\infty$.

Every real-world signal is time limited because it must start and stop somewhere. Consequently its spectrum cannot be bandlimited. In other words, its nonzero magnitude spectrum will extend to $\pm\infty$. However the magnitude must approach zero because the signal has finite total energy.[3] If a magnitude is less than, say 10^{-10}, it is nonzero in mathematics but could be very well considered zero in engineering. Thus every real-world signal will be considered time limited and bandlimited. For examples, the signals in Figures 1.1 through 1.3 are time limited and their frequency spectra, as will be computed in the next chapter, are bandlimited. In fact, the bandlimited property was used in selecting a carrier frequency in the preceding subsection and will be used again in the next subsection and next chapter. Thus in engineering, there is no need to be mathematically exact.

---

[3]Excluding some mathematically constructed magnitude spectra such as the type shown in Figure 3.4, if a spectrum does not approach zero as $|\omega| \to \infty$, then the signal has infinite total energy.

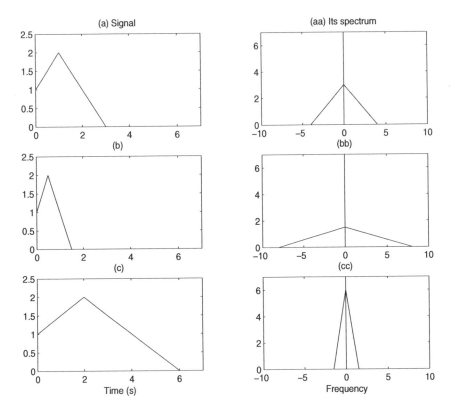

Figure 4.11: (a) Time signal $x(t)$. (aa) Its magnitude spectrum. (b) Time compression $x(2t)$. (bb) Frequency expansion. (c) Time expansion $x(t/2)$. (cc) Frequency compression.

4.4.3   Time duration and frequency bandwidth

Consider the signal $x(t)$ shown in Figure 4.11(a). It starts at $t = 0$ and ends at $t = 3$. Thus it has a time duration of 3 seconds. We plot $x(2t)$ in Figures 4.11(b) and $x(0.5t)$ in 4.11(c). We see that $x(2t)$ lasts only 1.5 seconds, half of the time duration of $x(t)$; whereas $x(0.5t)$ lasts six seconds, twice the time duration of $x(t)$. In general, we have

- $x(at)$ with $a > 1$: time compression or speed up

- $x(at)$ with $0 < a < 1$: time expansion or slow down

Now we study their effects on frequency spectra. By definition, we have

$$
\begin{aligned}
\mathcal{F}[x(at)] &= \int_{t=-\infty}^{\infty} x(at)e^{j\omega t}dt = \frac{1}{a}\int_{t=-\infty}^{\infty} x(at)e^{j(\omega/a)at}d(at) \\
&= \frac{1}{a}\int_{\tau=-\infty}^{\infty} x(\tau)e^{j(\omega/a)\tau}d\tau
\end{aligned}
$$

where we have used $\tau = at$ with $a > 0$. Thus we conclude

$$\mathcal{F}[x(at)] = \frac{1}{a}X\left(\frac{\omega}{a}\right) \tag{4.44}$$

See Section 3.4. Note that (4.44) holds only for a positive. If the spectrum of $x(t)$ is as shown in Figure 4.11(aa)[4], then the spectra of $x(2t)$ and $x(0.5t)$ are as shown in Figures 4.11(bb) and 4.11(cc). Thus we have

- Time compression $\Leftrightarrow$ frequency expansion

- Time expansion $\Leftrightarrow$ frequency compression

This is intuitively apparent. In time compression, the rate of change of the time signal increases, thus the spectrum contains higher-frequency components. In time expansion, the rate of change of the time signal decreases, thus the spectrum contains lower-frequency components. This fact can be used to explain the change of pitch of an audio tape when its speed is changed. When we increase the speed of the tape (time compression), its frequency spectrum is expanded and, consequently, contains higher-frequency components. Thus the pitch is higher. When we decrease the speed (time expansion), its spectrum is compressed and contains lower-frequency components. Thus its pitch is lower. This fact also has an implication in the transmission of a time signal. Speeding up the transmission (time compression), will increase the bandwidth. If the cost of transmission is proportional to bandwidth, then the cost will be increased.

If a signal lasts from t_1 to t_2 and is identically zero before t_1 and after t_2, then we may define its *time duration* as $L := t_2 - t_1$. If we use this definition, then the time signals $e^{-0.1t}$ and e^{-2t}, for $t \geq 0$, shown in Figures 2.15 have infinite time duration. Thus the time duration as defined is not very useful in practice. Instead we may define the time duration to be the width of the time interval in which

$$|x(t)| > ax_{max}$$

where x_{max} is the peak magnitude of $x(t)$, and a is some small constant. For example, if $a = 0.01$, then $e^{-0.1t}$ and e^{-2t} have, as discussed in Section 2.7, time duration 50 and 2.5 seconds, respectively. If we select a different a, then we will obtain different time durations. Another way is to define the time duration as the width of the smallest time interval that contains, say, 90% of the total energy of the signal. Thus there are many ways of defining the time duration of a signal.

Likewise, there are many ways of defining the bandwidth of the spectrum of a signal. We can define the frequency *bandwidth* as the width of the *positive* frequency range in which

$$|X(\omega)| > bX_{max}$$

where X_{max} is the peak magnitude of $X(\omega)$ and b is a constant such as 0.01. We can also define the frequency bandwidth as the width of the smallest positive frequency interval which contains, say, 48% of the total energy of the signal. The frequency bandwidth is

---

[4]The frequency spectrum of the time signal in Figure 4.11(a) is complex-valued and extends all the way to $\pm\infty$. See Problem 4.14 and the preceding subsection. It is assumed to have the form in Figure 4.11(aa) to simplify the discussion.

defined for ω in $[0, \infty)$. The frequency interval together with the corresponding interval in $(-\infty, 0]$ will contain 96% of the total energy. In any case, no matter how the time duration and frequency bandwidth are defined, generally we have

$$\text{time duration} \sim \frac{1}{\text{frequency bandwidth}} \tag{4.45}$$

It means that the larger the frequency bandwidth, the smaller the time duration and vice versa. For example, the time function $1.5\delta(t - 0.8)$ in Example 4.3.2 has zero time duration and its magnitude spectrum is 1.5, for all ω. Thus the signal has infinite frequency bandwidth. The time function $x(t) = 1$, for all t, has infinite time duration; its spectrum (as we will show in the next section) is $\mathcal{F}[1] = 2\pi\delta(\omega)$, which has zero frequency bandwidth. It is also consistent with our earlier discussion of time expansion and frequency compression.

A mathematical proof of (4.45) is difficult, if not impossible, if we use any of the aforementioned definitions. However, if we define the time duration as

$$L = \frac{\left(\int_{-\infty}^{\infty} |x(t)| dt\right)^2}{\int_{-\infty}^{\infty} |x(t)|^2 dt}$$

and the frequency bandwidth as

$$B = \frac{\int_{-\infty}^{\infty} |X(\omega)|^2 d\omega}{2|X(0)|^2}$$

with $|X(0)| \neq 0$, then we can show $BL \geq \pi$ which roughly implies (4.45). However, the physical meanings of L and B are not transparent. See also Reference [S7]. In any case, the inverse relationship in (4.45) is widely accepted in engineering.

A signal with its nonzero magnitude spectrum centered around $\omega = 0$ will be called a low-frequency or *baseband* signal. A modulated signal with its spectrum shown Figure 4.8 is not a baseband signal. If a baseband signal is time limited and bounded, then its nonzero magnitude spectrum must extend to $\pm\infty$. However it must also approach zero as $\omega \to \pm\infty$ because it has finite energy. The bandwidth of this type of signal can be defined simply as $B := \omega_{max}$ where

$$|X(\omega)| \approx 0 \quad \text{for all } |\omega| > \omega_{max}$$

or most energy of the signal is contained in $[-\omega_{max}, \omega_{max}]$. Clearly there is a large latitude in this definition. In addition, selecting a bandwidth may involve some compromise. A larger bandwidth will contain more energy (more fidelity) but will be more costly in transmission. In conclusion, even though the bandwidth of signals is widely used, its exact definition is not important or critical.[5]

4.5   Frequency spectra of CT pure sinusoids in $(-\infty, \infty)$ and in $[0, \infty)$

This section develops frequency spectra of sinusoidal signals. Direct computation is, if not impossible, difficult, thus we take an indirect approach.

---

[5]Bandwidth is also defined for filters (Subsection 9.7.2) and in other applications. See Wikipedia.

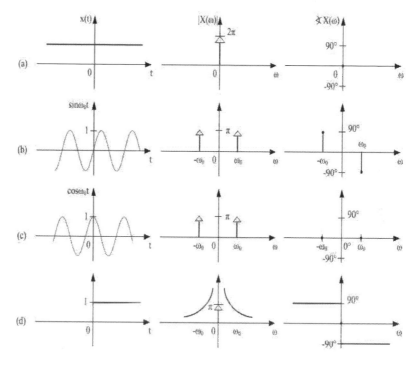

Figure 4.12: Frequency spectra of (a) $x(t) = 1$, for all t, (b) $\sin \omega_0 t$, (c) $\cos \omega_0 t$, and (d) $q(t)$.

Let us compute the inverse Fourier transform of a spectrum that consists of an impulse at $\omega = \omega_0$ or $A(\omega) := \delta(\omega - \omega_0)$. Using (4.22) and the sifting property in (2.14), we have

$$
\begin{aligned}
\mathcal{F}^{-1}[A(\omega)] &= \mathcal{F}^{-1}[\delta(\omega - \omega_0)] = \frac{1}{2\pi} \int_{\omega=-\infty}^{\infty} \delta(\omega - \omega_0) e^{j\omega t} d\omega \\
&= \frac{1}{2\pi} e^{j\omega t}\Big|_{\omega=\omega_0} = \frac{1}{2\pi} e^{j\omega_0 t} =: a(t)
\end{aligned} \tag{4.46}
$$

for all t in $(-\infty, \infty)$.

Because of the unique relationship between a time function and its Fourier transform, (4.46) implies

$$
\mathcal{F}[e^{j\omega_0 t}] = 2\pi \delta(\omega - \omega_0) \tag{4.47}
$$

which becomes, if $\omega_0 = 0$,

$$
\mathcal{F}[1] = 2\pi \delta(\omega) \tag{4.48}
$$

where 1 is the time function $x(t) = 1$ for all t in $(-\infty, \infty)$. Its frequency spectrum is an impulse at $\omega = 0$ with weight 2π as shown in Figure 4.12(a). Because $x(t) = 1$ contains no nonzero frequency component, it is called a dc signal.

Using the linearity property of the Fourier transform and (4.47), we have

$$\mathcal{F}[\sin \omega_0 t] = \mathcal{F}\left[\frac{e^{j\omega_0 t} - e^{-j\omega_0 t}}{2j}\right] = \frac{\pi}{j}[\delta(\omega - \omega_0) - \delta(\omega + \omega_0)]$$
$$= -j\pi\delta(\omega - \omega_0) + j\pi\delta(\omega + \omega_0) \qquad (4.49)$$

and

$$\mathcal{F}[\cos \omega_0 t] = \mathcal{F}\left[\frac{e^{j\omega_0 t} + e^{-j\omega_0 t}}{2}\right] = \pi\delta(\omega - \omega_0) + \pi\delta(\omega + \omega_0) \qquad (4.50)$$

The magnitude and phase spectra of (4.49) are shown in Figure 4.12(b). It consists of two impulses with weight π at $\omega = \pm\omega_o$ and phase $-\pi$ (rad) at $\omega = \omega_0$ and π at $\omega = -\omega_0$ as denoted by solid dots. The magnitude and phase spectra of (4.50) are plotted in Figure 4.12(c). The frequency spectra of $\sin \omega_0 t$ and $\cos \omega_0 t$ are zero everywhere except at $\pm\omega_0$. Thus their frequency spectra as defined are consistent with our perception of their frequency. Note that the sine and cosine functions are not absolutely integrable in $(-\infty, \infty)$ and their frequency spectra are still defined. Their spectra consist of impulses that are not bounded nor continuous.

We encounter in practice only positive-time signals. However, in discussing spectra of periodic signals, it is simpler to consider them to be defined for all t. For example, the frequency spectrum of a step function defined by $q(t) = 1$, for $t \geq 0$, and $q(t) = 0$, for $t < 0$, can be computed as

$$Q(\omega) = \mathcal{F}[q(t)] = \pi\delta(\omega) + \frac{1}{j\omega} \qquad (4.51)$$

The spectrum of $\cos \omega_0 t$, for $t \geq 0$, can be computed as

$$\mathcal{F}[\cos \omega_0 t \cdot q(t)] = 0.5\pi[\delta(\omega - \omega_0) + \delta(\omega + \omega_0)] + \frac{j\omega}{\omega_0{}^2 - \omega^2} \qquad (4.52)$$

Their derivations are fairly complex. See Reference [C8, 2nd ed.]. If the signals are extended to $-\infty$, then their spectra can be easily computed as in (4.49) and (4.50). This may be a reason of studying signals defined in $(-\infty, \infty)$. Moreover the spectra of $q(t)$ and $\cos \omega_0 t$, for $t \geq 0$, are nonzero for all ω, thus mathematically speaking they are not a dc or pure cosine function.

Many engineering texts call the set of Fourier coefficients the *discrete frequency spectrum*. For example, because of

$$\sin \omega_0 t = -0.5j e^{j\omega_0 t} + 0.5j e^{-j\omega_0 t}$$

the set of the Fourier coefficients $\mp 0.5j$ at $\pm\omega_0$ is called the discrete frequency spectrum of $\sin \omega_0 t$ in many texts. In this text, frequency spectra refer exclusively to Fourier transforms. Because

$$\mathcal{F}[\sin \omega_0 t] = -j\pi\delta(\omega - \omega_0) + j\pi\delta(\omega + \omega_0)$$

the frequency spectrum of $\sin \omega_0 t$ for t in $(-\infty, \infty)$ consists of impulses at $\pm\omega_0$ which are nonzero only at $\pm\omega_0$, thus we may call them discrete frequency spectrum.

To conclude this section, we discuss the term 'frequency' in frequency spectrum. A pure sinusoid has a well defined frequency. A periodic signal can be decomposed as a sum of pure sinusoids with discrete frequencies. Because the Fourier transform is developed from the Fourier series, an aperiodic signal may be interpreted as consisting of infinitely many pure sinusoids with various frequencies in a continuous range. This however is difficult to visualize. Alternatively, we may consider the frequency spectrum to indicate rates of changes of the time function. The ultimate rate of change is discontinuities. If a time function has a discontinuity anywhere in $(-\infty, \infty)$, then its nonzero spectrum, albeit very small, will extend to $\pm\infty$. For example, the step function has a discontinuity at $t = 0$ and its spectrum contains the factor $1/j\omega$ which is nonzero for all ω. Thus we may also associate the frequency spectrum of a signal with its rates of changes for all t.

4.6 DT pure sinusoids

Consider the CT pure sinusoids $e^{j\omega_0 t}$, $\sin \omega_0 t$, and $\cos \omega_0 t$. They are periodic for every ω_0 and repeat themselves every $P = 2\pi/\omega_0$ seconds. Sampling them with sampling period T yields $e^{j\omega_0 nT}$, $\sin \omega_0 nT$, and $\cos \omega_0 nT$. We call them *DT pure sinusoids*. Are DT pure sinusoids periodic for every ω_0? To answer this question, we must first define periodicity for DT signals.

Before proceeding, we mentioned in Section 2.6.1 that the processing of DT signals is independent of the sampling period T and T may be assumed to be 1. However, the frequency of DT pure sinusoids and, more generally, the frequency content of DT signals do depend on T. Thus in the remainder of this chapter, we do not assume $T = 1$. The only assumption is $T > 0$.

A DT signal $x[n] = x(nT)$ is defined to be periodic with period N (in samples) or NT (in seconds) if there exists an integer N so that

$$x[n] = x[n + N]$$

for all integers n. If $x[n]$ is periodic with period N, then it is periodic with period $2N, 3N, \ldots$. The smallest such N is called the *fundamental period*.

The CT $\sin \omega_0 t$ is periodic for every ω_0. Is the DT $\sin \omega_0 nT$ periodic for every ω_0? If so, then there exists an integer N such that

$$\sin \omega_0 nT = \sin \omega_0 (n + N)T = \sin \omega_0 nT \cos \omega_0 NT + \cos \omega_0 nT \sin \omega_0 NT$$

This holds for every integer n if and only if

$$\cos \omega_0 NT = 1 \quad \text{and} \quad \sin \omega_0 NT = 0$$

In order to meet these conditions, we must have

$$\omega_0 NT = k2\pi \quad \text{or} \quad N = \frac{2k\pi}{\omega_0 T} \tag{4.53}$$

for some integer k. In other words, the DT sinusoid $\sin \omega_0 nT$ is periodic if and only if there exists an integer k to make $2k\pi/\omega_0 T$ an integer. This is not always possible

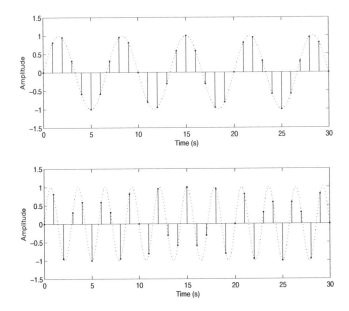

Figure 4.13: (a) $\sin 0.3\pi nT$ with $T = 1$. (b) $\sin 0.7\pi nT$ with $T = 1$. Note that each sequence repeats itself after every 20 samples.

for every $\omega_0 T$. For example, consider $\sin \omega_0 nT = \sin 2nT$ with $\omega_0 = 2$ and $T = 1$. Because $N = 2k\pi/\omega_0 T = 2k\pi/2 = k\pi$ and because π is an irrational number (cannot be expressed as a ratio of two integers), there exists no integer k to make $N = k\pi$ an integer. Thus $\sin 2nT$ with $T = 1$ is not periodic.

The condition for $\sin \omega_0 nT$ to be periodic is the existence of an integer k to make N in (4.53) an integer. In order for such k to exit, $\omega_0 T$ must be a rational-number multiple of π. For example, for $\sin 0.3\pi nT$ and $T = 1$, we have

$$N = \frac{2k\pi}{0.3\pi \cdot 1} = \frac{2k}{0.3} = \frac{20k}{3}$$

If $k = 3$, then $N = 20k/3 = 20$. Note that $k = 6$ and 9 will also make N an integer. The smallest such k will yield the fundamental period. Thus $\sin 0.3\pi nT$, with $T = 1$, is periodic with fundamental period 20 samples or $20T$ seconds as shown in Figure 4.13(a). In conclusion, $\sin \omega_0 nT$ may or may not be periodic.

4.6.1 Can we define the frequency of $\sin \omega_0 nT$ directly?

The frequency of CT sinusoid $\sin \omega_0 t$ equals $2\pi/P$, where P is its fundamental period. If $\sin \omega_0 nT$ is periodic with fundamental period $P = NT$, can we define its frequency as $2\pi/NT$? For example, $\sin 0.3\pi nT$ with $T = 1$ is periodic with fundamental period 20 (in seconds), can we define its frequency as $2\pi/20 = 0.314$ (rad/s)? The answer is negative. To see the reason, consider

$$\sin 0.7\pi nT$$

with $T = 1$ as shown in Figure 4.13(b) and consider

$$N = \frac{2k\pi}{\omega_0 T} = \frac{2k\pi}{0.7\pi} = \frac{2k}{0.7} = \frac{20k}{7}$$

The smallest integer k to make N an integer is 7 which yields $N = 20$. Thus $\sin 0.7\pi nT$ has the same fundamental period as $\sin 0.3\pi nT$ as we can also see from Figure 4.13. Thus if we define its frequency as $2\pi/NT$, then $\sin 0.3\pi nT$ and $\sin 0.7\pi nT$ have the same frequency. This is clearly unacceptable. Thus even if a DT pure sinusoid is periodic, its frequency should not be defined as $2\pi/P$ as in the CT case.

4.6.2 Frequency of DT pure sinusoids– Principal form

We discuss some properties of $\sin \omega_0 nT$ before defining its frequency. For the CT case, we have

$$\omega_1 \neq \omega_2 \quad \rightarrow \quad \sin \omega_1 t \neq \sin \omega_2 t \tag{4.54}$$

Does this hold for the DT case? We first use an example to illustrate the situation. Consider the following DT sinusoids:

$$\sin 1.2nT, \quad \sin 7.4832nT, \quad \sin 13.7664nT, \quad \sin(\; 5.0832nT) \tag{4.55}$$

for all integers n and $T = 1$. These four $\sin \omega_i nT$ with different ω_i appear to represent different DT sequences. Is this so? Let us use MATLAB to compute their values for n from 0 to 5:

n	0	1	2	3	4	5
$\sin 1.2nT$	0	0.9320	0.6755	-0.4425	-0.9962	-0.2794
$\sin(7.4832nT)$	0	0.9320	0.6754	-0.4426	-0.9962	-0.2793
$\sin(13.7664nT)$	0	0.9320	0.6754	-0.4426	-0.9962	-0.2793
$\sin(-5.0832nT)$	0	0.9320	0.6755	-0.4425	-0.9962	-0.2795

We see that other than some differences in the last digit due to rounding errors in computer computation, the four DT sinusoids appear to generate the same DT sequence. Actually this is the case as we show next.

Consider the DT complex exponential $e^{j\omega nT}$. We show that

$$e^{j\omega_1 nT} = e^{j\omega_2 nT} \quad \text{if } \omega_1 = \omega_2 \;(\text{mod } 2\pi/T) \tag{4.56}$$

for all integer n. Recall from (3.4) and (3.5) that if $\omega_1 = \omega_2 \;(\text{mod } 2\pi/T)$, then

$$\omega_1 = \omega_2 + k(2\pi/T)$$

for some integer k. Using $e^{a+b} = e^a e^b$, we have

$$e^{j\omega_1 nT} = e^{j(\omega_2 + k(2\pi/T))nT} = e^{j\omega_2 nT + jk(2\pi)n} = e^{j\omega_2 nT} e^{jk(2\pi)n} = e^{j\omega_2 nT}$$

for all integers n and k, where we have used $e^{jk(2\pi)n} = 1$. See Problem 3.6. This establishes (4.56). Note that (4.56) also holds if the complex exponentials are replaced by sine functions. In other words, we have

$$\sin \omega_1 nT = \sin \omega_2 nT \quad \text{if } \omega_1 = \omega_2 \;(\text{mod } 2\pi/T)$$

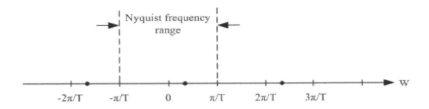

Figure 4.14: Representation of a DT frequency on a real number line.

for all n.

We now are ready to show that the four DT sinusoids in (4.55) denote the same DT signal. If $T = 1$ and if we use $\pi = 3.1416$, then we have $2\pi/T = 6.2832$ and

$$1.2 = 1.2 + 6.2832 = 1.2 + 2 \times 6.2832 = 1.2 - 6.2832 \quad (\text{mod } 6.2832)$$

or

$$1.2 = 7.4832 = 13.7664 = -5.0832 \quad (\text{mod } 6.2832)$$

Thus we have, for $T = 1$,

$$\sin 1.2nT = \sin 7.4832nT = \sin 13.7664nT = \sin(-5.0832nT)$$

for all n.

The preceding example shows that a DT sinusoid can be represented by many $\sin \omega_i nT$. If we plot ω_i on a real line as shown in Figure 4.14, then all dots which differ by $2\pi/T$ or its integer multiple denote the same DT sinusoid. Thus we may say that ω_0 in $\sin \omega_0 nT$ is periodic with period $2\pi/T$. In order to have a unique representation, we must select a frequency range of $2\pi/T$. The most natural one will be centered around $\omega = 0$. We cannot select $[-\pi/T, \pi/T]$ because the frequency π/T can also be represented by $-\pi/T$. Nor can we select $(-\pi/T, \pi/T)$ because it does not contain π/T. We may select either $(-\pi/T, \pi/T]$ or $[-\pi/T, \pi/T)$. We select the former and call $(-\pi/T, \pi/T]$ (in rad/s) the *Nyquist frequency range (NFR)*. We see that the frequency range is dictated by the sampling period T. The smaller T is, the larger the Nyquist frequency range.

The Nyquist frequency range can also be expressed in terms of sampling frequency. Let us define the *sampling frequency f_s* in Hz (cycles per second) as $1/T$ and the *sampling frequency ω_s* in rad/s as $\omega_s = 2\pi f_s = 2\pi/T$. Then the Nyquist frequency range can also be expressed as $(-0.5f_s, 0.5f_s]$ in Hz. Thus we have

$$
\begin{aligned}
\text{Nyquist frequency range (NFR)} \;=\;& (-\pi/T, \pi/T] = (-0.5\omega_s, 0.5\omega_s] \quad (\text{in rad/s}) \\
=\;& (-0.5f_s, 0.5f_s] \quad (\text{in Hz})
\end{aligned}
$$

Note that in mathematical equations, we *must* use frequencies in rad/s. But in text or statement, we often use frequencies in Hz. Note that if $T = 1$, then $f_s = 1/T = 1$ and the NFR becomes $(-\pi, \pi]$ (rad/s) or $(-0.5, 0.5]$ (Hz). This is the frequency range used in many texts on Digital Signal Processing. Note that the sampling frequency may be called, more informatively, the *sampling rate*, because it is the number of time samples in one second.

We are ready to give a definition of the frequency of DT sinusoids which include $e^{j\omega_0 nT}$, $\sin \omega_0 nT$, and $\cos \omega_0 nT$. Giving a DT sinusoid, we first compute its Nyquist frequency range. If ω_0 lies outside the range, we add or subtract $2\pi/T$, if necessary repeatedly, to shift ω_0 inside the range, called the frequency $\bar{\omega}$. Then the frequency of the DT sinusoid is defined as ω, that is

$$\text{Frequency of } e^{j\omega_0 nT} := \bar{\omega} \qquad \text{where } \bar{\omega} = \omega_0 \pmod{2\pi/T}$$
$$\text{and } -\pi/T < \bar{\omega} \leq \pi/T \qquad (4.57)$$

We first give an example and then give an interpretation of the definition.

Example 4.6.1 Consider the DT sinusoid $\sin 13.7664 nT$ with $T = 1$. According to our definition, its frequency is not necessarily 13.7664 rad/s. We first compute its Nyquist frequency range as $(-\pi/T, \pi/T] = (-\pi, \pi] = (-3.1416, 3.1416]$. The number 13.7664 is outside the range. Thus we must shift it inside the range. Subtracting $2\pi/T = 6.2832$ from 13.7664 yields 7.4832 which is still outside the range. Subtracting again 6.2832 from 7.4832 yields 1.2 which is inside the range $(-3.1416, 3.1416]$. Thus the frequency of $\sin 13.7664 nT$ with $T = 1$ is 1.2 rad/s. $\square$

Consider a DT sinusoid $\sin \omega_0 nT$. We call a CT sinusoid $\sin \hat{\omega} t$ an *envelope* of $\sin \omega_0 nT$ if

$$\sin \hat{\omega} t \big|_{t=nT} = \sin \omega_0 nT$$

for all n. Clearly $\sin \omega_0 t$ is an envelope of $\sin \omega_0 nT$. However, $\sin \omega_0 nT$ has many other envelopes. For example, because

$$\sin 1.2 nT = \sin 7.4832 nT = \sin 13.7664 nT = \sin(-5.0832 nT)$$

for all n, $\sin 13.7664 nT$ has envelopes $\sin 1.2t$, $\sin 7.4832t$, $\sin 13.7664t$, $\sin(-5.0832t)$ and many others. Among these envelopes, the one with the smallest $\hat{\omega}$ in magnitude will be called the *primary envelope*. Thus the primary envelope of $\sin 13.7664 nT$ is $\sin 1.2t$ which is a CT sinusoid with frequency 1.2 rad/s. Thus strictly speaking, the frequency of DT sinusoids is defined using the frequency of CT sinusoids. Moreover, the frequency of DT sinusoids so defined is consistent with our perception of frequency as we will see in the next subsection.

We call the DT sinusoids

$$e^{j\omega_0 nT}, \quad \sin \omega_0 nT, \quad \text{and} \quad \cos \omega_0 nT$$

in *principal form* if their frequencies equal ω_0. Clearly, in order for a DT sinusoid to be in principal form, ω_0 must lie inside its Nyquist frequency range. For example, $\sin 13.7664 nT$ and $\sin 7.4832 nT$ with $T = 1$ are not in principal form. Their principal form is $\sin 1.2 nT$. It is really meaningless to write $\sin 13.7664 nT$ or $\sin 7.4832 nT$. Thus from now on, when we are given a DT sinusoid, we must reduce it to its principal form by adding or subtracting $2\pi/T$, if necessary, repeatedly. If $\sin \omega_0 nT$ and $\cos \omega_0 nT$ are in principal form, then their primary envelopes are $\sin \omega_0 t$ and $\cos \omega_0 t$ which have frequency ω_0.

The frequency range of CT $e^{j\omega_0 t}$ is $(-\infty, \infty)$. The frequency range of DT $e^{j\omega_0 nT}$ however is limited to $(-\pi/T, \pi/T]$. This is the most significant difference between CT and DT sinusoids. The highest frequency of a CT sinusoid can be as large as desired. For a given T, the *highest* frequency of DT sinusoid is π/T (rad/s). If $T = 1$, the highest frequency is π (rad/s) or 0.5 (Hz).

4.7   Sampling of CT pure sinusoids – Aliased frequencies

We first use an example to discuss the effect of sampling. Consider $x(t) = \sin 5t$ as plotted in Figure 4.15(a) with solid line. It has frequency 5 rad/s. Its sampled sequence with sampling period T is $x[n] = \sin 5nT$. As discussed in the preceding subsection, the frequency of $\sin 5nT$ depends on T and may not equal 5. To find the frequency of $\sin 5nT$, we must compute the Nyquist frequency range (NFR) $(-\pi/T, \pi/T]$. For example, if $T = 0.9$, then the sampled sequence is as shown in Figure 4.15(a) with solid dots and its NFR is $(-\pi/T, \pi/T] = (-3.49, 3.49]$. The frequency 5 is outside the range and must be shifted inside the range by subtracting $2\pi/T = 6.98$ to yield $5 - 6.98 = -1.98$. In other words, $\sin 5nT$ with $T = 0.9$ has the principal form $\sin(-1.98)nT$ and frequency -1.98 rad/s. If we try to interpolate or construct a CT sine function from the dots in Figure 4.15(a), the most natural one is the primary envelope $\sin(-1.98)t$ as plotted with a dashed line. Thus the frequency of $\sin 5nT$ as defined is consistent with our perception. We call -1.98 an *aliased frequency*.

We now consider $x(t) = \sin \omega_0 t$. Its sampled sequence with sampling period T is $x[n] = x(nT) = \sin \omega_0 nT$. From the preceding discussion, we may conclude that the frequency of $\sin \omega_0 nT$ equals ω_0 if ω_0 lies inside the NFR $(-\pi/T, \pi/T]$. If ω_0 lies outside the NFR, then the frequency of $\sin \omega_0 nT$ is $\bar{\omega}$ which is obtained by shifting ω_0 inside the range by subtracting or adding $2\pi/T$, or is defined as in (4.57). Thus the frequency of $\sin \omega_0 t$ and the frequency of $\sin \omega_0 nT$ are related as shown in Figure 4.15(b). For example, if $\omega_0 = 5$ and $T = 0.9$, then $2\pi/T = 6.98$ and $\bar{\omega} = -1.98$ can be obtained from the plot as shown.

In digital processing of a CT signal $x(t)$ using its sampled sequence $x(nT)$, we must select a sampling period T so that all essential information of $x(t)$ is preserved in $x(nT)$. In the case of $\sin \omega_0 t$, we must preserve its frequency in $\sin \omega_0 nT$. From Figure 4.15(b), we conclude that T must be selected so that ω_0 lies inside the Nyquist frequency range $(-\pi/T, \pi/T]$. However if ω_0 is on the edge of the range or $T = \pi/\omega_0$, a nonzero CT sinusoid may yield a sequence which is identically zero. For example, if $x(t) = \sin 5t$ and if $5 = \pi/T$ or $T = \pi/5$, then $x(nT) = \sin 5nT = \sin \pi n = 0$, for all n. Thus we must exclude this case and require $\omega_0 = 5$ to lie inside the range $(-\pi/T, \pi/T)$. In conclusion, the sampled sequence of $\sin \omega_0 t$ has frequency ω_0 if $|\omega_0| < \pi/T$ or

$$T < \frac{\pi}{|\omega_0|} \qquad \text{or} \quad f_s > 2f_0 \tag{4.58}$$

where $f_s = 1/T$ and $f_0 = \omega_0/2\pi$. Otherwise the sampled DT sinusoid $\sin \omega_0 nT$ has a frequency different from ω_0, called an *aliased frequency*. Note that the (fundamental) frequency of $\sin \omega_0 t$ is $P_0 = 2\pi/|\omega_0|$. Thus $T < \pi/|\omega_0|$ implies $T < P_0/2$. In other words, the sampled sequence of $\sin \omega_0 t$ has frequency ω_0 if we take more than two

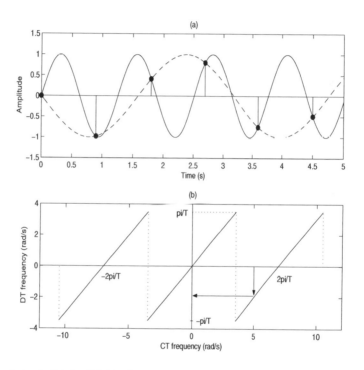

Figure 4.15: (a) $\sin 5t$ (solid line), its sampled sequence with $T = 0.9$ (solid dots) and its primary envelope or $\sin(-1.98t)$ (dashed line). (b) Relationship between the frequency of a CT sinusoid and that of its sampled sequence.

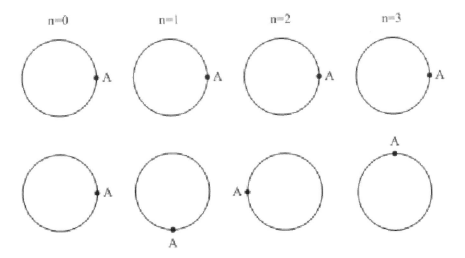

Figure 4.16: (Top) A spins complete cycles per second in either direction but appears stationary in its sampled sequence. (Bottom) A spins (3/4) cycle per second in the counterclockwise direction but appears to spin (1/4) cycle per second in the clockwise direction.

samples in one period and has an aliased frequency if we take two or less samples in one period.

We give one more explanation of aliased frequencies. Consider the wheel spinning with a constant speed shown in Figure 4.16(a). The spinning of point A on the wheel can be denoted by $e^{j\omega_0 t}$. Let us use a video camera to shoot, starting from $t = 0$, the spinning wheel. To simplify the discussion, we assume that the camera takes one frame per second. That is, we assume $T = 1$. Then point A on the n-th frame can be denoted by $e^{j\omega_0 n}$. If A spins in the counterclockwise direction one complete cycle every second, then we have $\omega_0 = 2\pi$ (rad/s) and A will appears as shown in Figure 4.16(top) for $n = 0 : 3$. Point A appears to be stationary or the sampled sequence appears to have frequency 0 even though the wheel spins one complete cycle every second. Indeed, if $T = 1$, then the Nyquist frequency range is $(-\pi, \pi]$. Because $\omega_0 = 2\pi$ is outside the range, we subtract $2\pi/T = 2\pi$ from 2π to yield 0. Thus the sampled sequence has frequency 0 which is an aliased frequency. This is consistent with our perception. In fact, the preceding perception still holds if point A spins complete cycles every second (no matter how many complete cycles) in either direction.

Now if point A rotates 3/4 cycle per second in the counterclockwise direction or with angular velocity $\omega_0 = 3\pi/2$ (rad/s), then the first four frames will be as shown in Figure 4.16(bottom). To our perception, point A is rotating in the clockwise direction 1/4 cycle every second or with frequency $-\pi/2$ rad/s. Indeed $\omega_0 = 3\pi/2$ is outside the Nyquist frequency range $(-\pi, \pi]$. Subtracting 2π from $3\pi/2$ yields $-\pi/2$. This is the frequency of the sampled sequence. Thus sampling $e^{j(3\pi/2)t}$ with sampling period $T = 1$ yields a DT sinusoid with frequency $-\pi/2$ rad/s which is an aliased frequency.

If $T = 1$, the *highest* frequency of the sampled sequence is π rad/s. This follows directly from Figure 4.16(b). If the frequency is slightly larger than π, then it appears to have a negative frequency whose magnitude is slightly less than π. Thus for a selected T, the highest frequency is π/T (rad/s).

To conclude this section, we mention that the plot in Figure 4.15(b) has an important application. A stroboscope is a device that periodically emits flashes of light and acts as a sampler. The frequency of emitting flashes can be varied. We aim a stroboscope at an object that rotates with a constant speed. We increase the strobe frequency and observe the object. The object will *appear* to increase its rotational speed and then suddenly reverse its rotational direction.[6] If we continue to increase the strobe frequency, the speed of the object will appear to slow down and then to stop rotating. The strobe frequency at which the object appears stationary is the rotational speed of the object. Thus a stroboscope can be used to measure rotational velocities. It can also be used to study vibrations of mechanical systems.

4.7.1 A sampling theorem

In this subsection, we first use an example to discuss the effect of sampling on a signal that consists of a couple of pure sinusoids and then develop a sampling theorem. Consider

$$x(t) = 2\sin 50t - 3\cos 70t$$

as plotted in Figure 4.17(a) with a solid line. Its sampled signal with sampling period T is given by

$$x(nT) = 2\sin 50nT - 3\cos 70nT \qquad (4.59)$$

We now discuss the effects of selecting various T.

If we select $T_1 = \pi/35$, then its Nyquist frequency range (NFR) is $(-\pi/T_1, \pi/T_1] = (-35, 35]$. Because both 50 and 70 are outside the range, they must be shifted inside the range by subtracting $2\pi/T_1 = 70$. Thus the resulting sampled signal in principal form is

$$x_1(nT_1) := 2\sin(50 - 70)nT_1 - 3\cos(70 - 70)nT_1 = -2\sin 20nT_1 - 3 \qquad (4.60)$$

where we have used $\sin(-\theta) = -\sin\theta$ and $\cos(0) = 1$. This DT signal contains the constant or dc (-3) which has frequency 0 and a sinusoid with frequency 20 rad/s. These two frequencies are different from the frequencies of $x(t)$ and are *aliased frequencies*. This phenomenon is called *frequency aliasing*. If we try to construct a CT signal from $x_1(nT_1)$, the most logical one is to use its primary envelopes or simply to replace nT_1 by t to yield $x_1(t)$. We plot $x_1(t)$ in Figure 4.17(a) with a dotted line. It is very different from $x(t)$.

Next we select $T_2 = \pi/60$. Then its NFR is $(-60, 60]$. The frequency 50 is inside the range but 70 is outside and must be shifted inside the range. Thus the resulting signal in principal form is

$$x_2(nT_2) := 2\sin 50nT_2 - 3\cos(70 - 120)nT_2 = 2\sin 50nT_2 - 3\cos 50nT_2 \qquad (4.61)$$

where we have used $\cos(-\theta) = \cos\theta$. This DT signal contains only one frequency 50 rad/s. The original 70 rad/s does not appear in (4.61). The most logically constructed

---

[6]This phenomenon may appear in a helicopter's rotor blades and a wagon's wheel in movies.

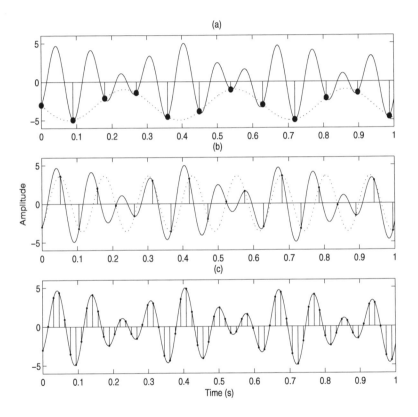

Figure 4.17: (a) $x(t)$ (solid line), its sampled sequence with $T_1 = \pi/35$ (solid dots), and reconstructed CT signal (dotted line). (b) With $T_2 = \pi/60$. (c) With $T_3 = \pi/200$.

CT signal of (4.61) is shown in Figure 5.1(b) with a dotted line. It is different from $x(t)$.

Finally we select $T_3 = \pi/200$. Then its NFR is $(-200, 200]$. Both 50 and 70 are inside the range and its resulting sampled signal in principal form is

$$x_3(nT_3) := 2\sin 50nT_3 - 3\cos 70nT_3 \qquad (4.62)$$

which contains all original frequencies and its reconstructed CT signal is identical to the original $x(t)$.

From the preceding discussion, we can now develop a sampling theorem. Consider a CT signal $x(t)$ that contains a finite number of sinusoids with various frequencies. Let $\omega_{max} = 2\pi f_{max}$ be the largest frequency in magnitude. If the sampling period $T = 1/f_s$ is selected so that

$$T < \frac{\pi}{\omega_{max}} \qquad \text{or} \qquad f_s > 2f_{max} \qquad (4.63)$$

then

$$(-\pi/T, \pi/T) \text{ contains all frequencies of } x(t)$$

and, consequently, the sampled sequence $x(nT)$ contains all frequencies of $x(t)$. If a sampling period does not meet (4.63), then the sampled sequence $x(nT)$ in principal form contains some aliased frequencies which are the frequencies of $x(t)$ lying outside the NFR shifted inside the NFR. This is called *frequency aliasing*. Thus *time sampling may introduce frequency aliasing*. If frequency aliasing occurs, it is not possible to reconstruct $x(t)$ from $x(nT)$.

On the other hand, if a sampling period meets (4.63), then $x(nT)$ in principal form contains all frequencies of $x(t)$. Furthermore $x(t)$ can be constructed exactly from $x(nT)$. This is called a *sampling theorem*. Note that $2f_{max}$ is often called the *Nyquist rate*. Thus if the sampling rate f_s is larger than the Nyquist rate, then there is no frequency aliasing.

The preceding sampling theorem is developed for CT signals that contain only sinusoids. Actually it is applicable to any CT signal $x(t)$, periodic or not, if ω_{max} can be defined for $x(t)$. This will be developed in the next chapter.

4.8 Frequency spectra of DT signals

As in the CT case, it is possible to develop the DT Fourier series and then DT Fourier transform. See Reference [C7]. In view of the discussion in Section 4.6, we will not do so because the frequency of DT sinusoids is defined from the the frequency of CT sinusoids. Thus we will use the (CT) Fourier transform to develop directly frequency spectra of DT signals.

Consider a DT signal $x(nT)$ which is zero for all $t \neq nT$. The application of the (CT) Fourier transform to $x(nT)$ yields

$$
\begin{aligned}
\mathcal{F}[x(nT)] &= \int_{t=-\infty}^{\infty} x(nT)e^{-j\omega t}dt = \cdots + \int_{0_-}^{0_+} x(0)e^{-j\omega t}dt \\
&+ \int_{t=T_-}^{T_+} x(T)e^{-j\omega t}dt + \int_{t=2T_-}^{2T_+} x(2T)e^{-j\omega t}dt + \cdots
\end{aligned}
$$

Because $x(nT)$ and $e^{jn\omega t}$ contain no impulses and because all integration intervals are practically zero, we have $\mathcal{F}[x(nT)] = 0$. See (2.11). In other words, the application of the (CT) Fourier transform to a DT signal directly will yield no information.

The DT signal $x(nT)$ consists of a sequence of numbers and can be expressed as in (2.26) or

$$x(nT) = \sum_{k=-\infty}^{\infty} x(kT)\delta_d(nT - kT)$$

Its value at kT is $x(kT)$. Now we replace the value by a CT impulse at kT with weight $x(kT)$ or replace $x(kT)\delta_d(nT - kT)$ by $x(kT)\delta(t - kT)$, where the impulse is defined as in (2.7). Let the sequence of impulses be denoted by $x_d(t)$, that is,

$$x_d(t) := \sum_{n=-\infty}^{\infty} x(nT)\delta(t - nT) \tag{4.64}$$

The function $x_d(t)$ is zero everywhere except at sampling instants where they consist of impulses with weight $x(nT)$. Thus $x_d(t)$ can be considered a CT representation of a DT signal $x(nT)$. The application of the Fourier transform to $x_d(t)$ yields

$$
\begin{aligned}
X_d(\omega) &:= \mathcal{F}[x_d(t)] = \int_{t=-\infty}^{\infty} \left[\sum_{n=-\infty}^{\infty} x(nT)\delta(t - nT) \right] e^{-j\omega t} dt \\
&= \sum_{n=-\infty}^{\infty} \left[x(nT) \int_{t=-\infty}^{\infty} \delta(t - nT)e^{-j\omega t} dt \right] = \sum_{n=-\infty}^{\infty} x(nT)\, e^{-j\omega t}\Big|_{t=nT}
\end{aligned}
$$

where we have interchanged the order of integration and summation and used (2.14), or

$$X_d(\omega) = \sum_{n=-\infty}^{\infty} x(nT)e^{-j\omega nT} \tag{4.65}$$

for all ω in $(-\infty, \infty)$. This is, by definition, the *discrete-time (DT) Fourier transform* of the sequence $x(nT)$. This is the counterpart of the (CT) Fourier transform of $x(t)$ defined in (4.21). Just as in the CT case, not every sequence has a DT Fourier transform. For example, if $x(nT)$ grows unbounded, then its DT Fourier transform is not defined. A sufficient condition is that $x(nT)$ is absolutely summable. See Section 3.5.[7] If $X_d(\omega)$ exists, it is called the *frequency spectrum* or, simply, *spectrum* of $x(nT)$. In general, $X_d(\omega)$ is complex-valued. We call its magnitude $|X_d(\omega)|$ the *magnitude spectrum* and its phase $\measuredangle X_d(\omega)$ the *phase spectrum*. Before giving examples, we discuss two formulas. First we have

$$\sum_{n=0}^{N} r^n = 1 + r + r^2 + \cdots + r^N = \frac{1 - r^{N+1}}{1 - r} \tag{4.66}$$

---

[7]This condition corresponds to that $x(t)$ is absolutely integrable. See the Dirichlet conditions for the CT case in Section 4.3. Note that every DT sequence is of bounded variation.

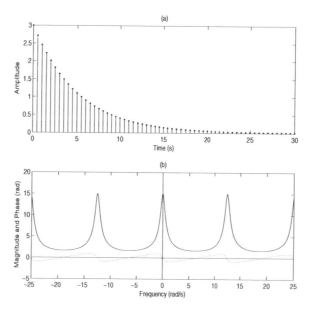

Figure 4.18: (a) Positive-time sequence $x(nT) = 3 \times 0.8^n$, for $T = 0.5$ and $n \geq 0$. (b) Its magnitude and phase spectra.

where r is a real or complex constant and N is any positive integer. See Problems 4.28 and 4.29. If $|r| < 1$, then $r^N \to 0$ as $N \to \infty$ and we have

$$\sum_{n=0}^{\infty} r^n = \frac{1}{1-r} \quad \text{for } |r| < 1 \tag{4.67}$$

The infinite summation diverges if $|r| > 1$ and $r = 1$. If $r = -1$, the summation in (4.67) is not well defined. See Section 3.5. Thus the condition $|r| < 1$ is essential in (4.67). Now we give an example.

Example 4.8.1 Consider the DT signal defined by

$$x(nT) = \begin{cases} 3 \times 0.8^n & \text{for } n \geq 0 \\ 0 & \text{for } n < 0 \end{cases} \tag{4.68}$$

with $T = 0.5$. It is plotted in Figure 4.18(a) for $n = 0 : 60$. Using (4.67), we can show that $x(nT)$ is absolutely summable. Its DT Fourier transform is

$$X_d(\omega) = \sum_{n=-\infty}^{\infty} x(nT)e^{-j\omega nT} = \sum_{n=0}^{\infty} 3 \times 0.8^n e^{-j\omega nT} = 3 \sum_{n=0}^{\infty} \left(0.8e^{-j\omega T}\right)^n$$

which reduces to, because $|0.8e^{-j\omega T}| = |0.8| < 1$,

$$X_d(\omega) = \frac{3}{1 - 0.8e^{-j\omega T}} \tag{4.69}$$

This is the frequency spectrum of the DT signal in (4.68).

The spectrum is complex-valued. Its value at $\omega = 0$ is $X_d(0) = 3/(1 - 0.8) = 15$. Its value at $\omega = \pi/2 = 1.57$ is

$$
\begin{aligned}
X_d(1.57) &= \frac{3}{1 - 0.8e^{-j\pi/4}} = \frac{3}{1 - 0.8 \times (0.7 - j0.7)} = \frac{3}{1 - 0.56 + j0.56} \\
&= \frac{3}{0.44 + j0.56} = \frac{3}{0.71e^{j0.9}} = 4.2e^{-j0.9}
\end{aligned}
$$

We see that its hand computation is tedious; it is better to delegate it to a computer. Typing in MATLAB

```
% Subprogram 4.2
w=-25:0.001:25;T=0.5;
Xd=3.0./(1-0.8*exp(-j*w*T));
plot(w,abs(Xd),w,angle(Xd),':')
```

will yield the magnitude (solid line) and phase (dotted line) spectra in Figure 4.18(b).□

Example 4.8.2 Consider the DT signal $x(0) = 1$, $x(T) = -2$, $x(2T) = 1$, $x(nT) = 0$, for $n < 0$ and $n \geq 3$, and $T = 0.2$. It is a finite sequence of length 3 as shown in Figure 4.19(a). Its frequency spectrum is

$$
X_d(\omega) = \sum_{n=-\infty}^{\infty} x(nT)e^{-jn\omega T} = 1 - 2e^{-j\omega T} + 1 \cdot e^{-j2\omega T} \tag{4.70}
$$

Even though (4.70) can be simplified (see Problem 4.34), there is no need to do so in using a computer to plot its spectrum. Typing in MATLAB

```
% Subprogram 4.3
w=-50:0.001:50;T=0.2;
Xd=1-2*exp(-j*w*T)+exp(-j*2*w*T);
plot(w,abs(Xd),w,angle(Xd),':')
```

will yield the magnitude (solid line) and phase (dotted line) spectra in Figure 4.19(b).□

If $x(nT)$ is absolutely summable, then its spectrum is bounded and is a continuous function of ω in $(-\infty, \infty)$.[8] Note that even if $x(nT)$ is defined for $n \geq 0$, its spectrum is defined for positive and negative frequencies.

Because $X(\omega)$ and $X_d(\omega)$ are the results of the same Fourier transform, all properties of $X(\omega)$ are directly applicable to $X_d(\omega)$. For example, if $x(nT)$ is real, its magnitude spectrum is even ($|X_d(\omega)| = |X_d(-\omega)|$) and its phase spectrum is odd

---

[8]There is a common misconception that the spectrum of $x(nT)$ is defined only at discrete frequencies because $x(nT)$ is defined at discrete time instants.

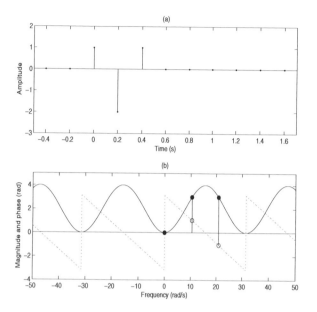

Figure 4.19: (a) Time sequence of length 3. (b) Its magnitude spectrum (solid line) and phase spectrum (dotted line).

$(\not{<} X_d(\omega) = -\not{<} X_d(-\omega))$ as shown in Figures 4.18(b) and 4.19(b). These properties are identical to those for the CT case. There is however an important difference between the CT and DT cases. Because of (4.56), we have

$$X_d(\omega_1) = X_d(\omega_2) \quad \text{if } \omega_1 = \omega_2 \ (\text{mod } 2\pi/T) \tag{4.71}$$

which implies that $X_d(\omega)$ is periodic with period $2\pi/T$. For example, the spectrum in Figure 4.18(b) is periodic with period $2\pi/0.5 = 12.6$ as shown and the spectrum in Figure 4.19(b) is periodic with period $2\pi/0.2 = 31.4$ as shown. This is similar to the frequencies of DT sinusoids discussed in Subsection 4.6.2. In order to have a unique representation, we restrict $X_d(\omega)$ to a range of $2\pi/T$. We can select the range as $(-\pi/T, \pi/T]$ or $[-\pi/T, \pi/T)$. As in Section 4.6.2, we select the former and call it the *Nyquist frequency range (NFR)*. Thus we have

$$\text{Frequency range of spectra of } x(t) \ = \ (\infty, \infty)$$
$$\text{Frequency range of spectra of } x(nT) \ = \ (-\pi/T, \pi/T] = (-0.5\omega_s, 0.5\omega_s] \quad (\text{in rad/s})$$
$$= \ (-0.5f_s, 0.5f_s] \quad (\text{in Hz})$$

where $f_s = 1/T$ and $\omega_s = 2\pi f_s = 2\pi/T$. This is the most significant difference between the spectra of $x(t)$ and $x(nT)$. Note that the highest frequency of a CT signal has no limit, whereas the highest frequency of a DT signal is limited to π/T (in rad/s) or $0.5f_s$ (in Hz). The DT signal in Example 4.8.1 is a low-frequency signal because the magnitudes of $X_d(\omega)$ in the neighborhood of $\omega = 0$ are larger than those in the neighborhood of the highest frequency $\pi/T = \pi/0.5 = 6.3$. For the DT signal in

Example 4.8.2, the highest frequency is $\pi/T = \pi/0.2 = 15.7$. It is a high-frequency signal because its magnitude spectrum around $\omega = 15.7$ is larger than the one around $\omega = 0$.

Even though a great deal more can be said regarding spectra of DT signals, all we need in this text is what has been discussed so far. Thus we stop its discussion here.

4.9   Concluding remarks

In a nutshell, this chapter deals only with the CT Fourier transform

$$X(\omega) = \int_{t=-\infty}^{\infty} x(t)e^{-j\omega t}dt \qquad (4.72)$$

Its application to a periodic or aperiodic CT signal yields the spectrum of the signal. The spectrum is generally complex valued. Its magnitude spectrum reveals the distribution of the energy of the signal in frequencies. This information is needed in defining for the time signal a frequency bandwidth which in turn is needed in selecting a carrier frequency and a sampling frequency, and in specifying a filter to be designed.

The application of the CT Fourier transform to a DT signal modified by CT impulses yields the spectrum of the DT signal. Because the spectra of CT and DT signals are obtained by applying the same transform, they have the same properties. The only difference is that spectra of CT signals are defined for all frequencies in $(-\infty, \infty)$; whereas spectra of DT signals are periodic with period $2\pi/T = \omega_s$ in rad/s or f_s in Hz and we need to consider them only in the Nyquist frequency range (NFR) $(-\pi/T, \pi/T] = (-\omega_s/2, \omega_s/2]$ in rad/s or $(-f_s/2, f_s/2]$ in Hz.

The equation in (4.72) is applicable to real- or complex-valued $x(t)$. The only complex-valued signal used in this chapter is the complex exponential $e^{j\omega t}$ defined for all t in $(-\infty, \infty)$. All other signals are real valued and positive time. Even though it is possible to define a Fourier transform for real-valued signals defined for $t \geq 0$, the subsequent development will be no simpler than the one based on (4.72). In fact, computing the spectrum of real-valued $\cos \omega t$ or $\sin \omega t$ will be more complex. Thus we will not introduce such a definition.

Fourier analysis of signals is important. Without it, modulation and frequency-division multiplexing in communication would not have been developed. On the other hand, the concept used was only the bandwidth of signals. Thus we must put the Fourier analysis in perspective.

Problems

4.1 Plot $2e^{j0.5t}$ on a circle at $t = k\pi$, for $k = 0 : 6$. What is its direction of rotation, clockwise or counterclockwise? What are its period and its frequency in rad/s and in Hz?

4.2 Plot $2e^{-j0.5t}$ on a circle at $t = k\pi$, for $k = 0 : 6$. What is its direction of rotation, clockwise or counterclockwise? What are its period and its frequency in rad/s and in Hz?

4.3 Is the signal $\sin 3t + \sin \pi t$ periodic? Give your reasons.

4.4 Is the signal $1.2 + \sin 3t$ periodic? Is yes, what is its fundamental period and fundamental frequency?

4.5 Repeat Problem 4.4 for the signal $1.2 + \sin \pi t$.

4.6 Is the signal $x(t) = -1.2 - 2\sin 2.1t + 3\cos 2.8t$ periodic? If yes, what is its fundamental frequency and fundamental period?

4.7 Express $x(t)$ in Problem 4.6 is complex Fourier series with its coefficients in polar form.

4.8 Use the integral formula

$$\int e^{at} dt = \frac{e^{at}}{a}$$

to verify that $x(t) = 3e^{-0.2t}$, for $t \geq 0$, is absolutely integrable in $[0, \infty)$.

4.9 What is the spectrum of $x(t) = 2e^{5t}$, for $t \geq 0$?

4.10 What is the spectrum of $x(t) = -2e^{-5t}$, for $t \geq 0$? Compute the spectrum at $\omega = 0, \pm 5$, and ± 100, and then sketch roughly its magnitude and phase spectra for ω in $[-100, 100]$. Is the magnitude spectrum even? Is the phase spectrum odd?

4.11 What is the spectrum of $x(t) = \delta(t - 0)$? Compute its total energy from its magnitude spectrum. Can an impulse be generated in practice?

4.12 Let $X(\omega)$ be the Fourier transform of $x(t)$. Verify that, for any $t_0 > 0$,

$$X_0(\omega) := \mathcal{F}[x(t - t_0)] = e^{-jt_0\omega} X(\omega)$$

Verify also $|X_0(\omega)| = |X(\omega)|$ and $\measuredangle X_0(\omega) = \measuredangle X(\omega) - t_0\omega$. Thus time shifting will not affect the magnitude spectrum but will introduce a linear phase into the phase spectrum.

4.13 Use Euler's formula to write (4.21) as

$$X(\omega) = \int_{t=-\infty}^{\infty} x(t)[\cos \omega t - j \sin \omega t] dt$$

and then use it to verify that if $x(t)$ is real, then the real part of $X(\omega)$ is even and the imaginary part of $X(\omega)$ is odd? Use the result to verify (4.29). This is the conventional way of verifying (4.29). Which method is simpler?

4.14 Show that if $x(t)$ is real and even $(x(t) = x(-t))$, then $X(\omega)$ is real and even. Can the spectrum of a positive-time signal be real? If we consider all real-world signals to be positive time, will we encounter real-valued spectra in practice?

4.15 Consider a signal $x(t)$. It is assumed that its spectrum is real valued and as shown in Figure 4.11(aa). What are the spectra of $x(5t)$ and $x(0.2t)$?

4.16 Consider a signal $x(t)$. It is assumed that its spectrum is real valued and of the form shown in Figure 4.11(aa). Plot the spectra of $x(t - 0.5)$, $x(0.5(t - 0.5))$, $x(2(t - 0.5))$, $x(t - 0.5)\cos 2t$, and $x(t - 0.5)\cos 10t$.

4.17* Consider the Fourier series in (4.14) and (4.15). Now if we use $P = 2P_0$ to develop its Fourier series as

$$x(t) = \sum_{m=-\infty}^{\infty} \bar{c}_m e^{jm\bar{\omega}_0 t}$$

with

$$\bar{c}_m = \frac{1}{P} \int_{t=-P/2}^{P/2} x(t)e^{-jm\bar{\omega}_0 t} dt = \frac{1}{P} \int_{t=0}^{P} x(t)e^{-jm\bar{\omega}_0 t} dt$$

where $\bar{\omega}_0 = 2\pi/P = \omega_0/2$. Verify

$$\bar{c}_m = \begin{cases} c_{m/2} & m \text{ even} \\ 0 & m \text{ odd} \end{cases}$$

We see that if we use $P = 2P_0$, then half of the computed Fourier coefficients will be zero. If we use $P = 3P_0$, then two third of the computed Fourier coefficients will be zero. In other words, the final result of the Fourier series remains the same if we use any period to carry out the computation. However, if we use the fundamental period, then the amount of computation will be minimum.

4.18* Verify (4.34) or

$$X(\omega) = \sum_{m=-\infty}^{\infty} c_m \frac{2\sin[(\omega - m\omega_0)L]}{\omega - m\omega_0}$$

4.19* Use the sinc function defined in (3.25) and (4.33) to express (4.34) as

$$X(\omega) = \sum_{m=-\infty}^{\infty} 2Lc_m \text{sinc}((\omega - m\omega_0)L) = \sum_{m=-\infty}^{\infty} X(m\omega_0)\text{sinc}((\omega - m\omega_0)L)$$

This relates the Fourier series and Fourier transform of a time signal of finite duration. This result is dual to the sampling theorem to be developed in the next chapter. See Problem 5.4.

4.20 Consider a signal $x(t)$. It is assumed that its spectrum $X(\omega)$ is real-valued and positive for all ω. Its squared value or $X^2(\omega)$ is plotted in Figure 4.20. What is the total energy of the signal? Plot its spectrum and the spectrum of its modulated signal $x_m(t) = x(t)\cos 10t$. What is the total energy of the modulated signal $x_m(t)$?

4.21 Verify that the total energy of a signal $x(t)$ is cut in half in its modulated signal $x_m(t) = x(t)\cos\omega_c t$ for any ω_c so long as $X(\omega - \omega_c)$ and $X(\omega + \omega_c)$ do not overlap.

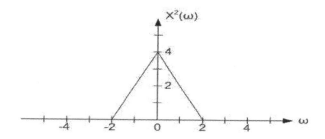

Figure 4.20: Squared frequency spectrum.

4.22 Consider the sequences $x_1(nT) = \cos 2\pi nT$, $x_2(nT) = \cos 4\pi nT$, and $x_3(nT) = \cos 6\pi nT$. If $T = 1$, do the three sequences denote the same sequence? If yes, what is its frequency? What is its principal form?

4.23 Consider the sequences $x_1(nT) = \cos 2\pi nT$, $x_2(nT) = \cos 4\pi nT$, and $x_3(nT) = \cos 6\pi nT$. If $T = 0.5$, do the three sequences denote the same sequence? If not, how many different sequences are there? what are their frequencies? What are their principal forms?

4.24 Repeat Problem 4.23 for $T = 0.1$.

4.25 If we sample $x(t) = \sin 10t$ with sampling period $T = 1$, what is its sampled sequence in principal form? Does the frequency of its sampled sinusoid equal 10 rad/s?

4.26 Repeat Problem 4.25 for $T = 0.5, 0.3$, and 0.1.

4.27 Consider $x(t) = 2 - 3\sin 10t + 4\cos 20t$. What is its sampled sequence in principal form if the sampling period is $T = \pi/4$? What are its aliased frequencies? Can you recover $x(t)$ from $x(nT)$?

4.28 Repeat Problem 4.27 for $T = \pi/5$, $\pi/10$, and $\pi/25$.

4.29 Verify (4.66) by multiplying its both sides by $(1 - r)$.

4.30 Verify that (4.66) holds for $N = 0$ and $N = 1$. Next assume that (4.66) holds for $N = k$ and then verify that it holds for $N = k + 1$. Thus (4.66) holds for any positive integer N. This process of proof is called mathematical induction.

4.31 What is the frequency spectrum of $x(nT) = 3^n$, for $n \geq 0$ and $T = 0.1$?

4.32 What is the frequency spectrum of $x(nT) = 0.3^n$, for $n \geq 0$ and $T = 0.1$?

4.33 Consider the sequence $x(nT) = 1.2$, for $n = 0, 2$, and $x(nT) = 0$, for all n other than 0 and 2, with $T = 0.2$. What is its frequency spectrum? Plot its magnitude and phase spectra for ω in its NFR.

4.34 Verify that the spectrum of $x(nT) = 3 \times 0.98^n$ with $T = 0.1$ and $n \geq 0$ is given

$$X_d(\omega) = \frac{3}{1 - 0.98e^{-j\omega T}}$$

4.35 Consider the finite sequence in Figure 4.19(a) of length $N = 3$ with $T = 0.2$. It spectrum is given in (4.70). Verify that (4.70) can be simplified as

$$X_d(\omega) = 2e^{-j(\omega T - \pi)}[1 - \cos(\omega T)]$$

What are its magnitude and phase spectra? Verify analytically the plots in Figure 4.19(b)? What are the magnitudes and phases of $X_d(\omega)$ at $\omega = m(2\pi/NT)$, for $m = 0 : 2$? Note that in computer computation, if a complex number is zero, then its phase is automatically assigned to be zero. In analytical computation, this may not be the case as shown in this problem. In practical application, there is no need to be concerned with this discrepancy.

4.36 Consider the sequence in Problem 4.35 with $T = 1$. Plot its magnitude and phase spectra of ω in $[-\pi, 2\pi]$. What are the magnitudes and phases of $X_d(\omega)$ at $\omega = m(2\pi/NT)$, for $m = 0 : 2$? Are those values the same as those in Problem 4.35?

4.37 Verify

$$\mathcal{F}[\text{sinc}] = \mathcal{F}\left[\frac{\sin t}{t}\right] = \begin{cases} \pi & \text{for } |\omega| \leq 1 \\ 0 & \text{for } |\omega| > 1 \end{cases}$$

Thus the spectrum of a sinc function is a rectangular box centered at $\omega = 0$ with height π and width 2. Note that the sinc function is not absolutely integrable but is squared absolutely integrable and its spectrum is still defined but is not continuous. [*Hint*: Use the inverse Fourier transform.]

4.38 Verify, for any integers n and m,

$$\int_{t=-P_0/2}^{P_0/2} e^{j(n-m)\omega_0 t} dt = \int_{t=0}^{P_0} e^{j(n-m)\omega_0 t} dt = \begin{cases} 0 & \text{if } n \neq m \\ P_0 & \text{if } n = m \end{cases}$$

where $\omega_0 = 2\pi/P_0$, and then use it to derive (4.15) from (4.14).

Chapter 5

Sampling theorem and spectral computation

5.1 Introduction

This chapter discusses computer computation of frequency spectra. It consists of three parts. The first part discusses the equation for such computation. We first derive it informally from the definition of integration and then formally by establishing the sampling theorem. This justifies mathematically the computation of the spectrum of a CT signal from its time samples. Even so, the computed phase spectrum may not be correct as we give the reason and then demonstrate it using an example. Thus we plot mostly magnitude spectra.

In practice, we can use only a finite number of data and compute only a finite number of frequencies. If their numbers are the same, then the DT Fourier transform (DTFT) reduces to the discrete Fourier transform (DFT). We then discuss how to use the fast Fourier transform (FFT), an efficient way of computing DFT, to compute spectra. We also discuss how to select a largest possible sampling period T by *downsampling*.

In the last part, we show that the FFT-computed spectrum of a step function of a finite duration looks completely different from the exact one. Even so, we demonstrate that the FFT-computed result is correct. We also discuss a paradox involving ∞. We then introduce *trailing zeros* in using FFT to complete the chapter.

Before proceeding, we mention that, in view of the discussion in Section 2.3.1, the spectrum of a CT signal $x(t)$ can be computed from its sampled sequence $x(nT)$ if T is selected to be sufficiently small. Clearly the smaller T is, the more accurate the computed spectrum. However it will be more difficult to design digital filters if the signal is to be processed digitally.[1] See Subsection 11.10.1. Thus the crux of this chapter is find a largest possible T or an upper bound of T. An *exact* upper bound of T exists only for bandlimited signals. For real-world signals, no such T exists. Thus

---

[1] A smaller T requires more storage and more computation. This however is no longer an important issue because of very inexpensive memory chips and very fast computing speed.

the selection of an upper bound will be subjective.

5.1.1   From the definition of integration

Consider a CT signal $x(t)$. Its frequency spectrum is defined as

$$X(\omega) = \int_{t=-\infty}^{\infty} x(t)e^{-j\omega t} dt \qquad (5.1)$$

If $x(t)$ is absolutely integrable, then its spectrum exists and is a bounded and continuous function of ω. If $x(t)$ is, in addition, bounded, then $x(t)$ has finite total energy (see Section 3.5.1). Every real-world signal, as discussed in Section 3.5.1, is absolutely integrable and bounded. In addition, it is time limited. Thus its nonzero spectrum will extend all the way to $\pm\infty$. However because it has finite total energy, the magnitude spectrum must approach zero, that is,

$$|X(\omega)| \to 0 \quad \text{as } |\omega| \to \infty \qquad (5.2)$$

See the discussion in Subsections 4.4.2 and 4.4.3. This condition will be used throughout this chapter.

For real-world signals, analytical computation of (5.1) is not possible. Thus we discuss its numerical computation. Using the following definition of integration

$$\int f(t)dt = \sum f(nT)T \;\; \text{as } T \to 0$$

we can write (5.1) as

$$X(\omega) = \sum_{n=-\infty}^{\infty} x(nT)e^{-j\omega nT} T = T\left[\sum_{n=-\infty}^{\infty} x(nT)e^{-j\omega nT}\right] \quad \text{as } T \to 0$$

where $x(nT)$ is the samples of $x(t)$ with sampling period T. The term inside the brackets is defined in (4.65) as the frequency spectrum of $x(nT)$ or

$$X_d(\omega) := \sum_{n=-\infty}^{\infty} x(nT)e^{-j\omega nT} \qquad (5.3)$$

Thus we have

$$X(\omega) = TX_d(\omega) \quad \text{as } T \to 0 \qquad (5.4)$$

for all ω in $(-\infty, \infty)$

Note that $T \to 0$ is a concept and we can never achieve it in actual computation. No matter how small T we select, there are still infinitely many T_i between T and 0. Thus for a selected T, the equality in (5.4) must be replaced by an approximation. Moreover, the approximation cannot hold for all ω in $(-\infty, \infty)$. Recall that we have assumed $|X(\omega)| \to 0$ as $|\omega| \to \infty$. Whereas $X_d(\omega)$ is periodic with period $2\pi/T = \omega_s$ for all ω in $(-\infty, \infty)$, and we consider $X_d(\omega)$ for ω only in the Nyquist frequency range $(-\pi/T, \pi/T] = (-\omega_s/2, \omega_s/2]$. Thus in actual computation, (5.4) must be replaced by

$$X(\omega) \approx \begin{cases} TX_d(\omega) & \text{for } |\omega| < \pi/T = \omega_s/2 \\ 0 & \text{for } |\omega| \geq \pi/T = \omega_s/2 \end{cases} \qquad (5.5)$$

for T sufficiently small. Clearly the smaller T, the better the approximation and the wider the applicable frequency range. Computer computation of spectra is based on this equation. The preceding derivation of (5.5) is intuitive. We develop it formally in the next section.

5.2   Relationship between spectra of $x(t)$ and $x(nT)$

Consider a CT signal $x(t)$ and its sampled sequence $x(nT)$. The frequency spectrum of $x(t)$ is defined in (4.21) as

$$X(\omega) = \mathcal{F}[x(t)] = \int_{t=-\infty}^{\infty} x(t)e^{-j\omega t}\, dt \tag{5.6}$$

The frequency spectrum of $x(nT)$ is defined in (4.65) as

$$X_d(\omega) = \mathcal{F}[x_d(t)] = \sum_{n=-\infty}^{\infty} x(nT)e^{-j\omega nT} \tag{5.7}$$

where

$$x_d(t) := \sum_{n=-\infty}^{\infty} x(nT)\delta(t - nT)$$

is the CT representation of the DT signal $x(nT)$ discussed in (4.64). Using $x(nT)\delta(t - nT) = x(t)\delta(t - nT)$ as shown in (2.13), we can write $x_d(t)$ as

$$x_d(t) := \sum_{n=-\infty}^{\infty} x(t)\delta(t - nT) = x(t) \sum_{n=-\infty}^{\infty} \delta(t - nT) \tag{5.8}$$

The infinite summation is the sampling function studied in (4.16) and is plotted in Figure 4.4. It is periodic with fundamental frequency $\omega_s := 2\pi/T$ and can be expressed in Fourier series as in (4.18). Substituting (4.18) into (5.8) yields

$$x_d(t) = x(t) \left[\frac{1}{T} \sum_{m=-\infty}^{\infty} e^{jm\omega_s t}\right] = \frac{1}{T} \sum_{m=-\infty}^{\infty} x(t)e^{jm\omega_s t} \tag{5.9}$$

Thus we have, using the linearity of the Fourier transform,

$$\begin{aligned}
X_d(\omega) &= \mathcal{F}[x_d(t)] = \mathcal{F}\left[\frac{1}{T}\sum_{m=-\infty}^{\infty} x(t)e^{jm\omega_s t}\right] \\
&= \frac{1}{T}\sum_{m=-\infty}^{\infty} \mathcal{F}\left[x(t)e^{jm\omega_s t}\right]
\end{aligned}$$

which becomes, using (4.38),

$$X_d(\omega) = \frac{1}{T}\sum_{m=-\infty}^{\infty} X(\omega - m\omega_s) \tag{5.10}$$

This equation relates the spectrum $X(\omega)$ of $x(t)$ and the spectrum $X_d(\omega)$ of $x(nT)$. We discuss in the next subsection the implication of (5.10).

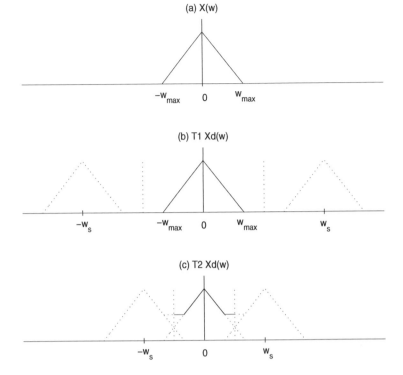

Figure 5.1: (a) Spectrum of $x(t)$. (b) Spectrum of $Tx(nT)$ with $\omega_s > 2\omega_{max}$. (c) Spectrum of $Tx(nT)$ with $\omega_s < 2\omega_{max}$.

5.2.1   Sampling theorem

A CT signal $x(t)$ is said to be *bandlimited* to ω_{max} if its spectrum $X(\omega)$ has the property

$$X(\omega) = 0 \quad \text{for all } |\omega| > \omega_{max} \tag{5.11}$$

In other words, all *nonzero* spectrum of $x(t)$ lies inside the range $[-\omega_{max}, \omega_{max}]$. Because a time-limited signal cannot be bandlimited, a bandlimited signal must extend to $-\infty$ and/or ∞ in time. The sinc function defined in (3.12) is bandlimited (see Problem 4.37). The sinusoids $\sin \omega_0 t$ and $\cos \omega_0 t$ defined for all t in $(-\infty, \infty)$ are bandlimited.[2] Other than these functions, we rarely encounter bandlimited signals.

For convenience of discussion, we assume that $X(\omega)$ is real-valued and is as shown in Figure 5.1(a). We write (5.10) as

$$TX_d(\omega) = \cdots + X(\omega + \omega_s) + X(\omega) + X(\omega - \omega_s) + \cdots$$

---

[2] The sinc and sinusoidal functions and their derivatives of all orders are continuous for all t in $(-\infty, \infty)$. Thus it is conjectured that if a signal or any of its derivative is not continuous in $(-\infty, \infty)$, then it cannot be bandlimited. For example, $\sin \omega_0 t$ for $t \geq 0$ is not bandlimited because its first derivative is not continuous at $t = 0$. $\cos \omega_0 t$ for $t \geq 0$ is not bandlimited because it is not continuous at $t = 0$.

Note that $X(\omega \pm \omega_s)$ are the shiftings of $X(\omega)$ to $\mp\omega_s$ as shown in Figure 5.1(b) with dotted lines. Thus the spectrum of $x(nT)$ multiplied by T is the sum of $X(\omega)$ and its repetitive shiftings to $m\omega_s$, for $m = \pm 1, \pm 2, \ldots$. Because $X_d(\omega)$ is periodic with period ω_s, we need to plot it only in the Nyquist frequency range $(-0.5\omega_s, 0.5\omega_s]$[3] bounded by the two vertical dotted lines shown. Furthermore, because $X(\omega)$ is even, the two shiftings immediately outside the range can be obtained by *folding* $X(\omega)$ with respect to $\pm 0.5\omega_s$. Now if the sampling frequency ω_s is selected to be larger than $2\omega_{max}$, then their shiftings will not overlap and we have

$$TX_d(\omega) = X(\omega) \quad \text{for } \omega \text{ in } (-0.5\omega_s, 0.5\omega_s) \tag{5.12}$$

Note that they are different outside the range because $X(\omega)$ is 0 and $X_d(\omega)$ is periodic outside the range. Now if $\omega_s < 2\omega_{max}$, then the repetitive shiftings of $X(\omega)$ will overlap as shown in Figure 5.1(c) and (5.12) does not hold. This is called *frequency aliasing*. In general, the effect of frequency aliasing is not as simple as the one shown in Figure 5.1(c) because it involves additions of complex numbers. See Problem 5.3.

In conclusion, if $x(t)$ is bandlimited to $\omega_{max} = 2\pi f_{max}$ and if $T = 1/f_s$ is selected so that

$$T < \frac{\pi}{\omega_{max}} \quad \text{or} \quad f_s > 2f_{max} \tag{5.13}$$

then we have

$$X(\omega) = \begin{cases} TX_d(\omega) & \text{for } |\omega| < \pi/T \\ 0 & \text{for } |\omega| \geq \pi/T \end{cases} \tag{5.14}$$

Thus the spectrum of $x(t)$ can be computed exactly from its samples $x(nT)$. Moreover, substituting (5.7) into (5.14) and then into the inverse Fourier transform in (4.22), we can express $x(t)$ in terms of $x(nT)$ as

$$x(t) = \sum_{n=-\infty}^{\infty} x(nT)\frac{T\sin[\pi(t - nT)/T]}{\pi(t - nT)} \tag{5.15}$$

See Problem 5.4. It means that for a bandlimited signal, the CT signal $x(t)$ can be recovered exactly from $x(nT)$. This is called the *Nyquist sampling theorem*.

Equation (5.15) is called the *ideal interpolator*. It is however not used in practice because it requires an infinite number of operations and, as seriously, cannot be carried out in real time. In practice we use a zero-order hold to construct a CT signal from a DT signal. In the remainder of this chapter, we focus on how to compute $X(\omega)$ from $x(nT)$.

All real-world signals have finite time duration, thus their spectra, as discussed in Section 4.4.3, cannot be bandlimited. However their spectra do approach zero as $|\omega| \to \infty$ because they have finite energy. Thus in practice we select ω_{max} so that

$$|X(\omega)| \approx 0 \quad \text{for all } |\omega| > \omega_{max} \tag{5.16}$$

Clearly, the selection is subjective. One possibility is to select ω_{max} so that $|X(\omega)|$ is less than 1% of its peak magnitude for all $|\omega| > \omega_{max}$. We then use (5.13) to select

---

[3]If $X(\omega)$ contains no impulse, there is no difference to plot it in $[-0.5\omega_s, 0.5\omega_s]$.

a T and compute the spectrum of $x(t)$ using $x(nT)$. In this computation, frequency aliasing will be small. The frequency aliasing can be further reduced if we design a CT lowpass filter to make the $|X(\omega)|$ in (5.16) even closer to zero. Thus the filter is called an *anti-aliasing* filter; it is the filter shown in Figure 2.12. Because frequency aliasing is not identically zero, we must replace the equality in (5.14) by an approximation as

$$X(\omega) \approx \begin{cases} TX_d(\omega) & \text{for } |\omega| < \pi/T = \omega_s/2 \\ 0 & \text{for } |\omega| \geq \pi/T = \omega_s/2 \end{cases} \qquad (5.17)$$

This is the same equation as (5.5) and will be the basis of computing the spectrum of $x(t)$ from $x(nT)$.

5.2.2   Can the phase spectrum of $x(t)$ be computed from $x(nT)$?

The spectrum of a real-world signal is generally complicated and cannot be described by a simple mathematical equation. Thus we plot it graphically. Because the spectrum is complex valued, its plotting against frequencies requires three dimension and is difficult to visualize. We may plot its real and imaginary parts which however have no physical meaning. Thus we plot its magnitude and phase against frequencies. The magnitude spectrum, as discussed in Section 4.4, will reveal the distribution of energy of the signal in frequencies.

Let us define

$$A(\omega) := X(\omega) - TX_d(\omega)$$

If $X(\omega) \to TX_d(\omega)$, then $A(\omega) \to 0$. Does this imply $|A(\omega)| \to 0$ and $\not\!\!\angle A(\omega) \to 0$? To answer this question, we consider $A = 0.00001 + j0.00001$ which is almost zero. Its magnitude is 0.000014 which is also almost zero, but its phase is $\pi/4 = 0.78$ which is quite different from zero. This can be explained graphically. Because A is complex or a point in a complex plane, it may approach 0 from any direction. Thus $A \to 0$ implies $|A| \to 0$ but does not imply $\not\!\!\angle A \to 0$. See also Problem 5.15. In conclusion, Equation (5.17) implies

$$|X(\omega)| \approx \begin{cases} |TX_d(\omega)| & \text{for } |\omega| < \pi/T = \omega_s/2 \\ 0 & \text{for } |\omega| \geq \pi/T = \omega_s/2 \end{cases} \qquad (5.18)$$

but may not imply

$$\not\!\!\angle X(\omega) \approx \begin{cases} \not\!\!\angle TX_d(\omega) & \text{for } |\omega| < \pi/T = \omega_s/2 \\ 0 & \text{for } |\omega| \geq \pi/T = \omega_s/2 \end{cases} \qquad (5.19)$$

This will be confirmed by the next example.

Example 5.2.1 Consider the CT signal $x(t) = 3e^{-0.2t}$, for $t \geq 0$. Its exact spectrum was computed in (4.25) as

$$X(\omega) = \frac{3}{j\omega + 0.2} \qquad (5.20)$$

and plotted in Figure 4.5(b). From the plot, we see that its magnitude spectrum is practically zero for $|\omega| > 20$. Thus we may select $\omega_{max} = 20$ and require $T < \pi/20 = 0.157$. Arbitrarily we select $T = 0.1$. The sampled sequence of $x(t)$ with $T = 0.1$ is

$$x(nT) = 3e^{-0.2nT} = 3\left(e^{-0.2T}\right)^n = 3 \times 0.98^n$$

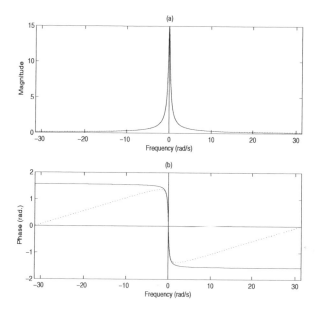

Figure 5.2: (a) Magnitude spectra of $x(t)$ (solid line) and $Tx(nT)$ (dotted line). (b) Corresponding phase spectra.

for $n \geq 0$. Its frequency spectrum can be computed as

$$X_d(\omega) = \frac{3}{1 - 0.98e^{-j\omega T}} \qquad (5.21)$$

(Problem 4.34). We plot in Figure 5.2(a) the magnitude spectrum of $x(t)$ (solid line) and the magnitude spectrum of $x(nT)$ multiplied by T (dotted line) for ω in the Nyquist frequency range $(-\pi/T, \pi/T] = (-31.4, 31.4]$ (rad/s) and in Figure 5.2(b) the corresponding phase spectra. From the plots, we see that their magnitude spectra are indistinguishable. However their phase spectra are different except in the immediate neighborhood of $\omega = 0$. This verifies the assertions regarding (5.18) and (5.19). □

In conclusion we can compute the magnitude spectrum but not the phase spectrum of $x(t)$ from $x(nT)$. For pure sinusoids and impulses, phase is related to time delay. But for general signals, the physical meaning of phase spectra is not transparent. Fortunately in most applications, we use only magnitude spectra because they reveal the distribution of energy in frequencies.

Because of evenness of magnitude spectra, we often plot spectra only for $\omega \geq 0$. In this case, (5.18) can be reduced to

$$|X(\omega)| \approx \begin{cases} |TX_d(\omega)| & \text{for } 0 \leq \omega < \pi/T = \omega_s/2 \\ 0 & \text{for } \omega \geq \pi/T = \omega_s/2 \end{cases} \qquad (5.22)$$

Before proceeding, we mention that in order for (5.18) and (5.22) to hold the computed $|TX_d(\omega)|$ must be practically zero in the neighborhood of $\pi/T = 0.5\omega_s$. If the computed $|TX_d(\omega)|$ is large at π/T, then $|X(\omega)|$ in (5.22) is discontinuous at $\omega = \pi/T$. This violates the fact that the spectrum of every real-world signal is continuous.

In application, we may be given a $x(t)$ and be asked to compute its spectrum. In this case, we cannot use (5.16) to select a T. We now discuss a procedure of selecting a T. We select arbitrarily a T and use the samples $x(nT)$ to compute $X_d(\omega)$ in (5.7). If the computed $X_d(\omega)$ multiplied by T at $\omega = \pi/T$ is significantly different from zero, the condition in (5.16) cannot be met and we must select a smaller T. We repeat the process until we find a T so that the computed $TX_d(\omega)$ is practically zero in the neighborhood of $\omega = \pi/T$, then we have (5.18) and (5.22).

The selection of T discussed so far is based entirely on the sampling theorem. In practice, the selection may involve other issues. For example, the frequency range of audio signals is generally limited to 20 Hz to 20 kHz, or $f_{max} = 20$ kHz. According to the sampling theorem, we may select f_s as 40 kHz. However, in order to design a less stringent and, consequently, less costly anti-aliasing analog filter, the sampling frequency is increased to 44 kHz. Moreover, in order to synchronize with the TV horizontal video frequency, the sampling frequency of CD was finally selected as 44.1 kHz. Thus the sampling period of CD is $T = 1/44100 = 2.2676 \cdot 10^{-5}$ s or 22.676 microsecond (μs). It is not determined solely by the sampling theorem.

5.2.3   Direct computation at a single frequency

We now discuss the actual computer computation of spectra of CT signals. As discussed in Chapter 2, we encounter only positive-time signals in practice. Moreover, practical signals cannot last forever. Thus we study in the remainder of this chapter signal $x(t)$ defined only for t in $[0, L)$ for some finite L. If we select N equally-spaced samples, then the sampling period T is given by $T = L/N$ and the N samples are $x(nT)$ for $n = 0 : N - 1$.[4] In this case, (5.3) reduces to

$$X_d(\omega) := \sum_{n=0}^{N-1} x(nT)e^{-j\omega nT} \tag{5.23}$$

where $T > 0$ and N is an integer. For a selected frequency, (5.23) can be readily computed using a computer. As an example, we compute $X_d(\omega)$ at $\omega = 128 \times 2\pi$ for the $x(nT)$ in Program 3.4. Recall that it is the samples of the sound generated by a 128-Hz tuning fork with sampling period $T = 1/8000$. We type in the control window of MATLAB the following

```
    %Program 5.1
>> x=wavread('f12.wav');
>> N=length(x),T=1/8000;n=0:N-1;
>> w=2*pi*128;e=exp(-j*w*n*T);
>> tic,Xd=sum(x'.*e),toc
```

---

[4]If $x(t)$ contains no impulses as is the case in practice, there is no difference in using $[0, L)$ or $[0, L]$. We adopt the former for convenience. See Problems 5.4 and 5.5.

The first two lines are the first two lines of Program 3.4. The third line specifies the frequency in radians per second and generates the $1 \times N$ row vector

$$e = [1 \quad e^{-j\omega T} \quad e^{-j\omega 2T} \quad \cdots \quad e^{j\omega(N-1)T}]$$

The x generated by wavread is an $N \times 1$ column vector. We take its transpose and multiply it with e term by term using '.*'. Thus x'.*e is a $1 \times N$ row vectors with its entry equal to the product of the corresponding entries of x' and e. The MATLAB function sum sums up all entries of x'.*e which is (5.23).[5] The function tic starts a stopwatch timer and toc returns the elapsed time. The result of the program is

$$N = 106000 \quad Xd = -748.64 - 687.72j \quad \text{elapsed\_time=0.0160}$$

We see that it takes only 0.016 second to compute $X_d(\omega)$, which involves over one hundred thousand additions and multiplications, at a single frequency. For some applications such as in DTMF (Dual-tone multi-frequency) detection, we need to compute $X_d(\omega)$ only at a small number of frequencies. In this case, we may compute (5.23) directly.

5.3 Discrete and fast Fourier transforms (DFT & FFT)

If we use (5.23) to compute the frequency spectrum of $x(nT)$, we must compute $X_d(\omega)$ at a large number of frequencies. We discuss in this section an efficient way of its computation.

Because $X_d(\omega)$ is periodic with period $2\pi/T$, we need to compute $X_d(\omega)$ for ω only in a frequency range of $2\pi/T = \omega_s$. The range consists of infinitely many ω, and we cannot compute them all. We can compute only a finite number of ω_m. Let us select the number of ω_m to be N, same as the number of samples $x(nT)$ used in (5.23). These N frequencies are to be equally spaced. If we select these N frequencies in the Nyquist frequency range $(-\pi/T, \pi/T] = (-\omega_s/2, \omega_s/2]$, then their indexing will be complicated. Because $X_d(\omega)$ is periodic with period ω_s, we can select the frequency range as $[0, \omega_s)$. By so doing, the N equally-spaced frequencies become

$$\omega_m = m\frac{2\pi}{NT} = m\frac{\omega_s}{N} \quad \text{for } m = 0 : N-1 \tag{5.24}$$

The first frequency is located at $\omega = 0$, and the last frequency is located at $(N-1)\omega_s/N$ which is slightly less than ω_s for all integer N. We call m the *frequency index* and

$$D := \frac{2\pi}{NT} = \frac{\omega_s}{N} \quad \text{(rad/s)} \tag{5.25}$$

the *frequency resolution*.

---

[5]Note that e is $1 \times N$ and x is $N \times 1$. Thus $e * x$ also yields (5.23). However if we type e*x in MATLAB, an error message will occur. This may be due to the limitation in dimension in matrix multiplication.

Substituting (5.24) into (5.23) yields

$$X_d(\omega_m) = \sum_{n=0}^{N-1} x(nT)e^{-jm(2\pi/NT)nT} = \sum_{n=0}^{N-1} x(nT)e^{-jmn(2\pi/N)} \qquad (5.26)$$

for $m = 0 : N - 1$. Let us define $X[m] = X_d(\omega_m)$, for $m = 0 : N - 1$, $x[n] := x(nT)$, for $n = 0 : N - 1$, and $W := e^{-j2\pi/N}$. Then (5.26) can be written as

$$X[m] = \sum_{n=0}^{N-1} x[n]W^{nm} \qquad (5.27)$$

for $m = 0 : N - 1$. This equation transforms N data $x[n] = x(nT)$, for $n = 0 : N - 1$ into N data $X[m]$, for $m = 0 : N - 1$ and is called the *discrete Fourier transform (DFT)*. It is possible to develop the inverse DFT and their general properties. These properties will not be used in this text and will not be discussed. The interested reader is referred to, for example, Reference [C7].

There are N terms in the summation of (5.27). Thus computing each $X[m]$ requires N multiplications and computing $X[m]$, for $m = 0 : N - 1$, requires a total of N^2 multiplications. To compute N data, if the number of operations is proportional to N^2, it is a curse because N^2 increases rapidly as N increases. See Reference [K3, p.294]. Thus direct computation of (5.27) for all m is not desirable. For example, for the problem discussed in the preceding subsection, we have $N = 106000$ and the direct computation at one frequency requires 0.016 second. If we compute N frequencies, then it requires $106000 \times 0.016 = 1696$ seconds or more than 28 minutes. It is much too long.

Now if we divide $x[n]$ into two subsequences of length $N/2$ and then carry out the computation of (5.27), the number of operations can be cut almost in half. If N is a power of 2, we can repeat the process $k = \log_2 N$ times. By so doing, the number of operations can be reduced to be proportional to $N \log_2 N$. Based on this idea, many methods, called collectively the *fast Fourier transform (FFT)*, were developed to compute (5.27) efficiently. They are applicable for any positive integer N. For example, to compute all $N = 106000$ data of the problem in the preceding subsection, FFT will take, as will be demonstrated in Section 5.4, only 0.078 second in contrast to 28 minutes by direct computation. Thus FFT is indispensable in computing frequency spectra of signals. For the development of one version of FFT, see Reference [C7].

We discuss only the use of FFT in this text. The use of FFT in MATLAB is very simple. Given a set of N data x expressed as a row vector. Typing Xd=fft(x) will yield N entries which are the values of $X[m]$, for $m = 0 : N - 1$, in (5.27). For example, typing in MATLAB

```
>> x=[1 -2 1];Xd=fft(x)
```

will yield

$$Xd = 0 \quad 1.5000 + 2.5981i \quad 1.5000 - 2.5981i$$

Note that i and j both stand for $\sqrt{-1}$ in MATLAB. What are the meanings of these two sets of three numbers? The first set [1 -2 1] denotes a time sequence of length $N = 3$. Note that the sampling period does not appear in the set, nor in (5.27). Thus

(5.27) is applicable for any $T > 0$. See Problems 4.34 and 4.35. If $T = 0.2$, then the sequence is as shown in Figure 4.19(a) and its spectrum is shown in Figure 4.19(b). Note that the spectrum is periodic with period $2\pi/T = 10\pi = 31.4$ and its plot for ω only in the Nyquist frequency range (NFR) $(-\pi/T, \pi/T] = (-15.7, 15.7]$ is of interest. The second set of numbers Xd is the three equally-spaced samples of the frequency spectrum of x for ω in $[0, 2\pi/T) = [0, 31.4)$ or at $\omega = 0, 10.47, 20.94$ Note that FFT computes the samples of $X_d(\omega)$ for ω in $[0, 2\pi/T)$, rather than in the NFR, for the convenience of indexing. Because the spectrum is complex, we compute its magnitude by typing abs(Xd) and its phase by typing angle(Xd) which yield, respectively,

$$\text{abs}(Xd) = 0 \quad 3 \quad 3$$

and

$$\text{angle}(Xd) = 0 \quad 1.0472 \quad -1.0472$$

They are plotted in Figure 4.19(b) with solid and hollow dots. They are indeed the three samples of $X_d(\omega)$ for ω in $[0, 31.4)$. See Problem 4.35.

In conclusion, if x is a row vector of N entries, then fft(x) will generate a row vector of N entries. The sampling period T comes into the picture only in placing the result of fft(x) at $\omega_m = m[2\pi/(NT)]$ for $m = 0 : N - 1$. This will be used to plot spectra in the next two subsections. Recall that $X_d(\omega)$ is a continuous function of ω and $X[m]$, $m = 0 : N - 1$, yield only samples of $X_d(\omega)$ at $\omega = mD$. Once $X[m]$ are computed, we must carry out interpolation using, for example, the MATLAB function plot to obtain $X_d(\omega)$ for ω in $[0, \omega_s)$.

5.3.1   Plotting spectra for ω in $[0, \omega_s/2]$

Based on the discussion in the preceding sections, we can now develop a procedure of computing and plotting the magnitude spectrum of $x(t)$ for t in $[0, L]$. Its phase spectrum will not be plotted, because it may not be a correct one. We list the procedure:

1. Let $x(t)$ be given analytically or graphically. We first select an L so that $x(t)$ is practically zero for $t \geq L$. We then select an integer N and compute $T = L/N$.[6]

2. Generate N samples of $x(t)$, that is, $x = x(nT)$, for $n = 0 : N - 1$.

3. Use FFT to generate N data and then multiply them by T as $X = T * fft(x)$. These are the N samples of $T X_d(\omega)$ at $\omega = m(2\pi/NT) = mD$ for $m = 0 : N - 1$. In MATLAB, if we type

```
% Subprogram A
X=T*fft(x);
m=0:N-1;D=2*pi/(N*T);
plot(m*D,abs(X))
```

---

[6] If $x(t)$ is to be measured and stored digitally, see Subsection 5.4.3.

then it will generate the magnitude of $TX_d(\omega)$ for ω in $[0,\omega_s)$. Note that the MATLAB function `plot` carries out linear interpolation as shown in Figure 2.8(c). Thus the magnitude so generated is a continuous function of ω in $[0,\omega_s)$. The program is not a complete one, thus we call it a subprogram.

4. If the generated magnitude plot of $TX_d(\omega)$ is significantly different from zero in the neighborhood of $\omega_s/2$ (the middle part of the plot), then the effect of frequency aliasing is significant. Thus we go back to step 1 and select a larger N or, equivalently, a smaller T. We repeat the process until the generated magnitude plot of $TX_d(\omega)$ is practically zero in the neighborhood of $\omega_s/2$ (the middle part of the plot). If so, then frequency aliasing is generally negligible. Subprogram A generates a plot for ω in $[0,\omega_s)$ in which the second half is not used in (5.22). We discuss next how to generate a plot for ω only in $[0,\omega_s/2]$.

5. If $m = 0 : N-1$, the frequency range of mD is $[0,\omega_s)$. Let us assume N to be even[7] and define $mp = 0 : N/2$. Because $(N/2)D = \pi/T = \omega_s/2$, the frequency range of $mp \times D$ is $[0,\omega_s/2]$. In MATLAB, if we type

```
% Subprogram B
X=T*fft(x);
mp=0:N/2;D=2*pi/(N*T);
plot(mp*D,abs(X(mp+1)))
```

then it will generate the magnitude of $TX_d(\omega)$ for ω in $[0,\omega_s/2]$. Note that MATLAB automatically assign the N entries of X by X(k) with $k = 1 : N$. In Subprogram B, if we type `plot(mp*D, abs(X))`, then an error message will appear because the two sequences have different lengths (mp*D has length $(N/2)+1$ and X has length N). Typing `plot(mp*D,abs(X(mp))` will also incur an error message because the internal index of X cannot be 0. This is the reason of using `X(mp+1)` in Subprogram B. The program generates the magnitude of $TX_d(\omega)$ for ω in $[0,\omega_s/2]$. If the plot decreases to zero at $\omega_s/2$, then frequency aliasing is generally negligible (see Problem 5.13) and we have

$$|X(\omega)| \approx \begin{cases} T|X_d(\omega)| & \text{for } 0 \le \omega \le \omega_s/2 \\ 0 & \text{for } \omega > \omega_s/2 \end{cases}$$

In most application, we use Subprogram B; there is no need to use Subprogram A.

6. In item (5) we assumed N to be even. If N is odd, Subprogram B is still applicable without any modification. The only difference is that the resulting plot is for ω in $[0,\omega_s/2)$ instead of $[0,\omega_s/2]$ as in item (5). See Problems 5.6 and 5.7.

Example 5.3.1 We use the preceding procedure to compute the magnitude spectrum of $x(t) = 3e^{-0.2t}\cos 10t$ for $t \ge 0$. It is a signal of infinite length. Thus we must

---

[7]This assumption will be removed in item (6).

truncate it to a finite length. Arbitrarily we select $L = 30$. Because $e^{-0.2t}$ decreases monotonically to 0 and $|\cos 10t| \leq 1$, we have

$$|x(t)| \leq |3e^{-0.2L}| = 3e^{-0.2 \times 30} = 0.0074$$

for all $t \geq 30$, which is practically zero. Thus we may use $x(t)$ for t in $[0, 30)$ to compute the spectrum of $x(t)$ for t in $[0, \infty)$. Next we select $N = 100$. We type in an edit window

```
%Program 5.2(f53.m)
L=30;N=100;T=L/N;n=0:N-1;
x=3*exp(-0.2*n*T).*cos(10*n*T);
X=T*fft(x);
m=0:N-1;D=2*pi/(N*T);
subplot(1,2,1)
plot(m*D,abs(X))
title('(a) Spectrum in [0, 2*pi/T)')
axis square,axis([0 2*pi/T 0 8])
xlabel('Frequency (rad/s)'),ylabel('Magnitude')

mp=0:N/2;
subplot(1,2,2)
plot(mp*D,abs(X(mp+1)))
title('(b) Spectrum in [0, pi/T]')
axis square,axis([0 pi/T 0 8])
xlabel('Frequency (rad/s)'),ylabel('Magnitude')
```

The first two lines generate the N samples of $x(t)$. The third line uses FFT to compute N frequency samples of $TX_d(\omega)$ in $[0, 2\pi/T) = [0, \omega_s)$ located at frequencies mD for $m = 0 : N - 1$. We use linear interpolation to plot in Figure 5.3(a) the magnitude of $TX_d(\omega)$ for ω in $[0, \omega_s) = [0, 20.9)$. We plot in Figure 5.3(b) only the first half of Figure 5.3(a). It is achieved by changing the the frequency indices from m=0:N-1 to mp=0:N/2. Let us save Program 5.2 as f53.m. If we type f53 in the command window, then Figure 5.3 will appear in a figure window.

Does the plot in Figure 5.3(b) yield the spectrum of $x(t)$ for ω in $[0, \omega_s/2]$ as in (5.22)? As discussed earlier, in order for (5.22) to hold , the computed $|TX_d(\omega)|$ must decrease to zero at $\pi/T = 10.45$. This is not the case. Thus frequency aliasing is appreciable and the computed $TX_d(\omega)$ will not approximate well the spectrum of $x(t)$ for ω in $[0, \pi/T] = [0, 10.5]$. In other words, the sampling period $T = L/N = 0.3$ used in Program 5.2 is too large and we must select a smaller one.$\square$

Example 5.3.2 We repeat Example 5.3.1 by selecting a larger N. Arbitrarily we select $N = 1000$. Then we have $T = L/N = 30/1000 = 0.03$ which is one tenth of $T = 0.3$. Program 5.2 with these N and T will yield the plots in Figures 5.4(a) and (b). We see that $|TX_d(\omega)|$ in Figure 5.4(a) is practically zero in the middle part of the frequency range or $|TX_d(\omega)|$ in Figure 5.5(b) decreases to zero at $\pi/T = 104.7$. Thus (5.22) is applicable and the magnitude spectrum of $x(t)$ for $0 \leq \omega \leq \pi/T = 104.7$ is as shown

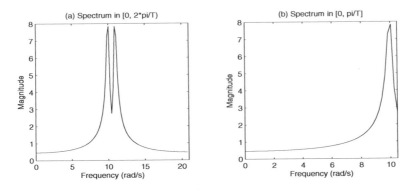

Figure 5.3: (a) Magnitude of $TX_d(\omega)$ for ω in $[0, \omega_s)$ with $T = 0.3$. (b) For ω in $[0, \omega_s/2]$.

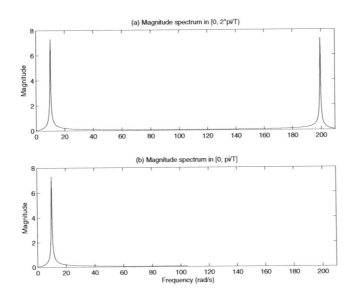

Figure 5.4: (a) Magnitude of $TX_d(\omega)$ for ω in $[0, 2\pi/T) = [0, \omega_s) = [0, 209.4)$ with $T = 0.03$. (b) For ω in $[0, \omega_s/2] = [0, 104.7]$ (rad/s).

in Figure 5.5(b) and is zero for $\omega > 104.7$. The FFT computed magnitude spectrum in Figure 5.4(b) is indistinguishable from the exact one shown in Figure 4.8(a) for $\omega \geq 0$. This shows that the magnitude spectrum of $x(t)$ can indeed be computed from its samples $x(nT)$ if T is selected to be sufficiently small.□

5.3.2 Plotting spectra for ω in $[-\omega_s/2, \omega_s/2)$

In this section,[8] we discuss how to plot $X(\omega)$ for ω in $[-\omega_s/2, \omega_s/2)$ from the FFT generated $TX_d(\omega)$ at

$$\omega = m(2\pi/T) \quad \text{for } m = 0 : N - 1$$

or for ω in $[0, \omega_s)$. We discuss in this subsection only for N even. For N odd, see Reference [C7].

Before proceeding, we introduce the MATLAB function `fftshift`. Let `a=[a1 a2]` be a row vector of even length with `a1` denotes the first half and `a2` the second half. Then the MATLAB function `fftshift(a)` will generate `[a2 a1]`, that is, interchange the positions of the two halves. For example, typing in MATLAB

```
>> fftshift([4 7 1 -2 -5 0])
```

generates

$$\text{ans} = -2 \quad -5 \quad 0 \quad 4 \quad 7 \quad 1$$

In order to plot $X(\omega)$ for ω in $[-\omega_s/2, \omega_s/2)$, we must shift the second half of the FFT generated $TX_d(\omega)$ and the corresponding frequency indices m to negative frequencies. We group m as $[m_1 \; m_2]$, with $m_1 = 0 : N/2 - 1$ and $m_2 = N/2 : N - 1$. Note that m_1 and m_2 have the same number of entries. Applying `fftshift` to m will yield $[m_2 \; m_1]$. Subtracting m_2 by N yields $-N/2 : -1$. Combining this and m_1, we define `mn=-N/2:N/2-1`. It is the frequency indices for ω in $[-\omega_s/2, \omega_s/2)$. Indeed, the first frequency of $mn * D$ is

$$\frac{-N}{2} \frac{2\pi}{NT} = \frac{-\pi}{T} = -0.5\omega_s$$

and the last frequency is

$$\frac{N-2}{2} \frac{2\pi}{NT} = \frac{(N-2)\pi}{NT} = \frac{N-2}{N} \frac{\pi}{T} < \frac{\pi}{T} = 0.5\omega_s$$

Thus the frequency range of $mn * D$ is $[-\omega_s/2, \omega_s/2)$.

Now if we modify Subprogram A as

```
% Subprogram C
X=T*fft(x);
mn=-N/2:N/2-1;D=2*pi/(N*T);
plot(mn*D,abs(fftshift(X)))
```

then the program will plot the magnitude of $TX_d(\omega)$ for ω in $[-\omega_s/2, \omega_s/2)$. It is the magnitude spectrum of $x(t)$ if the plot decreases to zero at $\pm\omega_s/2$.

Let us return to Example 5.3.2. Now we modify Program 5.2 as

---

[8]This section will be used only in generating Figure 5.12 in a later section and may be skipped.

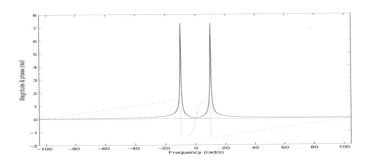

Figure 5.5: FFT computed magnitude (solid line) and phase (dotted line) spectra of $x(t)$.

```
%Program 5.3(f55.m)
L=30;N=1000;T=L/N;n=0:N-1;D=2*pi/(N*T);
x=3*exp(-0.2*n*T).*cos(10*n*T);
X=T*fft(x);
mn=-N/2:N/2-1;
plot(mn*D,abs(fftshift(X)),mn*D,angle(fftshift(X))),':')
```

This program will yield the plot in Figure 5.5. We see that by changing the frequency index to mn and using fftshift, we can readily plot spectra for positive and negative frequencies. Note that the last three lines of Program 5.3 can be replaced by

```
X=fftshift(T*fft(x));mn=-N/2:N/2-1;
plot(mn*D,abs(X),mn*D,angle(X)),':')
```

In application, it is however often sufficient to plot spectra for $\omega \geq 0$ by using Subprogram B.

The FFT generated magnitude spectrum (solid line) in Figure 5.5 is indistinguishable from the exact magnitude spectrum plotted in Figure 4.8(a). The FFT generated phase plot (dotted line), which decreases to zero for $|\omega|$ large, however is different from the exact phase spectrum plotted in Figure 4.8(b), which are constant for $|\omega|$ large. This confirms once again that we can compute the magnitude spectrum of $x(t)$ from $x(nT)$ but not phase spectrum.

5.4   Magnitude spectra of measured data

In this section, we use FFT to compute and to plot the frequency spectra of the signals shown in Figures 1.1 and 1.2. The signals were recorded as wave files in MS-Windows. The following

```
%Program 5.4(f56.m)
xb=wavread('f11.wav');
x=xb([2001:53500]);
```

```
N=length(x);fs=24000;T=1/fs;D=fs/N;
X=T*fft(x);
mp=0:N/2;
subplot(2,2,1)
plot(mp*D,abs(X(mp+1))),title('(a)')
subplot(2,2,2)
plot(mp*D,angle(X(mp+1))),title('(b)')
subplot(2,2,3)
plot(mp*D,abs(X(mp+1))),title('(c)')
axis([110 125 0 0.005])
subplot(2,2,4)
plot(mp*D,angle(X(mp+1))),title('(d)')
axis([110 125 -4 4])
```

is the main part of the program that generates Figure 5.6. The first two lines are those of Program 3.5. The sampling frequency is $f_s = 24{,}000$ Hz. Thus we have $T = 1/f_s$. The frequency resolution $D = 2\pi/NT$ defined in (5.25) is in the unit of rad/s. It becomes $D = 1/NT = f_s/N$ in the unit of Hz. Thus the frequency range of mp*D is $[0, 1/2T] = [0, f_s/2]$ in Hz. Figures 5.6(a) and (b) show the magnitude and phase spectra in $[0, f_s/2] = [0, 12000]$ (Hz) of the time signal shown in Figure 1.1. In order to see them better, we zoom in Figures 5.6(c) and (d) their segments in the frequency range $[110, 125]$ (Hz). This is achieved by typing axis([110 125 y_min y_max]) as shown. We see that the phase spectrum is erratic and is of no use.

Is the magnitude spectrum correct? Because the computed $|TX_d(\omega)|$ is not small in the neighborhood of π/T (rad/s) or $f_s/2 = 12000$ (Hz) as shown in Figure 5.6(a), the plot incurs frequency aliasing. Moreover it is known that spectra of speech are limited mostly to $[20\ 4000]$ (Hz). The magnitude spectrum shown in Figure 5.6(a) is not so limited. It has a simple explanation. The signal recorded contains background noises. The microphone and the cable used also generate noises. No wonder, recording requires a sound-proof studio and expensive professional equipment. In conclusion, the plot in Figure 5.6 shows the spectrum of not only the sound "signals and systems" uttered by the author but also noises.

We next compute the spectrum of the signal shown in Figure 1.2. We plot only its magnitude spectrum. Recall that its sampling frequency is 8,000 Hz. The program that follows

```
%Program 5.5 (f57.m)
x=wavread('f12.wav');
N=length(x);fs=8000;T=1/fs;D=fs/N;
tic,X=T*fft(x);toc
mp=0:N/2;
subplot(1,2,1)
plot(mp*D,abs(X(mp+1))),title('(a)')
axis square
ylabel('Magnitude'),xlabel('Freq. (Hz)')
subplot(1,2,2)
```

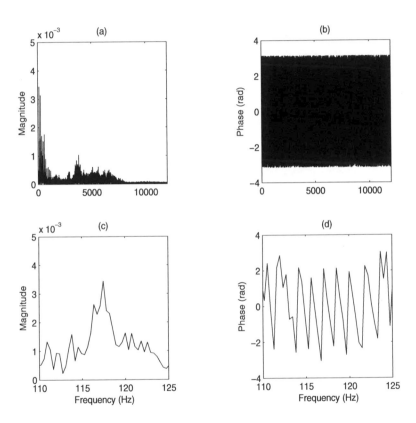

Figure 5.6: (a) Magnitude spectrum in $[0, 0.5f_s]$ (Hz) of the signal in Figure 1.1. (b) Corresponding phase spectrum. (c) Magnitude spectrum in $[110, 125]$ (Hz). (d) Corresponding phase spectrum.

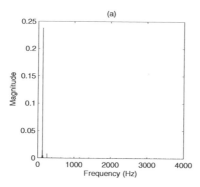 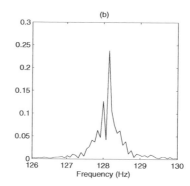

Figure 5.7: (a) Magnitude spectrum in $[0, 0.5 f_s] = [0, 4000]$ (Hz) of the signal in Figure 1.2 sampled with sampling frequency 8000 Hz. (b) In $[126, 130]$ (Hz).

```
plot(mp*D,abs(X(mp+1))),title('(b)')
axis square,axis([126 130 0 0.3])
xlabel('Freq. (Hz)')
```

will generate Figure 5.7 in a figure window and elapsed_time=0.078. Figure 5.7(a) shows the magnitude spectrum for the entire positive Nyquist frequency range $[0, f_s/2] = [0, 4000]$ (Hz); Figure 5.7(b) shows its small segment for frequencies in $[126, 130]$ (Hz). We see that most of the energy of the signal is concentrated around roughly 128 Hz. This is consistent with the fact that the signal is generated by a 128-Hz tuning fork. The peak magnitude however occurs at 128.15 Hz.

Because of the use of tic and toc, Program 5.5 also returns the elapsed time as 0.078 second. Direct computation of the spectrum will take over 28 minutes, FFT computation however takes only 0.078 second. The improvement is over twenty thousand folds. Thus FFT is indispensable in computing spectra of real-world signals.

To conclude this section, we mention an important difference between mathematics and engineering. The signal in Figure 1.2 is clearly time limited, thus its nonzero spectrum must extend to $\pm\infty$. However from Figure 5.7, the spectrum seems to be bandlimited to 1000 Hz or even 300 Hz. In fact, the FFT computed magnitude spectrum is in the order of 0.001 in the entire frequency range $[300, 4000]$. The value is less than 1% of the peak magnitude 0.24 or 0.0024 in Figure 5.7(a) and appears as zero as shown. This is consistent with the way of defining time constant in Section 2.7. Thus a zero in engineering is relative and is different from the zero in mathematics. In engineering, we often accept data or results which appear reasonable. Mathematical precision is not needed.

5.4.1 Downsampling

The magnitude spectrum of the signal in Figure 1.2, as we can see from Figure 5.7(a), is practically zero for frequencies larger than $f_{max} = 200$ Hz. According to the sampling theorem, any sampling frequency larger than $2f_{max} = 400$ Hz can be used to compute the spectrum of $x(t)$ from its sampled sequence $x(nT)$. Clearly the sampling frequency

$f_s = 8$ kHz used to generate the sampled sequence x is unnecessarily large or the sampling period T is unnecessarily small. This is called *oversampling*. We show in this subsection that the same magnitude spectrum of the signal can be obtained using a larger sampling period. In order to utilize the data stored in x, a new sampling period must be selected as an integer multiple of $T := 1/8000$. For example, if we select a new sampling period as $T_1 = 2T$, then its sampled sequence will consist of every other term of x. This can be generated as y=x([1:2:N]), where N is the length of x. Recall that the internal indices of x range from 1 to N and 1:2:N generates the indices 1, 3, 5, ..., up to or less than N. Thus the new sequence y consists of every other term of x. This is called *downsampling*. The program that follows

```
%Program 5.6 (f58.m)
x=wavread('f12.wav');
fs=8000;T=1/fs;N=length(x);a=2;T1=a*T;
y=x([1:a:N]);N1=length(y);
Y=T1*fft(y);
D1=1/(N1*T1);
mp=0:N1/2;
subplot(1,2,1)
plot(mp*D1,abs(Y(mp+1))),title('(a)')
axis square
ylabel('Magnitude'),xlabel('Frequency (Hz)')
subplot(1,2,2)
plot(mp*D1,abs(Y(mp+1))),title('(b)')
axis square,axis([126 130 0 0.3])
ylabel('Magnitude'),xlabel('Frequency (Hz)')
```

will generate Figure 5.8. Note that for $T_1 = 2T$, the new sampling frequency is $f_{s1} = f_s/2 = 4000$ (Hz) and its positive Nyquist frequency range is $[0, f_{s1}/2] = [0, 2000]$ (Hz) as shown in Figure 5.8(a). Its segment in $[128, 130]$ is shown in Figure 5.8(b). We see that the plot in Figure 5.8(b) is indistinguishable from the one in Figure 5.7(b) even though the sampling period of the former is twice of the sampling period of the latter.

If we change a=2 in Program 5.6 to a=10, then the program will yield the plots in Figure 5.9. The spectrum in Figure 5.9 is obtained by selecting its sampling period as $T_2 = 10T$ or its sampling frequency as 800 Hz. Thus it uses only about one tenth of the data in x. The result appears to be also acceptable.

5.4.2   Magnitude spectrum of middle-C sound

We now compute the magnitude spectrum of the signal shown in Figure 1.3. It is obtained using the sampling frequency $f_s = 22050$ Hz. We modify Program 5.5 as

```
%Program 5.7 (f510.m)
xb=wavread('f13.wav');
x=xb([30001:52000]);N=length(x);
```

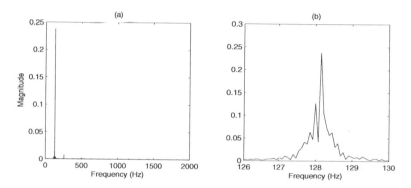

Figure 5.8: (a) Magnitude spectrum in [0, 2000] (Hz) of the signal in Figure 1.2 sampled with sampling frequency 4000 Hz. (b) In [126, 130] (Hz).

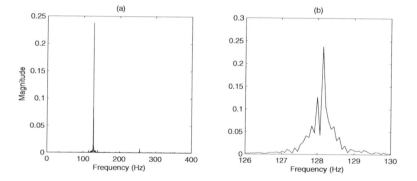

Figure 5.9: (a) Magnitude spectrum in [0, 400] (Hz) of the signal in Figure 1.2 sampled with sampling frequency 800 Hz. (b) In [126, 130] (Hz).

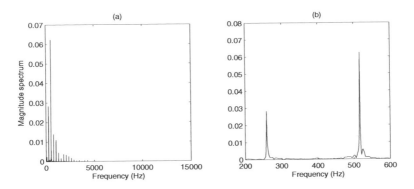

Figure 5.10: (a) Magnitude spectrum in $[0, 11025]$ (Hz) of the middle C in Figure 1.3 with sampling frequency 22050 Hz. (b) In $[200, 600]$ (Hz).

```
fs=22050;T=1/fs;D=fs/N;
X=T*fft(x);
mp=0:N/2;
subplot(1,2,1)
plot(mp*D,abs(X(mp+1))),title('(a)'),axis square
ylabel{'Magnitude spectrum'},xlabel('Frequency (Hz)')
subplot(1,2,2)
plot(mp*D,abs(X(mp+1))),title('(b)'),axis square
axis([200 600 0 0.08]),xlabel('Frequency (Hz)')
```

The first three lines are taken from Program 3.6. The rest is similar to Program 5.5. The program will generate in Figure 5.10(a) the magnitude spectrum for the entire positive Nyquist frequency range $[0, 0.5f_s] = [0, 11025]$ Hz and in Figure 5.10(b) for frequencies between 200 and 600 Hz. It is the magnitude spectrum of the middle C shown in Figure 1.3. Its spectrum has narrow spikes at roughly $f_c = 260$ Hz and kf_c, for $k = 2, 3, 4, \cdots$. Moreover the energy around $2f_c = 520$ is larger than the energy around 260 Hz.[9] In any case, it is incorrect to think that middle C consists of only a single frequency at 261.6 Hz as listed in *Wikipedia*.

From Figure 5.10(a) we see that the spectrum is identically zero for frequency larger than 2500 Hz. Thus we may consider middle C to be bandlimited to $f_{max} = 2500$. Clearly the sampling frequency $f_s = 22050$ used in obtaining Figure 1.3 is unnecessarily large. Thus we may select a smaller $\bar{f}$ which is larger than $2f_{max} = 5000$. In order to utilize the data obtained using $f_s = 22050$, we select $\bar{f}_s = f_s/4 = 5512.5$ or $\bar{T} = 4T = 4/22050$. Using a program similar to Program 5.6 we can obtain the magnitude spectrum shown in Figure 5.11. It is indistinguishable from the one in Figure 5.10(a).

5.4.3   Remarks for spectral computation

We give some remarks concerning FFT spectral computation of real-world signals.

---

[9]The grand piano I used may be out of tune.

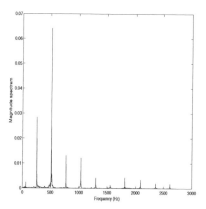

Figure 5.11: Magnitude spectrum in $[0, 0.5\bar{f}_s] = [0, 2756.25]$ (Hz) of the middle C in Figure 1.3 with sampling frequency $\bar{f}_s = f_s/4 = 5512.5$ (Hz).

1. A real-world signal is generally complicated and cannot be described by a simple mathematical equation. A 128-Hz tuning fork will not generate a pure sinusoid with frequency exactly 128 Hz; it generates only a signal with most of its energy residing in a narrow frequency interval centered around 128 Hz.

2. In recording real-world signals, they are often corrupted by noises due to devices used and other sources.

3. In digital recording of real-world signals, we should use the smallest possible sampling period or the highest possible sampling frequency, and hope that frequency aliasing due to time sampling will be small. If it is not small, there is nothing we can do because we already have used the smallest sampling period available.

4. Once the sampled sequence of a CT signal $x(t)$ is available, we can use Subprogram B or C to compute and to plot the magnitude spectrum of $x(t)$ for ω in $[0, \omega_s/2]$ or in $[-\omega_s/2, \omega_s/2)$ (rad/s) or for f in $[0, f_s/2]$ or in $[-f_s/2, f_s/2)$ (Hz). Even though the phase spectrum can also be generated; it may not be correct. Fortunately, we use mostly magnitude spectra in practice.

5. In digital processing of a CT signal, the smaller T used, the more difficult in designing digital filters. See Subsection 11.10.1. Thus it is desirable to use the largest possible T. Such a T can be found by downsampling as discussed in the preceding subsection.

5.5 FFT-computed magnitude spectra of step functions

The frequency spectrum of $x(t) = 1$, for all t in $(-\infty, \infty)$, as discussed in Section 4.5, is $2\pi\delta(\omega)$. The frequency spectrum of $x(t) = 1$, for t in $[0, \infty)$ is $\pi\delta(\omega) + 1/j\omega$. Can we induce these results from computed spectra?

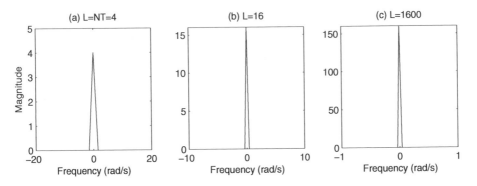

Figure 5.12: (a) FFT-computed magnitude spectrum of $x_L(t) = 1$ with $L = 4$ for ω in $[-20, 20]$. (b) With $L = 16$ for ω in $[-10, 10]$. (c) With $L = 160$ for ω in $[1, 1]$.

We consider

$$x_L(t) = \begin{cases} 1 & \text{for } t \text{ in } [0, L) \\ 0 & \text{for } t < 0 \text{ and } t \geq L \end{cases} \qquad (5.28)$$

for some finite $L > 0$. It is a constant signal of length L. Such a signal can be easily generated in practice. We now use FFT to compute the spectrum of $x_L(t)$ using its samples.

Arbitrarily we select the sampling period as 0.02. First we select $N = 200$. In other words, we compute the spectrum of $x_L(t) = 1$ of length $L = NT = 4$ using its 200 samples. Clearly all samples equal 1. The MATLAB function ones(n,m) generates an $n \times m$ matrix with all entries equal to 1. Thus the string of 200 ones can be generated in MATLAB as ones(1,200), a 1×200 row vector with all entries equal to 1. The program that follows

```
%Program 5.8 (f512.m)
T=0.02;N=200;L=N*T;
x=ones(1,N);
X=T*fft(x)
mn=-N/2:N/2-1;D=2*pi/(N*T);
plot(mn*D,abs(fftshift(X))),title('(a) L=NT=4')
ylabel('Magnitude'),xlabel('Frequency (rad/s)')
axis square,axis([-20 20 0 5])
```

will generate the magnitude spectrum of $x_L(t)$ for ω in $[-\pi/T, \pi/T] = [-157, 157]$ (rad/s). In order to see better the triangle at $\omega = 0$, we use Subprogram C to plot the spectrum for positive and negative frequencies. We plot in Figure 5.12(a) only for ω in $[-20, 20]$. This is achieved by typing the axis function axis([x_min x_max y_min y_max])=([-20 20 0 5]).

The spectrum in Figure 5.12(a) is a triangle. We explain how it is obtained. Recall that FFT generates N frequency samples of $X_L(\omega)$ at $\omega = mD = m(2\pi/NT) = m(2\pi/L)$, for $m = 0 : N - 1$. In our program, because no semicolon is typed after

`X=T*fft(x)`, the program will generate Figure 5.12(a) in a figure window as well as return in the command window 200 numbers. They are

$$4 \quad 0 \quad 0 \quad \cdots \quad 0 \quad 0 \tag{5.29}$$

That is, 4 followed by 199 zeros. If we shift the second half of (5.29) to negative frequencies and then connect neighboring points with straight lines, we will obtain the triangle shown in Figure 5.12(a). The height of the triangle is 4 or $X_L(0) = 4 = L$. Because $D = 2\pi/L = 1.57$, the base is located at $[-1.57, 1.57]$ or $[-2\pi/L, 2\pi/L]$. Thus the area of the triangle is 2π.

Next we select $N = 800$ and then $N = 8000$ or $L = NT = 16$ and then $L = 160$. Program 5.8 with these N and modified axis functions will yield the magnitude spectra in Figures 5.12(b) and (c). Each one consists of a triangle centered at $\omega = 0$ with height L and base $[-2\pi/L, 2\pi/L]$ and is zero elsewhere. As L increases, the height of the triangle becomes higher and its base width becomes narrower. However the area of the triangle remains as 2π. Thus we may induce from the FFT computed results that the magnitude spectrum of the step function $x_L(t) = 1$, with $L = \infty$ is $2\pi\delta(\omega)$, an impulse located at $\omega = 0$ and with weight 2π.

This conclusion is inconsistent with the mathematical results. The spectrum of $x(t) = 1$, for t in $[0, \infty]$ is $\pi\delta(\omega) + 1/j\omega$. The FFT computed spectrum yields an impulse at $\omega = 0$ but with weight 2π and does not show any sign of $1/j\omega$. Actually the FFT computed spectrum leads to the spectrum of $x(t) = 1$ for all t in $(-\infty, \infty)$. It is most confusing. Is it a paradox involving ∞ as in (3.16)? Or is there any error in the FFT computation?

We will show in the next section that the FFT computed spectrum of $x_L(t)$ is mathematically correct for any finite L, even though it leads to a questionable result for $L = \infty$. Fortunately, in engineering, there is no need to be concerned with ∞.

5.5.1 Comparison of FFT-computed and exact magnitude spectra

In this subsection we show that the FFT computed results in the preceding section is correct. In order to do so, we compute analytically the spectrum of $x_L(t) = 1$, for t in $[0, L)$. Its spectrum is, using (4.21),

$$
\begin{aligned}
X_L(\omega) &= \int_{t=-\infty}^{\infty} x_L(t)e^{-j\omega t}dt = \int_{t=0}^{L} e^{-j\omega t}dt = \left.\frac{e^{-j\omega t}}{-j\omega}\right|_{t=0}^{L} \\
&= \frac{e^{-j\omega L} - e^0}{-j\omega} = \frac{1 - e^{-j\omega L}}{j\omega}
\end{aligned}
\tag{5.30}
$$

Typing in MATLAB the following

```
%Program 5.9 (f513.m)
w=-10:0.01:10;L=4;
X=(1-exp(-j*w*L))./(j*w);
plot(w,abs(X))
xlabel('Frequency (rad/s)'),ylabel('Magnitude')
```

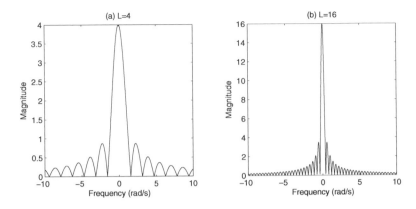

Figure 5.13: (a) Exact magnitude spectrum of $x_L(t)$ for $L = 4$. (b) $L = 16$.

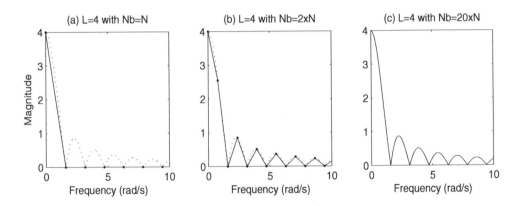

Figure 5.14: (a) Exact magnitude spectrum of $x_4(t)$ (dotted line), FFT-computed magnitudes (dots) using $N = 200$ samples of $x_4(t)$, and their linear interpolation (solid line). (b) With additional N trailing zeros. (c) With additional $19N$ trailing zeros.

yields in Figure 5.13(a) the magnitude spectrum of $x_L(t)$ for $L = 4$. Figure 5.13(b) shows the magnitude spectrum of $x_L(t)$ for $L = 16$.

Now we show that even though the FFT-generated magnitude spectrum in Figure 5.12 looks very different from the exact magnitude spectrum shown in Figure 5.13, the result in Figure 5.12 is in fact correct. Recall that FFT computes the samples of $X_L(\omega)$ at $\omega_m = 2m\pi/NT$. We superpose Figure 5.12(a) (solid line) and Figure 5.13(a) (dotted line) in Figure 5.14(a) for ω in $[0, 10]$. Indeed the FFT computed results are the samples of $X_L(\omega)$ which happen to be all zero except at $\omega = 0$.[10] In conclusion, the the FFT-generated magnitude spectrum in Figure 5.12 is correct. Figure 5.12 looks different from Figure 5.13 only because of the poor frequency resolution and the use of linear interpolation. This problem will be resolved in the next subsection.

---

[10]This can also be verified analytically.

5.5.2 Padding with trailing zeros

The FFT-generated magnitude spectrum in Figure 5.12(a) is different from the exact one in Figure 5.13(a). The question is then: Can we obtain the exact one using FFT? The answer is affirmative if we introduce *trailing zeros*. To be more specific, let us define

$$\bar{x}_L(t) = \begin{cases} x_L(t) & \text{for } t \text{ in } [0, L) \\ 0 & \text{for } t \text{ in } [L, M) \end{cases} \tag{5.31}$$

with $L = N \times T$, $M = \bar{N} \times T$, and $\bar{N} > N$. It is clear that the frequency spectrum of $x_L(t)$ equals the frequency spectrum of $\bar{x}_L(t)$. However if we use FFT to compute the spectrum of $\bar{x}_L(t)$, then its frequency resolution is

$$\bar{D} = \frac{2\pi}{\bar{N}T}$$

which is smaller than $D = 2\pi/(NT)$. Thus more frequency samples of $X_L(\omega)$ will be generated.

Let us use the samples of $\bar{x}_L(t)$ to compute the spectrum of $x_L(t)$. The first N samples of $\bar{x}_L(t)$ are 1 and the remainder are zero. The number of zeros is $\bar{N}$ N. Thus the sampled sequence of $x_L(t)$ can be expressed as

```
xb=[ones(1,N) zeros(1,Nb-N)],
```

where Nb stands for $\bar{N}$ and zeros(n,m) denotes an nxm matrix with all entries equal to zero. We see that padding trailing zeros to x yields xb. If we use xb to compute the spectrum of $x_L(t)$, then the frequency resolution can be improved.

If x has length N, then fft(x) computes N-point fast Fourier transform. The function fft(x,Nb) carries out $\bar{N}$-point fast Fourier transform by using the first N data of x if $N \geq \bar{N}$ or by using x padding with trailing zeros if $\bar{N} > N$. Now we modify Program 5.8 as

```
%Program 5.10 (f514.m)
T=0.02;N=200;Nb=2*N;L=N*T;
x=ones(1,N);
X=T*fft(x,Nb);
mp=0:Nb/2;D=2*pi/(Nb*T);
plot(mp*D,abs(X(mp+1)),'.',mp*D,abs(X(mp+1)))
title('L=4 with Nb=2xN')
ylabel('Magnitude'),xlabel('Frequency (rad/s)')
axis square,axis([0 10 0 L])
```

Then it will generate the solid dots and the solid line in Figure 5.14(b). Note that the solid dots are generated by plot(mp*D,abs(X(mp+1)),'.') and the solid line by plot(mp*D,abs(X(mp+1))). See Problem 3.29. Note that two or more plot functions can be combined as in the fifth line of the program. We plot in Figure 5.14(b) with dotted line also the exact magnitude spectrum for comparison. Program 5.10 uses Nb=2xN or adds 200 trailing zeros, thus it has a frequency resolution half of the one in Program 5.8. Consequently Program 5.10 generates one extra point between any two immediate points in Figure 5.14(a) as shown in Figure 5.14(b). We see that the exact and computed spectra become closer. If we use `Nb=20xN`, then the program generates extra 19 points between any two immediate points in Figure 5.14(a) and the computed magnitude spectrum (solid lines) is indistinguishable from the exact one (dotted lines) as shown in Figure 5.14(c). In conclusion, the exact magnitude spectrum in Figure 5.13(a) can also be obtained using FFT by introducing trailing zeros.

5.6   Magnitude spectra of dial and ringback tones

Our telephones use dual-tone multi-frequency signals to carry out all functions. When we pick up a handset, the dial tone will indicate that the telephone exchange is working and ready to accept a telephone number. We use an MS sound recorder (with audio format: PCM, 8.000 kHz, 8 Bit, and Mono) to record a dial tone over 6.5 seconds as shown in Figure 5.15(a). We plot in Figure 5.15(aa) its small segment for t in $[2, 2.05]$. We then use a program similar to Program 5.5 to plot the magnitude spectrum of the dial tone in Figure 5.15(b) for the entire positive Nyquist frequency range and in Figure 5.15(bb) its small segment for frequency in $[200, 600]$ (Hz). It verifies that the dial tone is generated by two frequencies 350 and 440 Hz.

We repeat the process by recording a ringback tone. Figure 5.16(a) shows a ringback tone for about 8 seconds. We plot in Figure 5.16(aa) its small segment for t in $[4, 4.1]$. We then use a program similar to Program 5.5 to plot its magnitude spectrum in Figure 5.16(b) for the entire positive Nyquist frequency range and in Figure 5.16(bb) its small segment for frequency in $[350, 550]$ (Hz). It verifies that the ringback tone is generated by two frequencies 440 and 480 Hz.

In a touch-tone telephone, each number generates a tone with two frequencies according to the following table:

	1209	1336	1477
697 Hz	1	2	3
770	4	5	6
852	7	8	9
941	*	0	#

In detecting these numbers, there is no need to compute their spectra. All that is needed is to detect the existence of such frequencies. Thus for each tone, we need to compute its spectrum at eight frequencies. The interested reader is referred to References [C4, M2].

5.7   Do frequency spectra play any role in real-time processing?

Consider a CT signal $x(t)$. If its Fourier transform or spectrum $X(\omega)$ can be computed *exactly*, then the time-domain description $x(t)$ and its frequency-domain description $X(\omega)$ are equivalent and either one can be computed from the other. Unfortunately, computing exact transforms is possible only for some mathematical functions. For a real-world signal, we can only compute approximately its magnitude spectrum (the

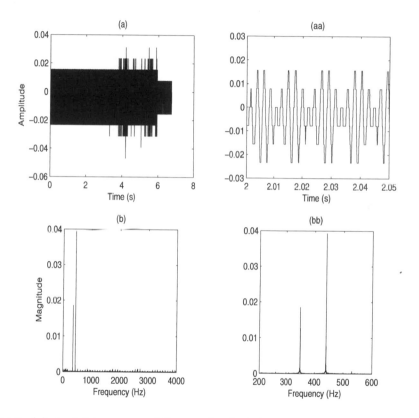

Figure 5.15: (a) A dial tone over 6.5 seconds. (aa) Its small segment. (b) Magnitude spectrum of (a). (bb) Its segment between $[200, 600]$.

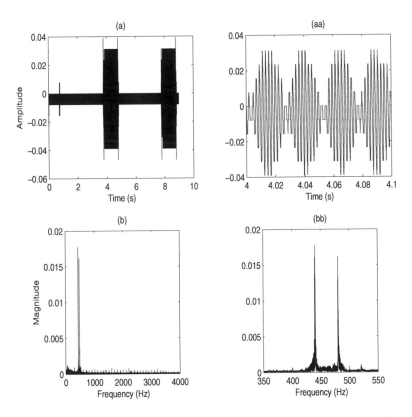

Figure 5.16: (a) A ringback tone over 8 seconds. (aa) Its small segment. (b) Magnitude spectrum of (a) in $[0, 4000]$ (Hz). (bb) Its segment in $[350, 550]$.

computed phase spectrum may not be correct). Fortunately, in practical application, we don't need the exact $X(\omega)$. For example, in determining a sampling period using $T < \pi/\omega_{max}$, we use only $|X(\omega)| \approx 0$ for $|\omega| > \omega_{max}$. Moreover what is approximately zero is open for argument. In communication and filter design, again we don't need exact $X(\omega)$, we need only the bandwidth of signals. In addition, we may have many different definitions of bandwidth. See Section 4.4.3. In engineering, the ultimate purpose is to design and build systems or devices which work. A precise definition of the bandwidth is not critical.

Once a system is designed and built, will the spectrum $X(\omega)$ be of any use in the processing of $x(t)$? When we speak into a telephone, the speech will appear almost instantaneously at the other end. When we turn on a CD player, music will appear as soon as the CD starts to spin. This is real-time processing. There is no need to wait until we complete a sentence or until the CD player completes the scanning of the entire CD. Thus all real-world processing is in the time domain and in real time. We can compute $X(\omega)$ only after the entire $x(t)$ is received. Thus its computation cannot be in real time. In conclusion, $X(\omega)$ plays no role in real-time processing.

5.7.1   Spectrogram

Even though we use only bandwidths of signals in this text, magnitude spectra of signals are used in the study of speech. Because the Fourier transform is a global property, it is difficult to subtract local characteristics of a time signal form its spectrum. Thus in speech analysis, we divide the time interval into small segments of duration, for example, 20 ms, and compute the spectrum of the time signal in each segment. We then plot the magnitude spectra against time and frequency. Note that the phase spectra are discarded. Instead of plotting the magnitude spectra using a third ordinate, magnitudes are represented by different shades or different colors on the time-frequency plane. A same shade or color is painted over the entire time segment and over a small frequency range of width, for example, 200 Hz. The shade is determined by the magnitude at, for example, the midpoint of the range and the magnitude is then quantized to the limited number of shades. The resulting two-dimensional plot is called a *spectrogram*. It reveals the change of magnitude spectrum with time. It is used in speech recognition and identification. See References [M4, S1].

Problems

5.1 Consider a CT signal $x(t)$. It is assumed that its spectrum is real and as shown in Figure 5.17. What are the spectra of $x(nT)$ and $Tx(nT)$ if $T = \pi/25$? Is there frequency aliasing?

5.2 Repeat Problem 5.1 for $T = \pi/15$ and $T = \pi/10$.

5.3 Consider a CT signal $x(t)$ with spectrum given by

$$X(\omega) = \begin{cases} e^{-j\omega t_0} & \text{for } |\omega| \leq 8 \\ 0 & \text{for } |\omega| > 8 \end{cases}$$

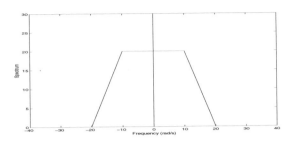

Figure 5.17: Spectrum of $x(t)$.

with $t_0 = \pi/8 = 0.39$.

1. Plot the magnitudes and phases of $X(\omega)$, $X(\omega - 8)$ and $X(\omega + 8)$, for ω in $[-20, 20]$. Can the magnitude spectra of $X(\omega-8)$ and $X(\omega+8)$ be obtained from the magnitude spectrum of $X(\omega)$ by folding with respect to ± 4? Can the phase spectra of $X(\omega - 8)$ and $X(\omega + 8)$ be obtained from the phase spectrum of $X(\omega)$ by folding with respect to ± 4?

2. Compute analytically the spectrum of $TX_d(\omega)$ with sampling period $T = \pi/4$ and then plot the magnitude and phase of $TX_d(\omega)$ for ω in the NFR $(-\pi/T, \pi/T] = (-4, 4]$. Can they be obtained from the plots in Part 1? Discuss the effect of t_0. Can you conclude that the effect of frequency aliasing is complicated if $X(\omega)$ is not real valued?

5.4* If $x(t)$ is bandlimited to ω_{max} and if $T < \pi/\omega_{max}$, then $X(\omega) = TX_d(\omega)$ for $|\omega| < \pi/T$ as shown in (5.14). Substitute (5.7) into (5.14) and then into (4.22) to verify

$$x(t) = \sum_{n=-\infty}^{\infty} x(nT)\frac{\sin[\pi(t - nT)/T]}{\pi(t - nT)/T} = \sum_{n=-\infty}^{\infty} x(nT)\text{sinc}(\pi(t - nT)/T)$$

Thus for a bandlimited signal, its time signal $x(t)$ can be computed exactly from its samples $x(nT)$. This is dual to Problem 4.19 which states that for a time-limited signal, its frequency spectrum (Fourier transform) can be computed exactly from its frequency samples (Fourier series).

5.5 Consider $x(t)$ defined for t in $[0, 5)$. If we select $N = 5$ equally spaced samples of $x(t)$, where are $x(nT)$, for $n = 0 : 4$? What is its sampling period T? Do we have $L = NT$?

5.6 Consider $x(t)$ defined for t in $[0, 5]$. If we select $N = 5$ equally spaced samples of $x(t)$, where are $x(nT)$, for $n = 0 : 4$? What is its sampling period T? Do we have $L = NT$?

5.7 Define $m = 0 : N - 1$, $mp = 0 : N/2$, and $D = 2\pi/NT$. If $N = 6$, what are m and mp? What are the frequency ranges of $m \times D$ and $mp \times D$?

5.8 Repeat Problem 5.7 for $N = 5$.

5.9 Use FFT to compute the magnitude spectrum of $a_L(t) = \cos 70t$, for t defined in $[0, L)$ with $L = 4$. Select the sampling period as $T = 0.001$. Plot the magnitude spectrum for ω in the positive Nyquist frequency range $[0, \pi/T]$ (rad/s) and then zoom it in the frequency range $[0, 200]$.

5.10 Consider Problem 5.9. From the magnitude spectrum obtained in Problem 5.9, can you conclude that $\cos 70t$, for t in $[0, 4]$, is roughly band-limited to 200 rad/s. Note that it is band-limited to 70 rad/s only if $\cos 70t$ is defined for all t in $(-\infty, \infty)$. Use $T = \pi/200$ to repeat Problem 5.9. is the result roughly the same as the one obtained in Problem 5.9? Compare the numbers of samples used in both computations.

5.11 Use FFT to compute and to plot the magnitude spectrum of

$$x(t) = \sin 20t + \cos 21t - \sin 40t$$

for t defined in $[0, L)$ with $L = 50$. Use $N = 1000$ in the computation. Are the peak magnitudes at the three frequencies 20, 21, and 40 rad/s the same? If you repeat the computation using $L = 100$ and $N = 2000$, will their peak magnitudes be the same?

5.12 Consider Problem 5.11 with $L = 50$ and $N = 1000$. Now introduce $4N$ trailing zeros into the sequence and use $5N$-point FFT to compute and to plot the magnitude spectrum of $x(t)$ for t in $[0, 50)$. Are the peak magnitudes at the three frequencies 20, 21, and 40 rad/s roughly the same? Note that this and the preceding problem compute spectra of the same signal. But the frequency resolution in this problem is $2\pi/5NT$, whereas the frequency resolution in Problem 5.11 is $2\pi/NT$. The former is five time smaller (or better) than the latter.

5.13 Consider the signal

$$x(t) = \sin 20t \cos 21t \sin 40t$$

defined for t in $[0, 50)$ and zero otherwise. It is time limited, therefore its spectrum cannot be band-limited. Use FFT and $N = 1000$ to compute and to plot its magnitude spectrum. At what frequencies are the spikes located? Use the formula

$$
\begin{aligned}
4 \sin A \cos B \sin C \quad = \quad & \cos(-A + B + C) - \cos(A - B + C) \\
& + \cos(A + B - C) - \cos(A + B + C)
\end{aligned}
$$

to verify the correctness of the result. Which spike is due to frequency aliasing? In general if the FFT-computed spectrum approaches zero in the neighborhood of π/T, then there is no frequency aliasing. This example shows an exception because the spectrum of $x(t)$ consists of only spikes. Repeat the computation using $N = 2000$. Where are the spikes located? Is frequency aliasing significant?

5.14 Consider the signal shown in Figure 5.18. The signal is defined for $t \geq 0$; it equals $x(t) = 0.5t$, for $0 \leq t < 2$ and then repeats itself every 2 seconds. Verify that the MATLAB program

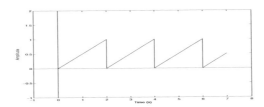

Figure 5.18: Periodic signal.

```
T=0.01;t=0:T:2-T;x=0.5*t;N=length(x);
n=0:N-1;D=2*pi/(N*T);X=T*fft(x);
mp=0:N/2;plot(mp*D,abs(X(mp+1)))
axis([0 50 0 1])
```

will yield the plot in Figure 5.19(a). It is the magnitude spectrum of $x(t)$ (one period). Verify that Figure 5.19(b) shows the magnitude spectrum of five periods and Figure 5.19(c) shows the magnitude spectrum of 25 periods. (Hint: Use x1=[x x x x x] and x2=[x1 x1 x1 x1 x1].)

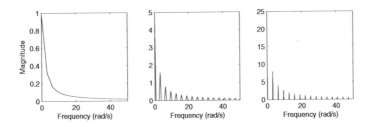

Figure 5.19: Magnitude spectra of (a) one period, (b) five periods, and (c) twenty five periods of the signal in Figure 5.18.

5.15 Let A be real and negative. What are its magnitude and phase? If $A \to 0$, will $|A|$ and $\measuredangle A$ approach zero?

Chapter 6

Systems – Memoryless

6.1 Introduction

Just like signals, systems exist everywhere. There are many types of systems. A system that transforms a signal from one form to another is called a *transducer*. There are two transducers in every telephone set: a microphone that transforms voice into an electrical signal and a loudspeaker that transforms an electrical signal into voice. Other examples of transducers are strain gauges, flow-meters, thermocouples, accelerometers, and seismometers. Signals acquired by transducers often are corrupted by noise. The noise must be reduced or eliminated. This can be achieved by designing a system called a *filter*. If a signal level is too small or too large to be processed by the next system, the signal must be amplified or attenuated by designing an *amplifier*. Motors are systems that are used to drive compact discs, audio tapes, or huge satellite dishes.

The temperature of a house can be maintained automatically at a desired temperature by designing a home heating system. By setting the thermostat at a desired temperature, a burner will automatically turn on to heat the house when the temperature falls below the desired temperature by a threshold and will automatically shut off when it returns to the desired temperature. The US economy can also be considered a system. Using fiscal policy, government spending, taxes, and interest rates, the government tries to achieve that the gross domestic product (GDP) grows at a healthy but not overheated rate, that the unemployment rate is acceptable and that inflation is under control. It is a complicated system because it is affected by globalization, international trades, and political crises or wars. Consumers' psychology and rational or irrational expectation also play a role. Such a system is very complicated.

To a user, an audio compact disc (CD) player is simply a system; it generates an audio signal from a CD. A CD player actually consists of many subsystems. A CD consists of one or more spiral tracks running from the inner rim to the outer rim. The spacing between tracks is 1.6 μm (10^{-6} meter). Roughly a human hair placing along the tracks can cover 40 tracks. There are pits and lands along the tracks with width 0.6 μm and various lengths. The transition from a pit to a land and vice versa denotes a 1 and no transition denotes a 0. A CD may contain over 650 million such binary digits or bits. To translate this stream of bits impressed on the disc's plastic substrate into an electrical form requires three control systems. A control system will spin the CD at various angular speeds from roughly 500 rpm (revolution per minute) at the inner rim to 200 rpm at the outer rim to maintain a constant linear velocity of 1.2 m/s; another control system will position the reading head; yet another control system will focus the laser beam on the track. The reflected beam is pick up by photodiodes to generate streams of 1s and 0s in electrical form. The stream of bits contains not only the 16-bit data words of an audio signal sampled at 44.1 kHz but also bits for synchronization, eight-to-fourteen modulation (EFM) (to avoid having long strings of 0s), and error correction codes (to recover data from scratches and fingerprints). A decoder will subtract from this stream of bits the original audio signal in binary form. This digital signal will go through a digital-to-analog converter (DAC) to drive a loudspeaker. Indeed a CD player is a very complicated system. It took the joint effort of Philips and Sony Corporations over a decade to bring the CD player to the market in 1982.

The word 'systems' appears in systems engineering, systems theory, linear systems, and many others. Although they appear to be related, they deal with entirely different subject areas. Systems engineering is a discipline dealing with macro design and management of engineering problems. Systems theory or general systems theory, according to Wikipedia, is an interdisciplinary field among ontology, philosophy, physics, biology, and engineering. It focuses on organization and interdependence of general relationship and is much broader than systems engineering. Linear systems, on the other hand, deal with a small class of systems which can be described by simple mathematical equations. The systems we will study in this text belong to the last group. The study of linear systems began in the early 1960 and the subject area has now reached its maturity. We study in this text only a small part of linear systems which is basic and relevant to engineering.

6.2   A study of CT systems

Physical systems can be designed empirically. Using available components and trial-and-error, physical systems have been successfully designed. For example, designing and installing a heating system is a long standing problem and much experience and data have been gathered. From the volume of the space to be heated and the geographical location of the house, we can select and install a satisfactory heating system. Indeed a large number of physical systems had been successfully designed even before the advent of engineering curriculums. See Section 9.10.1.

The empirical method relies heavily on past experience and is carried out by trial and error. The method may become unworkable if physical systems are complex or too expensive or too dangerous to be experimented on. In these cases, analytical methods become necessary. The analytical study of physical systems consists of four parts: modeling, development of mathematical descriptions, analysis, and design. We briefly introduce each of these tasks.

The distinction between a physical system and its model is basic in engineering. For example, circuits and control systems studied in any textbook are models of physical systems. A resistor with a constant resistance is a model; it will burn out if the applied voltage is over a limit. The power limitation is often disregarded in its analytical study.

An inductor with a constant inductance is also a model; in reality, the inductance may vary with the amount of current passing through it. Modeling is a very important problem, for the success of the design depends on whether the physical system is modeled properly.

A physical system may have different models depending on the questions asked. For example, a spaceship can be modeled as a particle in investigating its trajectory; but it must be modeled as a rigid body in maneuvering. A spaceship may even be modeled as a flexible body when it it is connected to the space station. In order to develop a suitable model for a physical system, a thorough understanding of the physical system and its operational range is essential. In this text, we will call a model of a physical system a *system*. Thus a physical system is a device or a collection of devices existing in the real world; a system is a model of a physical system.

A system or a model can be classified dichotomously as linear or nonlinear, time-invariant or time varying, and lumped or distributed. We study in this text mostly the class of systems which are linear (L), time-invariant (TI), and lumped. For this class of systems, we will develop four types of mathematical equations to describe them. They are convolutions, high-order differential (difference) equations, state-space (ss) equations, and transfer functions.

Once we have mathematical equations, we can carry out analysis and design. There are two types of analyses: quantitative and qualitative. Although we have four types of mathematical equations, we give reasons for using only ss equations in quantitative analysis and only transfer functions in qualitative analysis. We also give reasons for using only transfer functions in design.

Not every physical system can be approximated or modeled by an LTI lumped model. However a large number of physical systems can be so modeled under some approximation and limitation as we will discuss in the text. Using the models, we can then develop mathematical equations and carry out analysis and design. Note that the design is based on the model and often does not take the approximation and limitation into consideration. Moreover, the design methods do not cover all aspects of actual implementations. For example, they do not discuss, in designing a CD player, how to focus a laser beam on a spiral track of width 0.6 μm, how to detect mis-alignment, and how to maintain such a high precision control in a portable player. These issues are far beyond the design methods discussed in most texts. Solving these problems require engineering ingenuity, creativity, trial-and-error, and years of development. Moreover, textbooks' design methods do not concern with reliability and cost. Thus what we will introduce in the remainder of this text is to provide only the very first step in the study and design of physical systems.

6.2.1 What role do signals play in designing systems?

We studied time signals in the preceding chapters. We introduced the Fourier transform to develop the concept of frequency spectrum. Because most signals are complicated and cannot be described by exact mathematical equations, we can compute their spectra only numerically. Moreover we can compute only their magnitude spectra.

Many systems such as amplifiers and filters are designed to process signals. Once a system is designed, the system can be applied to any signal. For example, whoever talks into a telephone or even play into it a violin, the sound will be transmitted to the receiver. The transmission is carried out directly in the time domain and in real time. The concept of frequency spectrum does not play any role in the processing. It is used only in the specification of the system to be designed. For example, most frequency spectra of human speech are limited to 4 kHz. Thus the telephone system is designed to process signals whose spectra are limited to the range [0, 4000] Hz. The spectrum of the sound generated by a violin can go outside the range, but the telephone will still transmit the sound. The only problem is the poor quality of the received sound. In conclusion, frequency spectra are used only in the specification of the system to be designed. Moreover, we do not need exact frequency spectra, all we need is their frequency bandwidths or frequency ranges in which most of the energy of signals reside. Once a system is designed, it will process whatever signals it receives without regarding their spectra. However, the system will function as designed only if the spectrum of the applied signal lies inside the specified region.

Most signals are generated naturally and are processed directly, thus we did not discuss much their properties and manipulation in the preceding chapters. On the other hand, systems are to be designed and built, thus they require a great deal of study as mentioned in the preceding section. Because the Fourier transform introduced for studying signals can also be applied to study systems, many texts combine the study of signals and systems. Indeed the mathematical conditions studied in Section 3.5 are directly applicable to signals and systems. But all real-world signals automatically meet the condition, whereas systems to be designed may or may not meet the condition. Thus the study of systems will involve many concepts which have no counterpart in the study of signals, and should be different from the latter. They intersects only in specifications of systems to be designed.

6.3   Systems modeled as black box

This text studies only *terminal properties* of systems. Thus a system will be modeled as a *black box* with at least one input terminal and at least one output terminal as shown in Figure 6.1.[1] A terminal does not necessarily mean a physical wire sticking out of the box. It merely means that a signal can be applied or measured at the terminal. The signal applied at the input terminal is called an *input* or an *excitation*. The signal at the output terminal is called an *output* or a *response*. The only condition for a black box to qualify as a system is that every input excites a unique output. For example, consider the RC circuit shown in Figure 6.2(a). It can be modeled as a black box with the voltage source as the input and the voltage across the capacitor as the output. Figure 5.2(b) shows a block with mass m connected to a wall through a spring. The connection of the two physical elements can also be so modeled with the applied force $u(t)$ as the input and the displacement $y(t)$ measuring from the equilibrium position [2] as the output. It is called a black box because we are mainly concerned with its terminal properties. This modeling concept is a powerful one because it can be applied to any system, be it electrical, mechanical or chemical.

---

[1]It is not the same as the black box on an airplane which is a flight data recorder enclosed in an indestructible case. The black box is often painted in orange or red color for easy spotting.

[2]It is the position where the block is stationary before the application of an input.

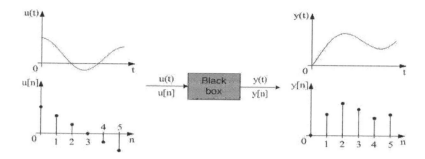

Figure 6.1: System.

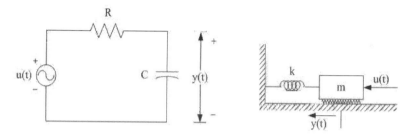

Figure 6.2: (a) RC circuit. (b) Mechanical system.

A system is called a single-input single-output (SISO) system if it has only one input terminal and only one output terminal. It is called a multi-input multi-output (MIMO) system if it has two or more input terminals and two or more output terminals. Clearly, we can have SIMO or MISO systems. We study in this text mostly SISO systems. However most concepts and computational procedure are directly applicable to MIMO systems.[3]

A SISO system is called a continuous-time (CT) system if the application of a CT signal excites a CT signal at the output terminal. It is a discrete-time (DT) system if the application of a DT signal excites a DT signal at the output terminal. The concepts to be introduced are applicable to CT and DT systems.

6.4 Causality, time-invariance, and initial relaxedness

For a CT system, we use $u(t)$ to denote the input and $y(t)$ to denote the output. In general, the output at time t_0 may depend on the input applied for all t. If we call $y(t_0)$ the *current* output, then $u(t)$, for $t < t_0$, is the past input; $u(t_0)$, the current input; and $u(t)$, for $t > t_0$, the future input. It is simple to define mathematically a system whose current output depends on past, current, and future inputs. See Problem 6.1.

---

[3]However the design procedure is much more complex. See Reference [C6].

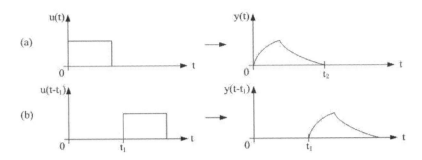

Figure 6.3: (a) An input-output pair. (b) If the input is shifted by t_1, then the output is shifted by t_1.

A system is defined to be *causal*,[4] if its current output depends on past and/or current input but not on future input. If a system is not causal, then its current output will depend on a future input. Such a system can *predict* what will be applied in the future. No physical system has such capability. Thus *causality is a necessary condition for a system to be built in the real world.* This concept will be discussed further in Section 7.2. This text studies only causal systems.[5]

If the characteristics of a system do not change with time, then the system is said to be *time invariant (TI)*. Otherwise it is *time varying*. For the RC circuit in Figure 6.2(a), if the resistance R and the capacitance C are constant, independent of t, then the circuit is time-invariant. So is the mechanical system if the mass m and the spring constant k are independent of time. For a time-invariant system, no matter at what time we apply an input, the output waveform will always be the same. For example, suppose the input $u(t)$ shown on the left hand side of Figure 6.3(a) excites the output $y(t)$ shown on its right hand side and denoted as

$$u(t), \ t \geq 0 \ \rightarrow \ y(t), \ t \geq 0$$

Now if we apply the same input t_1 seconds later as shown in Figure 6.3(b), and if the system is time invariant, then the same output waveform will appear as shown in Figure 6.3(b), that is,

$$u(t - t_1), \ t \geq t_1 \ \rightarrow \ y(t - t_1), \ t \geq t_1 \quad \text{(time-shifting)} \qquad (6.1)$$

Note that $u(t - t_1)$ and $y(t - t_1)$ are the shifting of $u(t)$ and $y(t)$ by t_1, respectively. Thus a time-invariant system has the time-shifting property.

For a system to be time-invariant, the time-shifting property must hold for all t_1 in $[0, \infty)$. Most real-world systems such as TVs, automobiles, computers, CD and DVD players will break down after a number of years. Thus they cannot be time invariant in the mathematical sense. However they can be so considered before their breakdown. We study in this text only time-invariant systems.

---

[4]Not to be confused with casual.

[5]Some texts on signals and systems call positive-time signals *causal signals*. In this text, causality is defined only for systems.

The preceding discussion in fact requires some qualification. If we start to apply an input at time t_0, we require the system to be *initially relaxed* at t_0. By this, we mean that the output $y(t)$, for $t \geq t_0$, is excited *exclusively* by the input $u(t)$, for $t \geq t_0$. For example, when we use the odometer of an automobile to measure the distance of a trip, we must reset the meter. Then the reading will yield the distance traveled in that trip. Without resetting, the reading will be incorrect. Thus when we apply to a system the inputs $u(t)$ and $u(t - t_1)$ shown in Figure 6.3, the system is required to be *initially relaxed* at $t = 0$ and $t = t_1$ respectively. If the output $y(t)$ lasts until t_2 as shown in Figure 6.3(a), then we require $t_1 > t_2$. Otherwise the system is not initially relaxed at t_1. In general, if the output of a system is identically zero before we apply an input, the system is initially relaxed. Resetting a system or turning a system off and then on will make the system initially relaxed.

We study in this text only time-invariant systems. For such a system, we may assume, without loss of generality, the initial time as $t_0 - 0$. The initial time is not an absolute one; it is the instant we start to study the system. Thus the time interval of interest is $[0, \infty)$

6.4.1   DT systems

We discuss the DT counterpart of what has been discussed so far for CT systems. A system is a DT system if the application of a DT signal excites a unique DT signal. We use $u[n] := u(nT)$ to denote the input and $y[n] := y(nT)$ to denote the output, where n is the time index and can assume only integers and T is the sampling period. As discussed in Section 2.6.1, processing of DT signals is independent of T so long as T is large enough to carry out necessary computation. Thus in the study of DT systems, we may assume $T = 1$ and use $u[n]$ and $y[n]$ to denote the input and output.[6] Recall that variables inside a pair of brackets must be integers.

The current output $y[n_0]$ of a DT system may depend on past input $u[n]$, for $n < n_0$, current input $u[n_0]$, and future input $u[n]$, for $n > n_0$. A DT system is defined to be *causal* if the current output does not depend on future input; it depends only on past and/or current inputs. If the characteristics of a DT system do not change with time, then the system is time invariant. For a time-invariant DT system, the output sequence will always be the same no matter at what time instant the input sequence is applied. This can be expressed as a terminal property using input-output pairs. If a DT system is time invariant, and if $u[n]$, $n \geq 0 \rightarrow y[n]$, $n \geq 0$, then

$$u[n - n_1], \ n \geq n_1 \ \rightarrow \ y[n - n_1], \ n \geq n_1 \quad \text{(time-shifting)} \quad (6.2)$$

for any $u[n]$ and any $n_1 \geq 0$. As in the CT case, the system must be initially relaxed when we apply an input. For a DT time-invariant system, we may select the initial time as $n_0 = 0$; it is the time instant we start to study the system.

6.5   Linear time-invariant (LTI) memoryless systems

In the remainder of this chapter we study memoryless systems. A system is *memoryless* if its output $y(t_0)$ depends only on the input applied at the same time instant t_0 or

---

[6]In practice, T rarely equals 1. See Section 11.10 and its subsection to resolve this problem.

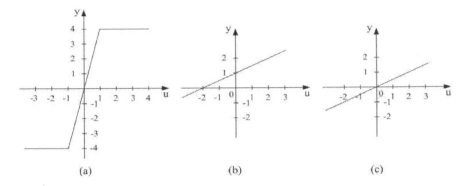

Figure 6.4: (a) Nonlinear. (b) Nonlinear. (c) Linear.

$u(t_0)$; it does not depend on $u(t)$, for $t < t_0$ and $t > t_0$. Because $y(t_0)$ does not depend on $u(t)$, for $t > t_0$, every memoryless system is causal.

The input and output of a memoryless system can be described by

$$y(t_0) = f(u(t_0), t_0) \quad \text{or} \quad y(t) = f(u(t), t)$$

if the system is time varying or by

$$y(t_0) = f(u(t_0)) \quad \text{or} \quad y(t) = f(u(t)) \tag{6.3}$$

if the system is time invariant. The function f in (6.3) is independent of t. In this case, we can drop t and write (6.3) as $y = f(u)$. In this equation, every u, a real number will generate a unique real number as its output y. Such a function can be represented graphically as shown in Figure 6.4. For each u, we can obtain graphically the output y. We use $u_i \to y_i$, for $i = 1, 2$, to denote that u_i excites y_i.

A time-invariant memoryless system is defined to be *linear* if for any $u_i \to y_i$, $i = 1, 2$, and for any real constant β, the system has the following two properties

$$u_1 + u_2 \quad \to \quad y_1 + y_2 \quad \text{(additivity)} \tag{6.4}$$
$$\beta u_1 \quad \to \quad \beta y_1 \quad \text{(homogeneity)} \tag{6.5}$$

If a system does not have the two properties, the system is said to be *nonlinear*. In words, if a system is linear, then the output of the system excited by the sum of any two inputs should equal the sum of the outputs excited by individual inputs. This is called the *additivity* property. If we increase the amplitude of an input by a factor β, the output of the system should equal the output excited by the original input increased by the same factor. This is called the *homogeneity* property. For example, consider a memoryless system whose input u and output y are related by $y = u^2$. If we apply the input u_i, then the output is $y_i = u_i^2$, for $i = 1, 2$. If we apply the input $u_3 = u_1 + u_2$, then the output y_3 is

$$y_3 = u_3^2 = (u_1 + u_2)^2 = u_1^2 + 2u_1 u_2 + u_2^2$$

which is different from $y_1 + y_2 = u_1^2 + u_2^2$. Thus the system does not have the additivity property and is nonlinear. Its nonlinearity can also be verified by showing that the system does not have the homogeneity property.

Let us check the memoryless system whose input and output are related as shown in Figure 6.4(a). From the plot, we can obtain $\{u_1 = 0.5 \to y_1 = 2\}$ and $\{u_2 = 2 \to y_2 = 4\}$. However, the application of $u_1 + u_2 = 2.5$ yields the output 4 which is different from $y_1 + y_2 = 6$. Thus the system does not have the additivity property and is nonlinear. Note that we can also reach the same conclusion by showing that $5u_1 = 2.5$ excites 4 which is different from $5y_1 = 10$. The nonlinearity in Figure 6.4(a) is called *saturation*. It arises often in practice. For example, opening a valve reaches saturation when it is completely open. The speed of an automobile and the volume of an amplifier will also saturate.

The system specified by Figure 6.4(b) is nonlinear as can be verified using the preceding procedure. We show it here by first developing an equation. The straight line in Figure 6.4(b) can be described by

$$y = 0.5u + 1 \tag{6.6}$$

Indeed, if $u = 0$, then $y = 1$ as shown in Figure 6.4(b). If $u = -2$, then $y = 0$. Thus the equation describes the straight line. If $u = u_1$, then $y_1 = 0.5u_1 + 1$. If $u = u_2$, then $y_2 = 0.5u_2 + 1$. However, if $u_3 = u_1 + u_2$, then

$$y_3 = 0.5(u_1 + u_2) + 1 \neq y_1 + y_2 = 0.5u_1 + 1 + 0.5u_2 + 1 = 0.5(u_1 + u_1) + 2$$

and the additivity property does not hold. Thus the system specified by Figure 6.4(b) or described by the equation in (6.6) is not a linear system.[7]

Consider now the memoryless system specified by Figure 6.6(c). To show a system not linear, all we need is to find two specific input-output pairs which do not meet the additivity property. However to show a system linear, we must check all possible input-output pairs. There are infinitely many of them and it is not possible to check them all. Thus we must use a mathematical equation to verify it. Because the slope of the straight line in Figure 6.4(c) is 0.5, the input and output of the system are related by

$$y(t) = 0.5u(t) \quad \text{or simply} \quad y = 0.5u \tag{6.7}$$

for all t. Using this equation, we can show that the system is linear. Indeed, for any $u = u_i$, we have $y_i = 0.5u_i$, for $i = 1, 2$. If $u_3 = u_1 + u_2$, then $y_3 = 0.5(u_1 + u_2) = y_1 + y_2$. If $u_4 = \beta u_1$, then $y_4 = 0.5(\beta u_1) = \beta(0.5u_1) = \beta y_1$. Thus the system satisfies the additivity and homogeneity properties and is linear. In conclusion, a CT time-invariant memoryless system is linear if and only if it can be described by

$$y(t) = \alpha u(t) \tag{6.8}$$

for some real constant α and for all t. Such a system can be represented as a *multiplier* or an *amplifier* with *gain* α as shown in Figure 6.5.

---

[7]It is a common misconception that (6.6) is linear. It however can be transformed into a linear equation. See Problem 6.10.

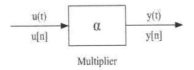

Multiplier

Figure 6.5: Output$= \alpha \times$ Input.

All preceding discussion is directly applicable to the DT case. A DT time-invariant memoryless system is linear if and only if it can be described by

$$y[n] = \alpha u[n] \tag{6.9}$$

for some real constant α and for all time index n. Such a system can also be represented as shown in Figure 6.5.

To conclude this section, we mention that the additivity and homogeneity properties can be combined as follows. A memoryless system is linear if for any $\{u_i \to y_i\}$ and any real constants β_i, for $i = 1, 2$, we have

$$\beta_1 u_1 + \beta_2 u_2 \to \beta_1 y_1 + \beta_2 y_2$$

This is called the *superposition property*. Thus the superposition property consists of additivity and homogeneity properties.

6.6   Op amps as nonlinear memoryless elements

The operational amplifier (op amp) is one of the most important electrical elements. It is usually represented as shown in Figure 6.6(a) and, more often, as in Figure 6.6(b). The op amp has two input terminals and one output terminal. The input terminal with a plus sign is called the *non-inverting terminal* and its voltage relative to the ground is denoted by $e_+(t)$ and the current entering the terminal is denoted by $i_+(t)$. The input terminal with a minus sign is called the *inverting terminal* with voltage $e_-(t)$ and current $i_-(t)$ as shown. The voltage at the output terminal is denoted by $v_o(t)$. In addition to the input and output terminals, there are terminals to be connected to dc power supplies. The power supplies could be ± 15 volts or as low as ± 3 volts. These terminals are usually not shown as in Figure 6.6(b). The resistors R_i and R_o shown in Figure 6.6(a) are called respectively *input resistance* and *output resistance*. The input resistance is usually very large, larger than $10^4 \Omega$ and the output resistance is very small, less than 50Ω.

If an op amp is *modeled* as time-invariant (TI) and memoryless, then its inputs and output can be described by

$$v_o(t) = f(e_+(t) - e_-(t)) \tag{6.10}$$

where f is as shown in Figure 6.7. The output voltage is a function of the difference of the two input voltages or $e_d(t) := e_+(t) - e_-(t)$. If the difference between e_+ and e_- lies within the range $[-a, a]$ as indicated in Figure 6.7, then v_o is a linear function of

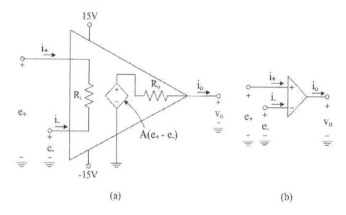

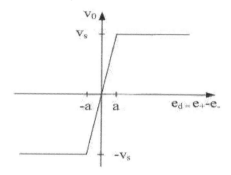

Figure 6.6: (a) Op-amp model. (b) Its simplified representation.

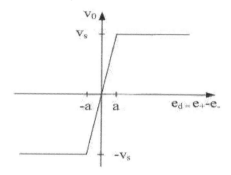

Figure 6.7: Characteristics of an op-amp.

e_d. If $|e_d|$ is larger than a, v_o reaches the positive or negative saturation region. The level of saturation, denoted by $\pm v_s$, is determined by the power supplies, usually one or two volts below the power supplies. Clearly an op amp is a nonlinear system.

In Figure 6.7, if $e_d = e_+ - e_-$ is limited to $[-a, a]$, then we have

$$v_o(t) = Ae_d(t) = A(e_+(t) - e_-(t)) \tag{6.11}$$

where A is a constant and is called the *open-loop gain*. It is the slope of the straight line passing through the origin in Figure 6.7 and is in the range of $10^5 \sim 10^{10}$. The value a can be computed roughly from Figure 6.7 as

$$a \approx \frac{v_s}{A}$$

If $v_s = 14$ V and $A = 10^5$, then we have $a = 14 \times 10^{-5}$ or 0.14 mV (millivolt).

Because of the characteristics shown in Figure 6.7, and because a is in the order of millivolts, op amps have found many applications. It can be used as a *comparator*. Let us connect v_1 to the non-inverting terminal and v_2 to the inverting terminal. If v_1 is slightly larger than v_2, then v_o will reach the saturation voltage v_s. If v_2 is slightly

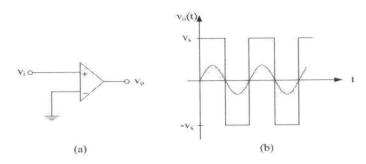

Figure 6.8: (a) Op amp. (b) Changing a sinusoid into a square wave.

larger than v_1, then v_o becomes $-v_s$. Thus from v_o, we know right away which voltage, v_1 or v_2, is larger. Comparators can be used to trigger an event and are used in building analog-to-digital converters.

Op amps can also be used to change a signal such as a sinusoidal, triangular, or other signal into a timing signal. For example, consider the op amp shown in Figure 6.8(a) with its inverting terminal grounded. If we apply a sine function to the non-inverting terminal, then $v_o(t)$ will be as shown in Figure 6.8(b). Indeed, the op amp is a very useful nonlinear element.

To conclude this section, we mention that an op amp is not a two-input one-output system even though it has two input terminals and one output terminal. To be so, we must have

$$v_o(t) = f_1(e_+) - f_2(e_-)$$

where f_1 and f_2 are some functions, independent of each other. This is not the case as we can see from (6.10). Thus the op amp is an SISO system with the single input $e_d = e_+ - e_-$ and the single output v_o.

6.7 Op amps as LTI memoryless elements

The op amp is a very useful nonlinear element. It turns out that it is also a workhorse for building linear time-invariant systems, as we discuss in this and later chapters. In order for an op amp to operate as a linear element, the voltage $e_d = e_+ - e_-$ must be limited to $[-a, a]$. This has an important implication. From Figure 6.6(a), we have

$$i_+ = -i_- = \frac{e_+ - e_-}{R_i} \le \frac{a}{R_i} \approx \frac{v_s}{AR_i}$$

which is practically zero because v_s is limited by the power supplies and A and R_i are very large. Thus we may assume

$$i_+(t) = 0 \quad \text{and} \quad i_-(t) = 0 \tag{6.12}$$

This is a standing assumption in the study of linear op-amp circuits.

It turns out that we *must* introduce feedback in order for an op amp to act as a linear element. We discuss in this section the simplest feedback. Consider the connection in

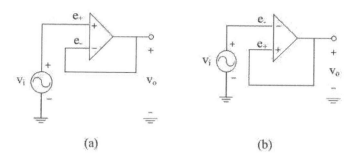

Figure 6.9: (a) Voltage follower. (b) Useless system.

Figure 6.9(a) in which the output is fed back directly to the inverting terminal. Because the terminal has a negative sign, it is called a *negative feedback*. Let us apply an input voltage $v_i(t)$ to the non-inverting terminal as shown. To develop an equation to relate the input $v_i(t)$ and the output $v_o(t)$, we substitute $e_+(t) = v_i(t)$ and $e_-(t) = v_o(t)$ into (6.11) to yield

$$v_o(t) = A(v_i(t) - v_o(t)) = Av_i(t) - Av_o(t)$$

which implies

$$(1 + A)v_o(t) - Av_i(t)$$

and

$$v_o(t) = \frac{A}{1 + A} v_i(t) =: \alpha v_i(t) \tag{6.13}$$

where $\alpha = A/(A + 1)$. This equation describes the negative feedback system in Figure 6.9(a). Thus the system is LTI memoryless and has gain α. For A large, α practically equals 1 and (6.13) becomes $v_o(t) = v_i(t)$ for all t. Thus the op-amp circuit is called a *voltage follower*.

Equation (6.13) is derived from (6.11) which holds only if $|e_d| \leq a$. Thus we must check whether or not $|e_d| \leq a \approx v_s/A$. Clearly we have

$$e_d = e_+ - e_- = v_i - v_o = v_i - \frac{A}{1 + A} v_i = \frac{1}{1 + A} v_i \tag{6.14}$$

If $|v_i| < v_s$, then (6.14) implies $|e_d| < |v_s/(1 + A)|$ which is less than $a = v_s/A$. Thus for the voltage follower, if the magnitude of the input $v_i(t)$ is less than v_s, then the op amp circuit functions as an LTI memoryless system and $v_o(t) = v_i(t)$, for all t.

An inquisitive reader may ask: If $v_o(t)$ equals $v_i(t)$, why not connect $v_i(t)$ directly to the output terminal? When we connect two electrical devices together, the so-called *loading* problem often occurs.[8] See Section 10.4. Because of loading, a measured or transmitted signal may be altered or distorted and will not represent the original signal. This is not desirable. This loading problem can be eliminated or at least reduced by

---

[8]Airplanes, elevators, and others are designed to carry loads (passengers and/or cargo). These loads must be included in design criteria. Except possibly in the last stage for driving a loudspeaker or a motor, electrical systems often involve only recording, transmission, or procession of signals. In these cases, characteristics of the signals must be preserved.

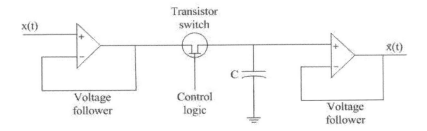

Figure 6.10: Sample-and-hold circuit.

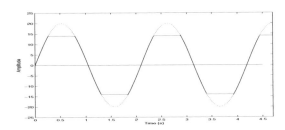

Figure 6.11: The input $v_i(t) = 20 \sin 3t$ (dotted line) and its output (solid line) of a voltage follower.

inserting a voltage follower. Thus a voltage follower is also called a *buffer* or an *isolating amplifier*. Figure 6.10 shows a sample-and-hold circuit in which two voltage followers are used to shield the electronic switch and the capacitor from outside circuits. Indeed, voltage followers are widely used in electrical systems.

To conclude this section, we discuss the effect of feedback. Consider the equation in (6.11). The equation is applicable only under the constraint $e_+(t) \approx e_-(t)$ for all t. The requirement is automatically met as shown in (6.14) if we introduce feedback and if $|v_i(t)| < v_s$. Consequently the op amp can function as a linear element. Without feedback, it will be difficult to meet the contraint. Thus all linear op-amp circuits involve feedback. Equation (6.13) also shows an important advantage of using feedback. The gain $\alpha = 1$ in (6.13) is independent of the exact value of A so long as A is large. It means that even if A changes its value, the gain α remains roughly 1. This is a general property of feedback systems: A feedback system is insensitive to its internal parameter variations.

6.7.1   Limitations of physical devices

Because the output voltage $v_o(t)$ of a voltage follower is limited to $\pm v_s$, if the magnitude of an input voltage is larger than v_s, we cannot have $v_o(t) = v_i(t)$. For example, if $v_s = 14$ and if $v_i(t) = 20 \sin 3t$ as shown in Figure 6.11 with a dotted line, then the output $v_o(t)$ will be as shown with a solid line. The input voltage is clipped by the circuit and we no longer have $v_o(t) = v_i(t)$, for all t. Thus the op-amp circuit ceases to function as a voltage follower.

Now if $v_i(t)$ is limited to v_s in magnitude, then we have $v_o(t) = v_i(t)$ for all t. Does

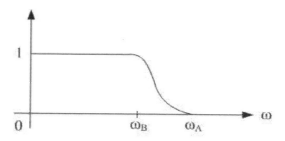

Figure 6.12: Operational frequency range of a voltage follower.

this hold for every $v_i(t)$ with its magnitude limited to v_s? Let us build an op amp circuit and carry out some experiments. If $v_s = 14$, and if $v_i(t) = 4\sin\omega_0 t$ with $\omega_0 = 10$ rad/s, then we can measure $v_o(t) = 4\sin 10t$, for all t. If $\omega_0 = 1000$, we can still measure $v_o(t) = 4\sin 1000t$. However, if $\omega_0 = 10^6$, we can measure roughly $v_o(t) = 2\sin 10^6 t$, its amplitude is only about half of the input amplitude. If $\omega_0 = 10^{10}$, no appreciable output can be measured or $v_o(t) \approx 0$. It means that the voltage followers cannot track any input even if its magnitude is limited to v_s.

Every device has an operational frequency range or a bandwidth. For example, there are low-frequency, intermediate-frequency (IF), or high-frequency amplifiers. A dc motor is to be driven by direct current or battery. An ac motor may be designed to be driven by a 120-volt, 60-Hz household electric power supply. The operational frequency range of a voltage follower, as we will discuss in Subsection 10.2.1, is of the form shown in Figure 6.12. We call $[0, \omega_B]$ the operational frequency range. If ω_0 of $v_i(t) = \sin\omega_0 t$ lies inside the range $[0, \omega_B]$ shown, then we have $v_o(t) = v_i(t)$. If $\omega_0 > \omega_A$, then $v_o(t) = 0$. If ω_0 lies roughly in the middle of ω_B and ω_A, then we have $v_o(t) = 0.5v_i(t)$. Real-world signals such as the ones shown in Figures 1.1 and 1.2 are rarely pure sinusoids. Thus it is more appropriate to talk about frequency spectra. If the frequency spectrum of a signal lies inside the operational frequency range, then the op-amp circuit can function as a voltage follower.

In conclusion, in order for a voltage follower to function as designed, an input signal must meet the following two limitations:

1. The magnitude of the signal must be limited.

2. The magnitude spectrum of the signal must lie inside the operational frequency range of the voltage follower.

If an input signal does not meet the preceding two conditions, the voltage follower will still generates an output which however may not equal the input. Note that *these two limitations on input signals apply not only to every voltage follower but also to every physical device.*

6.7.2 Limitation of memoryless model

Consider the op amp circuit shown in Figure 6.9(b) in which the output is fed back to the non-inverting terminal. Note that the positions of the two input terminals in Figure

6.9(b) are reversed from those in Figure 6.9(a). Because the non-inverting terminal has a positive sign, it is called a *positive-feedback* system.

Let us apply a voltage signal $v_i(t)$ to the inverting terminal. To develop an equation to relate the input $v_i(t)$ and the output $v_o(t)$, we substitute $e_+(t) = v_o(t)$ and $e_-(t) = v_i(t)$ into (6.11) to yield

$$v_o(t) = A(v_o(t) - v_i(t)) = Av_o(t) - Av_i(t)$$

which implies

$$(1 - A)v_o(t) = -Av_i(t)$$

and

$$v_o(t) = \frac{-A}{1 - A}v_i(t) = \frac{A}{A - 1}v_i(t) =: \alpha v_i(t) \qquad (6.15)$$

where $\alpha = A/(A - 1)$. This equation describes the positive feedback system in Figure 6.9(b). Thus the system is LTI memoryless and has gain α. For A large, α practically equals 1 and (6.15) becomes $v_o(t) = v_i(t)$ for all t. Can the op-amp circuit in Figure 6.9(b) be used as a voltage follower?

Before answering the question we check first the validity of (6.11) or $|e_d| \leq a$. Clearly we have

$$e_d = e_+ - e_- = v_o - v_i = \frac{A}{A - 1}v_i - v_i = \frac{1}{A - 1}v_i \qquad (6.16)$$

If $|v_i| < [(A - 1)/A]v_s$, then (6.16) implies $|e_d| < |v_s/A| = a$. Thus for the circuit in Figure 6.9(b), if $|v_i(t)| < [(A-1)/A]v_s \approx v_s$, then (6.11) holds and we have $v_o(t) = v_i(t)$, for all t.

Let us build a positive-feedback op-amp circuit shown in Figure 6.9(b) with ± 15V power supplies. If we apply $v_i(t) = 4\sin 3t$ to the circuit, then the output becomes $v_o(t) = -v_s$. It is different from $v_o(t) = 4\sin 3t$ as implied by (6.15). Thus the actual op-amp circuit cannot be used as a voltage follower.

Why does the actual op-amp circuit behave differently from the one derived mathematically? There is no error in the derivation. Thus the only explanation is that the memoryless model used or (6.11) cannot be used here. Indeed, this is the case as we will show in Section 10.2. If we use a model that has memory, then the circuit in Figure 6.9(b) can be shown to be not stable (to be introduced in Chapter 9), as confirmed by the actual circuit. Using a model with memory to verify the stability of an op-amp circuit is complicated and unnecessary. It is simpler to build an actual circuit and then to check its stability. See Section 9.5.2. Thus in practice, we use a memoryless model or even a simpler model (ideal model to be discussed next) to carry out the design.

6.8   Ideal op amps

The op amp studied in the preceding section is assumed to be describable by (6.11), that is, it is LTI memoryless with a finite open-loop gain A. It is called *ideal* if $A = \infty$, $R_i = \infty$, and $R_o = 0$. If $A = \infty$, then $|e_+(t) - e_-(t)| = |v_o(t)|/A \leq v_s/A$ implies

$$e_+(t) = e_-(t) \quad \text{(virtually short)} \qquad (6.17)$$

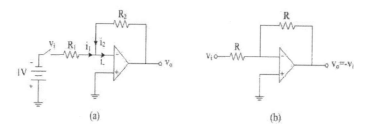

Figure 6.13: (a) Inverting amplifier with gain R_2/R_1. (b) Inverter.

Because of $i_+ = -i_- = (e_+ - e_-)/R_i$, we also have

$$i_+(t) = 0 \quad \text{and} \quad i_-(t) = 0 \quad \text{(virtually open)} \tag{6.18}$$

for all $t \geq 0$. This is the standing assumption in (6.12). For an ideal op amp, its two input terminals are virtually open and virtually short at the same time. Using these two properties, analysis and design of op-amp circuits can be greatly simplified. For example, for the voltage follower in Figure 6.9(a), we have $v_i(t) = e_+(t)$ and $v_o(t) = e_-(t)$. Using (6.17), we have immediately $v_o(t) = v_i(t)$. This can be confirmed by building an actual circuit. For the op-amp circuit in Figure 6.9(b), we can also obtain $v_o(t) = v_i(t)$ immediately. The equation however is meaningless as discussed in the preceding subsection.

We use the ideal model to analyze the op-amp circuit shown in Figure 6.13(a). The input voltage $v_i(t)$ is connected to the inverting terminal through a resistor with resistance R_1 and the output voltage $v_o(t)$ is fed back to the inverting terminal through a resistor with resistance R_2. The non-inverting terminal is grounded as shown. Thus we have $e_+(t) = 0$.

Let us develop the relationship between $v_i(t)$ and $v_o(t)$. Equation (6.17) implies $e_-(t) = 0$. Thus the currents passing through R_1 and R_2 are, using Ohm's law,

$$i_1(t) = \frac{v_i(t) - e_-(t)}{R_1} = \frac{v_i(t)}{R_1}$$

and

$$i_2(t) = \frac{v_o(t) - e_-(t)}{R_2} = \frac{v_o(t)}{R_2}$$

respectively. Applying Kirchoff's current law, we have

$$i_1(t) + i_2(t) - i_-(t) = 0$$

which together with (6.18) imply $i_2(t) = -i_1(t)$ or

$$\frac{v_o(t)}{R_2} = -\frac{v_i(t)}{R_1}$$

Thus we have

$$v_o(t) = -\frac{R_2}{R_1} v_i(t) \tag{6.19}$$

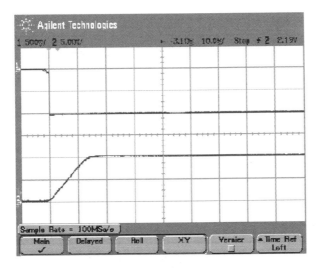

Figure 6.14: Response of an inverting amplifier with gain 10.

This equation describes the op-amp circuit in Figure 6.13(a). The circuit is an LTI memoryless system. It is called an *inverting amplifier* with gain R_2/R_1 or an *amplifier* with gain $-R_2/R_1$. It is called an *inverter* if $R_1 = R_2 = R$ as shown in Figure 6.13(b). An inverter changes only the sign of the input.

6.8.1   Response time

We use an LM 741 operational amplifier, $R_1 = 1k\Omega$ and $R_2 = 10k\Omega$ to build an inverting amplifier with gain $R_2/R_1 = 10$.[9] We use a dual-channel oscilloscope to measure and record the input $v_i(t)$ and the output $v_o(t)$. The top plot above the dotted line in Figure 6.14 records the input. Its vertical division denotes 0.5 V and its horizontal division denotes 0.1 second. If $v_i(t) = -1$, for $t \geq 0$, by closing the switch in Figure 6.13(a), then the input reaches two divisions or -1 V almost instantaneously as shown with a solid line. The excited output is plotted with a solid line in the bottom plot below the dotted line with its vertical division denoting 5 V. From the plot, we see that the output reaches 10 V in about 0.14 second. In conclusion, if we apply $v_i(t) = -1$, then the output reach almost immediately 10 V. Thus the op-amp circuit in Figure 6.13(a) with $R_1 = 1k\Omega$ and $R_2 = 10k\Omega$ is an amplifier with gain -10 as derived in (6.19).

Equation (6.19) implies $v_o(t) = -10v_i(t)$, for all $t \geq 0$. Thus if we apply $v_i(t) = -1$, for $t \geq 0$, then the output should jump from 0 to 10 volts instantaneously. In reality, the output increases continuously from 0 to 10 in roughly 0.14 second as shown in Figure 6.14. This discrepancy is due to the use of the memoryless model. If we use a model with memory to carry out the analysis, then it will show that the output takes some finite time, called the *response time*, to reach 10 V. Because the response time is often very small in practice, we often pay no attention to it. In conclusion, most

---

[9]This experiment was carried out by Anthony Olivo.

physical systems have memory. However we often model them as memoryless to simply analysis and design.

To conclude this section, we mention that values or numbers associated with physical elements or devices are often not precise. For example, the actual resistances of the 1kΩ- and 10kΩ-resistors are respectively 948Ω and 9789Ω. Thus the amplifier should have a gain of $-9789/948 = -10.3$ rather than -10. In some applications, we often accept a device so long as it works reasonable. Mathematical precision is not needed.

6.9 Concluding remarks

We start to study systems in this chapter. After introducing the concepts of causality and time-invariance, we focus on the study of memoryless systems. The linearity of such systems can be determined graphically and the equation describing such systems is very simple. It is $y(t) = \alpha u(t)$ or $y[n] = \alpha u[n]$.

Most physical systems have memory. However they are often modeled as memoryless to simplify analysis and design. We used op amps as an example to discuss the issues involved. An op amp is a very useful nonlinear element. However, after feedback, it operates only in its linear range under some constraints. Thus op amps can be used to build linear time-invariant systems. In design, we use its ideal model (LTI memoryless with an infinite open-loop gain). Because the design is based on a (mathematical) model, we must build an actual op-amp circuit to verify the result. If the actual circuit does not function as expected as in the case of the circuit in Figure 6.9(b), we must redesign. In engineering, no design is completed without building a working physical system.

Problems

6.1 Consider a CT system whose input $u(t)$ and output $y(t)$ are related by

$$y(t) = \int_{\tau=0}^{t+1} u(\tau)d\tau$$

for $t \geq 0$. Does the current output depend on future input? Is the system causal?

6.2 Consider a CT system whose input $u(t)$ and output $y(t)$ are related by

$$y(t) = \int_{\tau=0}^{t} u(\tau)d\tau$$

for $t \geq 0$. Does the current output depend on future input? Is the system causal?

6.3 Consider a DT system whose input and output are related by

$$y[n] = 3u[n-2] - 0.5u[n] + u[n+1]$$

for $n \geq 0$. Is the system causal?

6.4 Consider a DT system whose input and output are related by

$$y[n] = 3u[n-2] - 0.5u[n-1] + u[n]$$

for $n \geq 0$. Is the system causal?

6.5 Is a CT system described by

$$y(t) = 1.5[u(t)]^2$$

memoryless? time-invariant? linear?

6.6 Is a CT system described by

$$y(t) = (2t+1)u(t)$$

memoryless? time-invariant? linear?

6.7 Consider $y(t) = (\cos \omega_c t)u(t)$. It is the modulation discussed in Section 4.4.1. Is it memoryless? linear? time-invariant?

6.8 Is a DT system described by

$$y[n] = 1.5 \left(u[n] \right)^2$$

memoryless? time-invariant? linear?

6.9 Is a DT system described by

$$y[n] = (2n+1)u[n]$$

memoryless? time invariant? linear?

6.10 Is the equation $y = \alpha u + 1$ linear? If we define $\bar{y} := y - 1$ and $\bar{u} := u$, what is the equation relating $\bar{u}$ and $\bar{y}$? Is it linear?

6.11 Consider a memoryless system whose input u and output y are related as shown in Figure 6.15. If the input u is known to limit to the immediate neighborhood of u_0 shown, can you develop a linear equation to describe the system? It is assumed that the slope of the curve at u_0 is R. The linear equation is called a small signal model. It is a standard technique in developing a linear model for electronic circuits.

6.12 Consider the op amp shown in Figure 6.8(a). If the applied input $v_i(t)$ is as shown in Figure 6.16, what will be the output?

6.13 Use the ideal op-amp model to find the relationship between the input signal $v_i(t)$ and the output signal $v_o(t)$ shown in Figure 6.17. Is it the same as (6.19)? The circuit however cannot be used as an inverting amplifier. See Problem 10.2.

6.14 Verify that the op-amp circuit in Figure 6.18 is an amplifier with gain $1 + R_2/R_1$.

6.15 Verify that a multiplier with gain α can be implemented as shown in Figure 6.19(a) if $\alpha > 0$ and in Figure 6.19(b) if $\alpha < 0$.

6.16 Consider the op-amp circuit shown in Figure 6.20 with three inputs $u_i(t)$, for $i = 1:3$ and one output $v_o(t)$. Verify $v_o(t) = u_1(t) + u_2(t) + u_3(t)$. The circuit is called an *adder*.

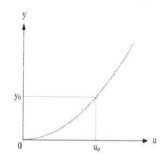

Figure 6.15: Nonlinear system.

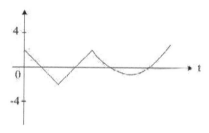

Figure 6.16: Input.

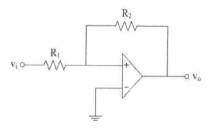

Figure 6.17: Positive-feedback op-amp circuit.

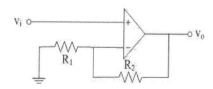

Figure 6.18: Noninverting amplifier.

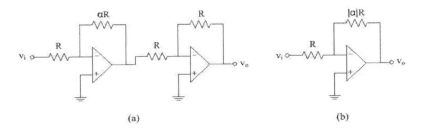

Figure 6.19: Multiplier with gain (a) $\alpha > 0$ and (b) $\alpha < 0$.

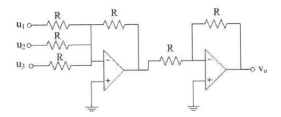

Figure 6.20: Adder.

Chapter 7

DT LTI systems with finite memory

7.1 Introduction

In the preceding chapter, we studied linear time-invariant (LTI) memoryless systems. The input u and output y of such a system can be related graphically. Moreover its mathematical description is very simple. It is $y(t) = \alpha u(t)$ for the CT case and $y[n] = \alpha u[n]$ for the DT case, where α is a real constant.

In the remainder of this text, we study systems with memory. There are two types of memory: finite and infinite. Thus we will encounter four types of systems:

- DT systems with finite memory

- DT systems with infinite memory

- CT systems with finite memory

- CT systems with infinite memory

For a system with memory, it is not possible to relate its input u and output y graphically. The only way is to use mathematical equations to relate them. Moreover, its study requires some new concepts. These concepts can be most easily explained using DT systems with finite memory. Thus we study first this class of systems.

After introducing the concepts of initial conditions, forced and natural responses, we extend the concepts of linearity (L) and time-invariance (TI) to systems with memory. We then introduce the following four types of mathematical equations to describe DT LTI systems:

1. Discrete convolutions

2. High-order difference equations

3. State-space (ss) equations (sets of first-order difference equations)

4. Rational transfer functions

The first description is developed for DT systems with finite and infinite memory. The remainder however are developed only for finite memory. The introduction is only a preview, the four types of equations will be further discussed in the next chapter and chapter 11.

As discussed in Section 6.4.1, in the study of DT systems we may assume the sampling period to be 1. This will be the standing assumption throughout this text unless stated otherwise.

7.2   Causal systems with memory

A DT system is memoryless if its current output $y[n_0]$ depends only on current input $u[n_0]$. It has memory if its current output $y[n_0]$ depends on current input $u[n_0]$, past input $u[n]$ with $n < n_0$, and future input $u[n]$ with $n > n_0$. A system is causal if its current output does not depend on future input. Thus *the current output of a causal system depends only on current and/or past inputs.* A non-causal system can predict what will be applied in the future. No physical system has such capability. Thus we study in this text only causal systems.[1]

If $y[n_0]$ depends on current input $u[n_0]$ and past N inputs $u[n_0-1], u[n_0-2], \ldots, u[n_0-N]$, the system is defined to have a *memory* of N samples. It has finite memory if N is finite, and infinite memory if N is infinite. The input and output of a DT time-invariant causal system with a memory of N samples can be expressed as

$$y[n] = f(u[n], u[n-1], u[n-2], \ldots, u[n-N]) = f(u[k], n - N \leq k \leq n) \qquad (7.1)$$

where f is some function and is independent of n. The equation reduces to $y[n] = f(u[n])$ or $y = f(u)$ if a system is memoryless or $N = 0$.

Equation (7.1) holds for all integers n. Replacing n by $n - 2$, $n - 1$, $n + 1$, and so forth, we have

$$\vdots$$

$$y[n-2] \;=\; f(u[n-2], u[n-3], u[n-4], \ldots, u[n-N-2])$$
$$y[n-1] \;=\; f(u[n-1], u[n-2], u[n-3], \ldots, u[n-N-1])$$
$$y[n] \;=\; f(u[n], u[n-1], u[n-2], \ldots, u[n-N])$$
$$y[n+1] \;=\; f(u[n+1], u[n], u[n-1], \ldots, u[n-N+1])$$

$$\vdots$$

$$y[n+N] \;=\; f(u[n+N], u[n+N-1], u[n+N-2], \ldots, u[n])$$
$$y[n+N+1] \;=\; f(u[n+N+1], u[n+N], u[n+N-1], \ldots, u[n+1])$$

We see that the current input $u[n]$ does not appear in $y[n-1]$, $y[n-2]$, and so forth. Thus the current input of a causal system will not affect past output. On the other

---

[1]The concept of causality is associated with time. In processing of signals such as pictures which have two spatial coordinates as independent variables, causality is not an issue.

hand, $u[n]$ appears in $y[n+k]$, for $k = 0 : N$. Thus the current input will affect the current output and the next N outputs. Consequently even if we stop to apply an input at n_1, that is, $u[n] = 0$ for $n \geq n_1$, the output will continue to appear N more samples. If a system has infinite memory, then the output will continue to appear forever even if we stop to apply an input. In conclusion, if a DT system is causal and has a memory of N samples, then

- its current *output* does not depend on future *input*; it depends on current and past N input samples,

or, equivalently,

- its current *input* has no effect on past *output*; it affects current and next N output samples.

Example 7.2.1 Consider a savings account in a bank. Let $u[n]$ be the amount of money deposited or withdrawn on the n-th day and $y[n]$ be the total amount of money in the account at the end of the n-th day. By this definition, $u[n]$ will be included in $y[n]$. The savings account can be considered a DT system with input $u[n]$ and output $y[n]$. It is a causal system because the current output $y[n]$ does not depend on future input $u[m]$, with $m > n$. The savings account has infinite memory because $y[n]$ depends on $u[m]$, for all $m \leq n$ or, equivalently, $u[n]$ affects $y[m]$, for all $m \geq n$. □

7.2.1 Forced response, initial conditions, and natural response

For a time-invariant (TI) memoryless system, the input $u[n]$ will excite a unique output $y[n]$. Thus we can use $\{u[n] \to y[n]\}$ to denote an input-output pair. Such a notation is not acceptable for a system with memory because $y[n]$ depends not only on $u[n]$ but also on $u[k]$, for $k < n$. Thus we first discuss a notation to denote input-output pairs for systems with memory.

If the characteristics of a system do not change with time, then the system is, as defined in Section 6.4, time invariant. For such a system, the output waveform will always be the same no matter at what time instant, called the *initial time*, we start to apply an input. Thus we may assume the initial time to be $n_0 = 0$ in studying time-invariant systems. This will be the standing assumption throughout this text. Note that the initial time $n_0 = 0$ is selected by us; it is the time instant we start to study the system.

When we start to study a system with memory at $n_0 = 0$, we will be interested in its input $u[n]$ and output $y[n]$ only for $n \geq 0$. Thus we may use

$$u[n], \quad n \geq 0 \quad \to \quad y[n], \quad n \geq 0$$

to denote an input-output pair. This notation however requires some qualification. If the system has memory, then the output $y[n]$, for $n \geq 0$, also depends on the input applied before $n = 0$. Thus if $u[n] \neq 0$, for some $n < 0$, then the input $u[n]$, for $n \geq 0$, may not excite a unique $y[n]$, for $n \geq 0$, and the notation is not well defined. Thus when we use the notation, we must assume $u[n] = 0$, for all $n < 0$. Under this assumption, the notation denotes a unique input-output pair and the output is called a *forced response*, that is,

- Forced response: Output $y[n]$, for $n \geq 0$, excited by $\begin{cases} u[n] = 0 & \text{for all } n < 0 \\ u[n] \neq 0 & \text{for some or all } n \geq 0 \end{cases}$

As in the CT case defined in Section 6.4, a DT system is defined to be *initially relaxed* at $n_0 = 0$ if the output $y[n]$, for $n \geq 0$, is excited *exclusively* by the input $u[n]$, for $n \geq 0$. Thus a sufficient condition for a system to be initially relaxed at $n_0 = 0$ is that no input is applied before $n_0 = 0$ or $u[n]$ is identically zero for all $n < 0$. In conclusion, if a system is initially relaxed, the excited output is a forced response.

If $u[n] = 0$, for all $n < 0$, then any DT system is initially relaxed at $n_0 = 0$. This condition can be relaxed for some systems as we discuss next. Consider a DT system with a memory of past $N = 3$ samples. Then the input $u[n]$ affects the output $y[n + k]$, for $k = 0 : 3$. Thus the output $y[n]$, for $n \geq 0$ is excited not only by $u[n]$, for $n \geq 0$, but also by $u[-k]$, for $k = 1 : 3$. Consequently, if $u[-k] = 0$, for $k = 1 : 3$, then the output $y[n]$, for $n \geq 0$, is excited exclusively by $u[n]$, for $n \geq 0$ and the system is initially relaxed at $n_0 = 0$. We call $u[-k]$, for $k = 1 : 3$ the *initial conditions*. The set of three initial conditions can be expressed as a 3×1 column vector $\mathbf{x}[0] := [u[-1] \ \ u[-2] \ \ u[-3]]'$, where the prime denotes the transpose. We call $\mathbf{x}[0]$ the *initial state* of the system at $n_0 = 0$. If the initial state is zero, the system is initially relaxed at $n_0 = 0$ even if $u[n] \neq 0$, for all $n \leq -4$. For the savings account studied in Example 7.2.1, if $y[-1] = 0$, then the output $y[n]$ for $n \geq 0$ will be excited exclusively by $u[n]$, for $n \geq 0$, even if $u[n] \neq 0$ for some or all $n < 0$. Thus the savings account has only one initial condition $y[-1]$. We see that introducing the set of initial conditions or initial state simplifies greatly the verification of initial relaxedness. In this text, we study, as we will discuss in the next chapter, only systems which have finite numbers of initial conditions. For this class of systems, if the initial state is zero or $\mathbf{x}[0] = \mathbf{0}$, then the excited response is a forced response. Thus a forced response is also called a *zero-state* response and we have

- Forced response = Zero-state response
    = Output $y[n]$, for $n \geq 0$, excited by $\begin{cases} \text{initial state } \mathbf{x}[0] = \mathbf{0} \text{ (initially relaxed)} \\ u[n] \neq 0 \text{ for some or all } n \geq 0 \end{cases}$

Before proceeding, we introduce *zero-input responses* defined by

- Zero-input response: Output $y[n]$, for $n \geq 0$, excited by $\begin{cases} u[n] \neq 0 & \text{for some or all } n < 0 \\ u[n] = 0 & \text{for all } n \geq 0 \end{cases}$

It is so named because the input is identically zero for all $n \geq 0$. In this case, the response is determined entirely by the characteristics or the *nature* of the system. Thus it is also called a *natural response*. If a system has a finite number of initial conditions, then we have

- Natural response = Zero-input response
    = Output $y[n]$, for $n \geq 0$, excited by $\begin{cases} \text{initial state } \mathbf{x}[0] \neq \mathbf{0} \\ u[n] = 0 \text{ for all } n \geq 0 \end{cases}$

The classification of outputs into forced or natural responses will simplify the development of mathematical equations for the former. Note that natural responses relate inputs applied *before* $n_0 = 0$ and outputs appeared *after* $n_0 = 0$. Because their time intervals are different, developing equations relating such inputs and outputs will be difficult. Whereas forced responses relate the input and output in the same time interval $[0, \infty)$. Moreover, under the relaxedness condition, every input excites a unique output. Thus developing equations relating them will be comparatively simple. In the remainder of this text, we study mainly forced responses. This is justified because in designing systems, we consider only forced responses.

7.3 Linear time-invariant (LTI) systems

A system can be classified dichotomously as causal or noncausal, time-invariant or time varying, linear or nonlinear. As discussed earlier, a system is causal if its current output does not depend on future input. A system is time invariant if its characteristics do not change with time. The output waveform of such a system will always be the same no matter at what time instant the input sequence is applied. We express this as a terminal property. Let

$$u[n], \ n \geq 0 \ \rightarrow \ y[n], \ n \geq 0$$

be any input-output pair and let n_1 be any positive integer. If the system is time invariant, then we have

$$u[n - n_1], \ n > n_1 \ \rightarrow \ y[n - n_1], \ n \geq n_1 \quad \text{(time shifting)}$$

This is the *time-shifting* property discussed in (6.2).

We next classify a system to be linear or nonlinear. Consider a DT system. Let

$$u_i[n], \ \text{for } n \geq 0 \ \rightarrow \ y_i[n], \ \text{for } n \geq 0$$

for $i = 1, 2$, be any two input-output pairs, and let β be any real constant. Then the system is defined to be *linear* if it has the following two properties

$$u_1[n] + u_2[n], \ n \geq 0 \ \rightarrow \ y_1[n] + y_2[n], \ n \geq 0 \ \text{(additivity)} \qquad (7.2)$$

$$\beta u_1[n], \ n \geq 0 \ \rightarrow \ \beta y_1[n], \ n \geq 0 \ \text{(homogeneity)} \qquad (7.3)$$

If not, the system is *nonlinear.* These two properties are stated in terms of inputs and outputs without referring to the internal structure of a system. Thus it is applicable to any type of system.[2]

Consider a DT system. If we can find one or two input-output pairs which do not meet the homogeneity or additivity property, then we can conclude that the system is not linear. However to conclude a system to be linear, the two properties must hold for all possible input-output pairs. There are infinitely many of them and there are no way to check them all. Moreover we require the system to be initial relaxed before we apply an input. Thus, it is not possible to check the linearity of a system from measurements at its input-output terminals. However using the conditions, we can develop a general mathematical equation to describe such systems.

---

[2]Additivity almost implies homogeneity but not conversely. See Problems 7.3 and 7.4. Thus we often check only the additivity property.

7.3.1   Finite and infinite impulse responses (FIR and IIR)

Consider a DT TI system with a memory of $N = 3$ past samples. Then its output $y[n]$ depends on $u[n], u[n-1], u[n-2]$, and $u[n-3]$. Suppose the output can be expressed as

$$y[n] = 3(u[n])^2 + 0 \times u[n-1] + u[n-2] \times u[n-3] \qquad (7.4)$$

Then we can show that the system is not linear. Indeed, if $u_i[n]$, $i = 1, 2$, then we have

$$y_i[n] = 3(u_i[n])^2 + u_i[n-2] \times u_i[n-3]$$

If $u_3[n] = u_1[n] + u_2[n]$, then we have

$$
\begin{aligned}
y_3[n] &= 3(u_1[n] + u_2[n])^2 + (u_1[n-2] + u_2[n-2]) \times (u_1[n-3] + u_2[n-3]) \\
&= 3(u_1[n])^2 + 3(u_2[n])^2 + 6u_1[n]u_2[n] + u_1[n-2]u_1[n-3] \\
&\quad + u_1[n-2]u_2[n-3] + u_2[n-2]u_1[n-3] + u_2[n-2]u_2[n-3] \\
&\neq y_1[n] + y_2[n]
\end{aligned}
$$

Thus the DT system described by (7.4) does not have the additivity property and is therefore not linear.

It turns out that the DT TI system is linear if and only if the output can be expressed as a *linear combination* of the inputs with constant coefficients such as

$$y[n] = 3u[n] - 2u[n-1] + 0 \times u[n-2] + 5u[n-3] \qquad (7.5)$$

Indeed, by direct substitution, we can verify that (7.5) meets the additivity and homogeneity properties. Note that (7.5) is a causal system because its output $y[n]$ does not depend on any future input $u[m]$ with $m > n$.

We define an important response, called the *impulse response*, for a system. It is the output of the system, which is initially relaxed at $n = 0$, excited by the input $u[0] = 1$ and $u[n] = 0$, for all $n > 0$. If the system is initially relaxed at $n = 0$, we may assume $u[n] = 0$ for all $n < 0$. In this case, the $u[n]$ becomes $u[0] = 1$ and $u[n] = 0$ for all $n \neq 0$. It is the impulse sequence defined in (2.24) with $n_0 = 0$. Thus the impulse response of a system can be defined as the output excited by $u[n] = \delta_d[n]$ where

$$\delta_d[n] = \begin{cases} 1 & \text{for } n = 0 \\ 0 & \text{for } n \neq 0 \end{cases} \qquad (7.6)$$

is the impulse sequence at $n = 0$. We now compute the impulse response of the system described by (7.5). Substituting $u[n] = \delta_d[n]$ into (7.5) yields

$$y[n] = 3\delta_d[n] - 2\delta_d[n-1] + 0 \times \delta_d[n-2] + 5\delta_d[n-3]$$

By direct substitution, we can obtain

$$
\begin{aligned}
y[-1] &= 3\delta_d[-1] - 2\delta_d[-2] + 0 \times \delta_d[-3] + 5\delta_d[-4] = 0 \\
y[0] &= 3\delta_d[0] - 2\delta_d[-1] + 0 \times \delta_d[-2] + 5\delta_d[-3] = 3 \\
y[1] &= 3\delta_d[1] - 2\delta_d[0] + 0 \times \delta_d[-1] + 5\delta_d[-2] = -2 \\
y[2] &= 3\delta_d[2] - 2\delta_d[1] + 0 \times \delta_d[0] + 5\delta_d[-1] = 0 \\
y[3] &= 3\delta_d[3] - 2\delta_d[2] + 0 \times \delta_d[1] + 5\delta_d[0] = 5 \\
y[4] &= 3\delta_d[4] - 2\delta_d[3] + 0 \times \delta_d[2] + \delta_d[1] = 0
\end{aligned}
$$

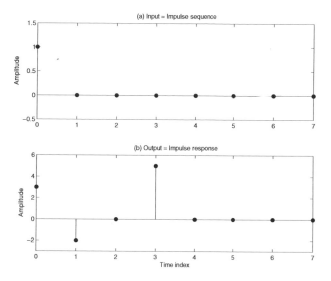

Figure 7.1: (a) Impulse sequence $\delta[n-0]$. (b) Impulse response $h[n]$.

and $y[n] = 0$ for all $n \geq 4$ and for all $n < 0$. This particular output sequence is the impulse response and will be denoted by $h[n]$. Thus we have $h[0] = 3$, $h[1] = -2$, h[2]=0, h[3]=5, and $h[n] = 0$, for $n < 0$ and $n \geq 4$. They are simply the coefficients of the equation in (7.5). We plot in Figure 7.1(a) the impulse sequence $\delta_d[n]$ and in Figure 7.1(b) the impulse response $h[n]$, for $n \geq 0$. We see that after the input is removed, the output continues to appear for $N = 3$ more sampling instants.

The impulse response is defined as the output excited by an impulse sequence applied at $n = 0$. If a system is initially relaxed and causal, the output must be zero before the application of an input. Thus we have $h[n] = 0$ for all $n < 0$. In fact, $h[n] = 0$, *for all $n < 0$, is a necessary and sufficient condition for a system to be causal.* This is an important condition and will be used later.

Example 7.3.1 (Moving average) Consider a DT system defined by

$$y[n] = \frac{1}{5}\left(u[n] + u[n-1] + u[n-2] + u[n-3] + u[n-4]\right)$$

Its current output is a linear combination of its current input and past four inputs with the same coefficient 0.2. Thus the system is linear (L) and time-invariant (TI), and has a memory of 4 samples. From the equation, we see that $y[n]$ is the average of the five inputs $u[n-k]$, for $k = 0 : 4$, thus it is called a 5-point *moving average*. If n denotes trading days of a stock market, then it will be a 5-day moving average. Its impulse response is

$$h[n] = 1/5 = 0.2 \quad \text{for } n = 0 : 4$$

and $h[n] = 0$ for $n < 0$ and $n \geq 5$. Note that if $u[n]$ is erratic, then $y[n]$ will be less erratic. Thus a moving average is used widely to smooth a curve or to generate a trend. For example, the smoother curve in Figure 1.5(d) is obtained using a 90-day moving average.□

Example 7.3.2 Consider the savings account studied in Example 7.2.1. It is a
DT system. If the interest rate changes with time, the system is time-varying. If the
interest rate depends on the total amount of money in the account, such as, the interest
rate is 0 if $y[n] < 500.00$, 3% if $500.00 \le y[n] < 10,000.00$, and 5% if $y[n] \ge 10,000.00$,
then the system is nonlinear. If the interest rate is independent of time and of the total
amount of money in the account, then the savings account is an LTI system.

To simplify the discussion, we assume that the interest rate is 0.015% per day and
compounded daily.[3] If we deposit one dollar the first day ($n = 0$) and none thereafter
(the input $u[n]$ is an impulse sequence), then we have

$$y[0] = 1$$
$$y[1] = y[0] + y[0] \times 0.00015 = (1 + 0.00015)y[0] = 1.00015$$

and

$$y[2] = y[1] + y[1] \times 0.00015 = (1 + 0.00015)y[1] = (1.00015)^2 y[0]$$
$$= (1.00015)^2$$

In general, we have $y[n] = (1.00015)^n$. This particular output is the impulse response,
that is,

$$h[n] = (1.00015)^n \tag{7.7}$$

for $n = 0, 1, 2, \ldots$, and $h[n] = 0$ for $n < 0$. □

A DT system is defined to be an *FIR (finite impulse response)* system if its impulse
response has a finite number of nonzero entries. It is called an *IIR (infinite impulse
response)* system if its impulse response has infinitely many nonzero entries. An FIR
system has finite memory and an IIR system has infinite memory. The system described
by (7.5) and the moving average in Example 7.3.1 are FIR and the savings account in
Example 7.3.2 is IIR.

The impulse response can be defined for any linear or nonlinear system. However,
the impulse response of a nonlinear system is completely useless because it cannot
be used to predict or compute the output excited by any other input. On the other
hand, if we know the impulse response of an LTI system, then we can use the impulse
response to predict or compute the output excited by any input. This is shown in the
next subsection.

7.3.2   Discrete convolution

Consider a DT LTI system. We show that if its impulse response $h[n]$ is known, then
we can develop a general equation to relate its input $u[n]$ and output $y[n]$. Let $u[n]$ be
an arbitrary input sequence defined for $n \ge 0$. Then it can be expressed as

$$u[n] = \sum_{k=0}^{\infty} u[k]\delta_d[n-k] \tag{7.8}$$

---

[3]The annual interest rate is 5.63%.

See (2.25). Let $h[n]$ be the impulse response of a DT system. If the system is linear and time invariant, then we have

$$
\begin{aligned}
\delta_d[n] &\rightarrow h[n] \quad \text{(definition)} \\
\delta_d[n-k] &\rightarrow h[n-k] \quad \text{(time-shifting)} \\
u[k]\delta_d[n-k] &\rightarrow u[k]h[n-k] \quad \text{(homogeneity)} \\
\sum_{k=0}^{\infty} u[k]\delta_d[n-k] &\rightarrow \sum_{k=0}^{\infty} u[k]h[n-k] \quad \text{(additivity)}
\end{aligned}
\tag{7.9}
$$

Because the left-hand side of (7.9) is the input in (7.8), the output excited by the input $u[n]$ is given by the right-hand side of (7.9), that is,

$$
y[n] = \sum_{k=0}^{\infty} h[n-k]u[k]
\tag{7.10}
$$

This is called a *discrete convolution*. The equation will be used to introduce the concept of transfer functions and a stability condition. However once the concept and the condition are introduced, the equation will not be used again. Thus we will not discuss further the equation. We mention only that if an impulse response is given as in Figure 7.1(b), then (7.10) reduces to (7.5) (Problem 7.9). Thus *a discrete convolution is simply a linear combination of the current and past inputs with constant coefficients.*

7.4   Some difference equations

Consider the equation

$$
\begin{aligned}
y[n] + a_1 y[n-1] &+ a_2 y[n-2] + \cdots + a_N y[n-N] \\
&= b_0 u[n] + b_1 u[n-1] + b_2 u[n-2] + \cdots + b_M u[n-M]
\end{aligned}
\tag{7.11}
$$

where N and M are positive integers, and a_i and b_i are real constants. Note that the coefficient a_0 associated with $y[n]$ has been normalized to 1. It is called an LTI *difference equation* of order $\max(N, M)$. It is called a *non-recursive* difference equation if $a_i = 0$ for all i and a *recursive* difference equation if one or more a_i are different from 0.

Consider the DT system described by (7.5) or

$$
y[n] = 3u[n] - 2u[n-1] + 0 \times u[n-2] + 5u[n-3]
$$

It is a special case of (7.11) with $N = 0$ and $M = 3$. Thus it is an LTI non-recursive difference equation of order 3. In fact, *for a DT LTI system with finite memory, there is no difference between its discrete convolution and its non-recursive difference equation description.*

A moving average is an FIR and can be described by a convolution which is also a non-recursive difference equation. However, if the moving average is very long, we can also develop a recursive difference equation to describe it as the next example shows.

Example 7.4.1 Consider a twenty-point moving average defined by

$$y[n] = 0.05 \, (u[n] + u[n-1] + \cdots + u[n-18] + u[n-19]) \qquad (7.12)$$

It is a non-recursive difference equation of order 19. Its current output $y[n]$ is the sum of of its current and past nineteen inputs and then divided by 20 or multiplied by $1/20 = 0.05$. The equation is a non-recursive difference equation of order 19. Computing each $y[n]$ requires nineteen additions and one multiplication.

We next transform the non-recursive difference equation into a recursive one. Equation (7.12) holds for all n. It becomes, after subtracting 1 from all its indices,

$$y[n-1] = 0.05 \, (u[n-1] + u[n-2] + \cdots + u[n-19] + u[n-20])$$

Subtracting this equation from (7.12) yields

$$y[n] - y[n-1] = 0.05(u[n] - u[n-20]) \qquad (7.13)$$

or

$$y[n] = y[n-1] + 0.05(u[n] - u[n-20])$$

It is a recursive difference equation of order 20. Computing each $y[n]$ requires two additions (including subtractions) and one multiplication. Thus the recursive equation requires less computation than the non-recursive equation. □

This example shows that for a long moving average, its non-recursive difference equation can be changed to a recursive difference equation. By so doing, the number of operations can be reduced. This reduction is possible because a moving average has the same coefficients. In general, it is futile to change a non-recursive difference equation into a recursive one.

Example 7.4.2 Consider the savings account studied in Example 7.3.2. Its impulse response is $h[n] = (1.00015)^n$, for $n \geq 0$ and $h[n] = 0$, for $n < 0$. Its convolution description is

$$y[n] = \sum_{k=0}^{\infty} h[n-k]u[k] = \sum_{k=0}^{n} (1.00015)^{n-k} u[k] \qquad (7.14)$$

Note that the upper limit of the second summation is n not ∞. If it were ∞, then the second equality does not hold because it does not use the causality condition $h[n] = 0$, for $n < 0$. Thus care must be exercised in actual use of the discrete convolution in (7.10).

Using (7.14), we can develop a difference equation. See Problem 7.12. It is however simpler to develop the difference equation directly. Let $y[n-1]$ be the total amount of money in the account in the $(n-1)$th day. Then $y[n]$ is the sum of the principal $y[n-1]$, its one-day interest $0.00015y[n-1]$ and the money deposited or withdrawn on the nth day $u[n]$, that is,

$$y[n] = y[n-1] + 0.00015y[n-1] + u[n] = 1.00015y[n-1] + u[n] \qquad (7.15)$$

or

$$y[n] - 1.00015y[n-1] = u[n] \qquad (7.16)$$

This is a special case of (7.11) with $N = 1$ and $M = 0$ and is a first-order recursive difference equation. □

7.4.1   Comparison of convolutions and difference equations

The savings account discussed in Example 7.4.2 can be described by the convolution in (7.14) and the difference equation in (7.16). we now compare the two descriptions.

1. The convolution expresses the current output $y[n]$ as a linear combination of all past and current inputs; whereas the difference equation expresses the current output as a linear combination of one past output and current input. Thus the latter expression is simpler.

2. In using the convolution, we need $h[n]$ and all past inputs. In using the difference equation, we need only the coefficient 0.00015, $y[n-1]$, and current input. All past inputs are no longer needed and can be discarded. Thus the convolution requires more memory locations than the difference equation.

3. The summation in (7.14) consists of $(n+1)$ terms, thus computing $y[n]$ requires n additions. If we compute $y[n]$ from $n = 0$ to N, then the total number of additions is

$$\sum_{n=0}^{N} n = \frac{N(N+1)}{2} = 0.5N^2 + 0.5N$$

Computing $y[n]$ using (7.15) or (7.16) requires only one addition. Thus to compute $y[n]$, for $n = 0 : N$, requires only $(N+1)$ additions. This number is much smaller than $0.5N^2 + 0.5N$ for N large. For example, if $N = 1000$, then the convolution requires more than half a million additions; whereas the difference equation requires only 1001 additions. Similar remarks apply to the number of multiplications. Although the number of operations in computing (7.14) can be reduced using the FFT discussed in Section 5.4, its number of operations is still larger than the one using (7.16). Moreover, FFT will introduce some additional numerical errors due to changing real-number computation into complex-number computation.

4. The difference equation can be easily modified to describe the time-varying (interest rate changes with time) or nonlinear case (interest rate depends on $y[n]$). This will be difficult in using the convolution.

In conclusion, a difference equation is simpler, requires less memory and less computations, and is more flexible than a convolution. Thus if a system can be described by both descriptions, we prefer the former. Although all DT LTI system with infinite memory can be described by a convolution, only a small number of them can also be described by difference equations, as we will discuss in Chapter 11. This text, as every other text on signals and systems, studies only this small class of DT LTI systems. Thus we will downplay convolutions in this text.

7.5   DT LTI basic elements and basic block diagrams

A system is often built up using simple components. If every component is linear and time-invariant, then the system is automatically linear and time-invariant. It turns out

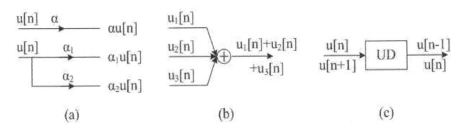

Figure 7.2: Time domain: (a) Multiplier. (b) Adder. (c) Unit-delay element.

that every DT LTI system with finite memory (FIR) can be built or simulated using the three types of basic elements shown in Figure 7.2. The element in Figure 7.2(a) denoted by a line with an arrow and a real number α is called a *multiplier* with gain α; its input and output are related by

$$y[n] = \alpha u[n] \quad \text{(multiplier)}$$

Note that α can be positive or negative. If $\alpha = 1$, it is direct transmission and the arrow and α may be omitted. If $\alpha = -1$, it changes only the sign of the input and is called an *inverter*. In addition, a signal may branch out to two or more signals with gain α_i as shown. The element denoted by a small circle with a plus sign as shown in Figure 7.2(b) is called an *adder*. It has two or more inputs denoted by entering arrows and one and only one output denoted by a departing arrow. Its inputs $u_i[n]$, for $i = 1 : 3$ and output $y[n]$ are related by

$$y[n] = u_1[n] + u_2[n] + u_3[n] \quad \text{(adder)}$$

The element in Figure 7.2(c) is a *unit-sample-delay element* or, simply, *unit-delay element*; its output $y[n]$ equals the input $u[n]$ delayed by one sample, that is,

$$y[n] = u[n - 1] \quad \text{or} \quad y[n + 1] = u[n] \quad \text{(unit delay)}$$

A unit-delay element will be denoted by *UD*.

The multiplier with gain α is, as discussed in the preceding chapter, linear and time-invariant. The adder is a multi-input single-output (MISO) system; its output is a linear combination of the inputs with constant coefficients (all equal 1). Thus it is also linear and time-invariant. See Problems 7.14 and 7.15. So is the unit-delay element. A diagram that consists of only these three types of elements is called a *basic block diagram*. Any system built using these types of elements is linear and time invariant.

The multiplier and adder are memoryless. The unit-delay element has a memory of one sample. The application of an input $u[n]$, for $n \geq 0$, to a unit-delay element will yield the output $y[n]$, for $n \geq 1$. The value $y[0] = u[-1]$ however is not specified. Thus we must specify $y[0]$ in order to have a unique output sequence $y[n]$, for $n \geq 0$. We call $y[0] = u[-1]$ the initial condition. Unless stated otherwise, the initial condition of every unit-delay element will be assumed to be zero.

We now show that the equation in (7.5) or

$$y[n] = 3u[n] - 2u[n - 1] + 5u[n - 3]$$

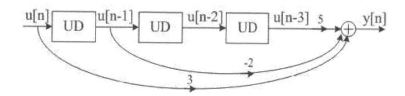

Figure 7.3: Time domain: Basic block diagram of (7.6).

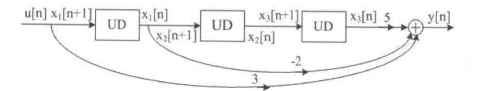

Figure 7.4: Basic block diagram.

can be implemented using basic elements. The equation has a memory of three samples. Thus its implementation requires three unit-delay elements as shown in Figure 7.3. They are connected in tandem as shown. If we assign the input of the left-most element as $u[n]$, then the outputs of the three unit-delay elements are respectively $u[n-1]$, $u[n-2]$, and $u[n-3]$ as shown. We then use three multipliers with gains $3, -2, 5$ and an adder to generate $y[n]$ as shown in Figure 7.3. Note that the coefficient of $u[n-2]$ is zero and no multiplier is connected to $u[n-2]$. Such a diagram that consists of only the three types of basic elements is called a *basic block diagram*. The procedure for developing such a diagram is simple and straightforward and is applicable to any DT LTI system with finite memory.

7.6   State-space (ss) equations

In this section we use a basic block diagram to develop a different mathematical equation. Consider the basic block diagram shown in Figure 7.3 and replotted in Figure 7.4. Now we assign the *output* of each unit-delay element as a variable, called a *state variable*. The three state variables $x_i[n]$, for $i = 1 : 3$, are assigned as shown in Figure 7.4. Note that we may select a different assignment, but all discussion will remain the same. Now if the output of a unit-delay element is assigned as $x_i[n]$, then its input is $x_i[n+1]$ as shown in Figure 7.4. From Figure 7.4, we can readily write down the following set of equations:

$$
\begin{aligned}
x_1[n+1] &= u[n] \\
x_2[n+1] &= x_1[n] \\
x_3[n+1] &= x_2[n] \\
y[n] &= 3u[n] - 2x_1[n] + 5x_3[n]
\end{aligned}
\tag{7.17}
$$

This set of equations is called a *state-space (ss)* equation. State-space equations are customarily expressed in matrix form. Let us write (7.17) as

$$
\begin{aligned}
x_1[n+1] &= 0 \times x_1[n] + 0 \times x_2[n] + 0 \times x_3[n] + 1 \times u[n] \\
x_2[n+1] &= 1 \times x_1[n] + 0 \times x_2[n] + 0 \times x_3[n] + 0 \times u[n] \\
x_3[n+1] &= 0 \times x_1[n] + 1 \times x_2[n] + 0 \times x_3[n] + 0 \times u[n] \\
y[n] &= -2 \times x_1[n] + 0 \times x_2[n] + 5 \times x_3[n] + 3 \times u[n]
\end{aligned}
$$

Then they can be expressed in matrix form as, see Section 3.3,

$$
\begin{bmatrix} x_1[n+1] \\ x_2[n+1] \\ x_3[n+1] \end{bmatrix}
=
\begin{bmatrix} 0 & 0 & 0 \\ 1 & 0 & 0 \\ 0 & 1 & 0 \end{bmatrix}
\begin{bmatrix} x_1[n] \\ x_2[n] \\ x_3[n] \end{bmatrix}
+
\begin{bmatrix} 1 \\ 0 \\ 0 \end{bmatrix} u[n]
\tag{7.18}
$$

$$
y[n] = \begin{bmatrix} -2 & 0 & 5 \end{bmatrix}
\begin{bmatrix} x_1[n] \\ x_2[n] \\ x_3[n] \end{bmatrix}
+ 3u[n]
\tag{7.19}
$$

and, using matrix symbols,

$$
\mathbf{x}[n+1] = \mathbf{A}\mathbf{x}[n] + \mathbf{b}u[n] \tag{7.20}
$$

$$
y[n] = \mathbf{c}\mathbf{x}[n] + du[n] \tag{7.21}
$$

where

$$
\mathbf{x} = \begin{bmatrix} x_1 \\ x_2 \\ x_3 \end{bmatrix}, \quad
\mathbf{A} = \begin{bmatrix} 0 & 0 & 0 \\ 1 & 0 & 0 \\ 0 & 1 & 0 \end{bmatrix}, \quad
\mathbf{b} = \begin{bmatrix} 1 \\ 0 \\ 0 \end{bmatrix}
$$

$$
\mathbf{c} = \begin{bmatrix} -2 & 0 & 5 \end{bmatrix}, \quad d = 3
$$

Note that both $\mathbf{A}\mathbf{x}[n]$ and $\mathbf{b}u[n]$ are 3×1 and their sum equals $\mathbf{x}[n+1]$. Both $\mathbf{c}\mathbf{x}[n]$ and $du[n]$ are 1×1 and their sum yields $y[n]$.

The set of two equations in (7.20) and (7.21) is called a (DT) state-space (ss) equation of *dimension* 3. We call (7.20) a *state equation* and (7.21) an *output equation*. The vector $\mathbf{x}$ is called the *state* or *state vector*; its components are called *state variables*. The scalar d is called the *direct transmission* part; it connects $u[n]$ directly to $y[n]$ without passing through any unit-delay element. Note that the state equation consists of a set of *first-order* difference equations.

From (7.21) we see that if the state $\mathbf{x}$ at time instant n_1 is available, all we need to determine $y[n_1]$ is the input $u[n_1]$. Thus the state $\mathbf{x}[n_1]$ summarizes the effect of $u[n]$, for all $n < n_1$, on $y[n_1]$. Indeed comparing Figures 7.3 and 7.4, we have $x_1[n_1] = u[n_1 - 1]$, $x_2[n_1] = u[n_1 - 2]$, and $x_3[n_1] = u[n_1 - 3]$. For this example, the state vector consists of simply the past three input samples and is consistent with the fact that the system has a memory of three samples. Note that the output equation in (7.21) is memoryless. The evolution or dynamic of the system is described by the state equation in (7.20).

7.6.1 Computer computation and real-time processing using ss equations

In quantitative analysis of a system, we are interested in the response of the system excited by some input. If the system can be described by an ss equation, then such computation is straightforward. Consider the ss equation in (7.20) and (7.21) or

$$
\begin{aligned}
y[n] &= \mathbf{c}\mathbf{x}[n] + du[n] \\
\mathbf{x}[n+1] &= \mathbf{A}\mathbf{x}[n] + \mathbf{b}u[n]
\end{aligned}
\tag{7.22}
$$

To compute the output $y[n]$, for $n \geq 0$, excited by an input $u[n]$, for $n \geq 0$, we must specify $\mathbf{x}[0]$, called the *initial state*. The initial state actually consists of the set of initial conditions discussed in Section 7.2.1. For the system in Figure 7.4, the initial state is $\mathbf{x}[0] = [u[-1] \ u[-2] \ u[-3]]'$, where the prime denotes the transpose. Unlike discrete convolutions which can be used only if systems are initially relaxed or all initial conditions are zero, ss equations can be used even if initial conditions are different from zero.

Once $\mathbf{x}[0]$ and $u[n]$, for $n \geq 0$ are given, we can compute $y[n]$, for $n \geq 0$, *recursively* using (7.22). We use $\mathbf{x}[0]$ and $u[0]$ to compute $y[0]$ and $\mathbf{x}[1]$ in (7.22). We then use the computed $\mathbf{x}[1]$ and the given $u[1]$ to compute $y[1]$ and $\mathbf{x}[2]$. Proceeding forward, we can compute $y[n]$, for $n = 0 : n_f$, where n_f is the final time index. The computation involves only additions and multiplications. Thus ss equations are most suited for computer computation.

As discussed in Subsection 2.6.1, there are two types of processing: real time and non-real time. We discuss next that if the sampling period T is large enough, ss equations can be computed in real time. We first assume $\mathbf{x}[0] = \mathbf{0}$ and $d = 0$. In this case, we have $y[0] = 0$ no matter what $u[0]$ is. Thus we can send out $y[0] = 0$ when $u[0]$ arrives at $n = 0$. We can also start to compute $\mathbf{x}[1]$ using $\mathbf{x}[0] = \mathbf{0}$ and $u[0]$ and then $y[1] = \mathbf{c}\mathbf{x}[1]$. If this computing time is less than the sampling period T, we store $\mathbf{x}[1]$ and $y[1]$ in memory. We then deliver $y[1]$ at the same time instant as $u[1]$ arrives. We also start to compute $\mathbf{x}[2]$ using $u[1]$ and stored $\mathbf{x}[1]$ and then $y[2]$. We deliver $y[2]$ at the time instant $n = 2$ and use just arrived $u[2]$ and stored $\mathbf{x}[2]$ to compute $\mathbf{x}[3]$ and $y[3]$. Proceeding forward, the output sequence $y[n]$ can appear at the same time instant as $u[n]$.

If $d \neq 0$, we can compute $y[0] = \mathbf{c}\mathbf{x}[0] + du[0]$ only after $u[0]$ arrives. The multiplication of d and $u[0]$ will take some finite time, albeit very small, thus $y[0]$ cannot appear at the same time instant as $u[0]$. In this case, if we delay $y[0]$ by one sampling period, then the preceding discussion can be applied. Because the sampling period is very small in practice, this delay can hardly be detected. In conclusion, ss equations can be used in real-time processing if the sampling period T is large enough to complete the necessary computation.

In the remainder of this subsection, we use an example to discuss the use of MATLAB functions to compute system responses. Consider a DT system described by (7.22) with

$$
\mathbf{A} = \begin{bmatrix} -0.18 & -0.8 \\ 1 & 0 \end{bmatrix} \quad \mathbf{b} = \begin{bmatrix} 1 \\ 0 \end{bmatrix}
\tag{7.23}
$$

$$
\mathbf{c} = [2 \quad -1] \quad d = 1.2
$$

We use the MATLAB function lsim, abbreviation for *linear simulation*, to compute the response of the system excited by the input $u[n] = \sin 1.2n$, for $n \geq 0$ and the initial state $\mathbf{x}[0] = [-2 \ \ 3]'$, where the prime denotes the *transpose*. Up to this point, nothing is said about the sampling period T and we can select any $T > 0$. On the other hand, if $u[n]$ is given as $x[n] = x(nT) = \sin 2.4nT$ with a specified T, say $T = 0.5$, then we must select $T = 0.5$. We type in an edit window the following:

```
% Program 7.1 (f75.m)
a=[-0.18 -0.8;1 0];b=[1;0];c=[2 -1];d=1.2;T=0.5;
pig=ss(a,b,c,d,T);
n=0:30;t=n*T;
u=sin(1.2*n);x0=[-2;3];
y=lsim(pig,u,t,x0);
subplot(3,1,1)
stem(t,y),title('(a)')
ylabel('Amplitude')
subplot(3,1,2)
stairs(t,y),title('(b)')
ylabel('Amplitude')
subplot(3,1,3)
lsim(pig,u,t,x0)
```

The first line expresses $\{\mathbf{A}, \mathbf{b}, \mathbf{c}, d\}$ in MATLAB format. In MATLAB, a matrix is expressed row by row separated by semicolons and bounded by a pair of brackets as shown. Note that $\mathbf{b}$ has two rows, thus there is a semicolon in b=[1;0]. Whereas $\mathbf{c}$ has only one row, and there is no semicolon in c=[2 -1]. Note also that for a scalar, the pair of brackets may be omitted as in d=1.2. The second line defines the system. We call the system 'pig', which is defined using the state-space model, denoted by ss. It is important to have the fifth argument T inside the parentheses. *Without a T, it defines a CT system.* If no T is specified, we may set $T = 1$. The third line denotes the number of samples to be computed and the corresponding time instants. The fourth line is the input and initial state. We then use lsim to compute the output. Note that the T in defining the system and the T in defining t must be the same, otherwise an error message will appear. The output is plotted in Figure 7.5(a) using the function stem and in Figure 7.5(b) using the function stairs. The actual output of the DT system is the one shown in Figure 7.5(a). The output shown in Figure 7.5(b) is actually the output of the DT system followed by a zero-order hold. A zero-order hold holds the current value constant until the arrival of the next value. If it is understood that the output of the DT system consists of only the values at sampling instants, then Figure 7.5(b) is easier for viewing than Figure 7.5(a), especially if the sampling period is very small. Thus all outputs of DT systems in MATLAB are plotted using the function stairs instead of stem. If the function lsim does not have the left-hand argument as shown in the last line of Program 7.1, then MATLAB automatically plots the output as shown in Figure 7.5(c).[4] It automatically returns the title and x- and y-labels. We save Program 7.1 as an m-file named f75.m. Typing in the command window >>

---

[4]It also generates plot(t,u)

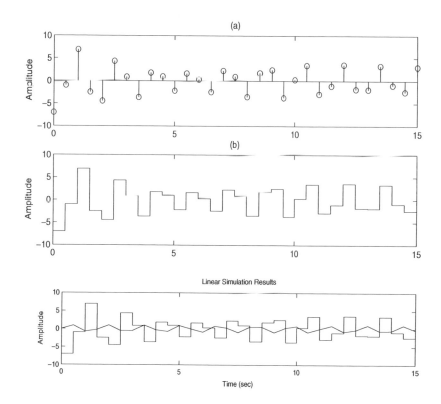

Figure 7.5: (a) Response of (7.23). (b) Response of (7.23) followed by a zero-order hold. (c) Output of `lsim(pig,u,t,x0)`
without left-hand argument.

f75 will generate Figure 7.5 in a figure window. Note that if the initial state $\mathbf{x}[0]$ is zero, there is no need to type x0 in Program 7.1 and the last argument of lsim may be omitted. Note that we can also run the program by clicking 'Debug' on the edit window.

MATLAB contains the functions impulse and step which compute the impulse and step responses of a system. The impulse response is the output excited by an impulse sequence ($u[n] = \delta_d[n]$) and the step response is the output excited by a step sequence ($u[n] = 1$, for all $n \geq 0$). In both responses, the initial state is assumed to be zero. The program that follows

```
%Program 7.2 (f76.m)
a=[-0.18 -0.8;1 0];b=[1;0];c=[2 -1];d=1.2;T=0.5;
pig=ss(a,b,c,d,T);
subplot(2,1,1)
impulse(pig)
subplot(2,1,2)
step(pig)
```

will generate the impulse and step responses in Figure 7.6. Note that in using lsim, we must specify the number of samples to be computed. However impulse and step will select automatically the number of samples to be computed or, more precisely, they will automatically stop computing when the output hardly changes its values. They also automatically generate titles and labels. Thus their use is very simple.

7.7   Transfer functions – z-transform

In this section, we introduce a different type of equation to describe DT LTI systems.

Consider a DT signal $x[n]$ defined for all n in $(-\infty, \infty)$. Its z-transform is defined as

$$X(z) := \mathcal{Z}[x[n]] := \sum_{n=0}^{\infty} x[n]z^{-n} \tag{7.24}$$

where z is a complex variable. Note that it is defined only for the positive-time part of $x[n]$ or $x[n]$, for $n \geq 0$. Because its negative-time part ($x[n]$, for $n < 0$) is not used, we often assume $x[n] = 0$, for $n < 0$, or $x[n]$ to be positive time. Before proceeding, we mention that the z-transform has the following linearity property:

$$\mathcal{Z}[\beta_1 x_1[n] + \beta_2 x_2[n]] = \beta_1 \mathcal{Z}[x_1[n]] + \beta_2 \mathcal{Z}[x_2[n]]$$

for any constants β_i and any sequences $x_i[n]$, for $i = 1, 2$. This is the same superposition property discussed at the end of Section 6.5.

The z-transform in (7.24) is a power series of z^{-1}. In this series, z^{-n} indicates the nth sampling instant. For example, $z^0 = 1$ indicates the initial time instant $n = 0$, z^{-1} indicates the sampling instant $n = 1$, and so forth. In this sense, there is not much difference between a time sequence and its z-transform. For example, the z-transform

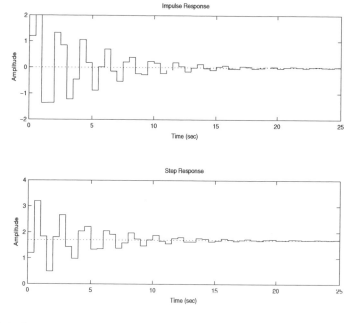

Figure 7.6: (Top) Impulse response of (7.23) followed by a zero-order hold. (Bottom) Step response of (7.23) followed by a zero-order hold.

of the DT signal shown in Figure 7.1(a) is $1z^0 = 1$. The z-transform of the DT signal shown in Figure 7.1(b) is

$$3z^0 + (-2)z^{-1} + 0 \cdot z^{-2} + 5z^{-3} = 3 - 2z^{-1} + 5z^{-3}$$

Let us apply the z-transform to $y[n]$ in (7.10). Then we have

$$
\begin{aligned}
Y(z) &= \mathcal{Z}[y[n]] = \sum_{n=0}^{\infty} y[n]z^{-n} = \sum_{n=0}^{\infty} \left[\sum_{k=0}^{\infty} h[n-k]u[k] \right] z^{-(n-k)-k} \\
&= \sum_{k=0}^{\infty} \left[\sum_{n=0}^{\infty} h[n-k]z^{-(n-k)} \right] u[k]z^{-k}
\end{aligned}
$$

where we have interchanged the order of summations. Let us define $l := n - k$. Then the preceding equation can be written as

$$Y(z) = \sum_{k=0}^{\infty} \left[\sum_{l=-k}^{\infty} h[l]z^{-l} \right] u[k]z^{-k}$$

which can be reduced as, using the causality condition $h[l] = 0$, for $l < 0$,

$$Y(z) = \sum_{k=0}^{\infty} \left[\sum_{l=0}^{\infty} h[l]z^{-l} \right] u[k]z^{-k} = \left[\sum_{l=0}^{\infty} h[l]z^{-l} \right] \left[\sum_{k=0}^{\infty} u[k]z^{-k} \right]$$

Thus we have

$$Y(z) = H(z) \times U(z) = H(z)U(z) \tag{7.25}$$

where

$$H(z) := \mathcal{Z}[h[n]] = \sum_{n=0}^{\infty} h[n]z^{-n} \tag{7.26}$$

It is called the *DT transfer function* of the system. It is, by definition, the z-transform of the impulse response $h[n]$. Likewise $Y(z)$ and $U(z)$ are respectively the z-transforms of the output and input sequences. We give an example.

Example 7.7.1 Consider the DT system described by the equation in (7.5). Its impulse response was computed as $h[0] = 3$, $h[1] = -2$, $h[2] = 0$, $h[3] = 5$ and $h[n] = 0$, for $n \geq 4$. Thus its transfer function is

$$\begin{aligned} H(z) &= 3z^0 + (-2)z^{-1} + 0\dot{z}^{-2} + 5z^{-3} = \frac{3 - 2z^{-1} + 5z^{-3}}{1} \tag{7.27} \\ &= \frac{3z^3 - 2z^2 + 5}{z^3} \tag{7.28} \end{aligned}$$

The transfer function in (7.27) is a polynomial of negative powers of z or a negative-power rational function of z with its denominator equal to 1. Multiplying z^3 to its numerator and denominator, we obtain a rational function of positive powers of z as in (7.28). They are called respectively a negative-power transfer function and a positive-power transfer function. Either form can be easily obtained from the other. $\square$

Although the transfer function is defined as the z-transform of the impulse response, it can also be defined without referring to the latter. Using (7.25), we can also define a transfer function as

$$H(z) := \frac{\mathcal{Z}[\text{output}]}{\mathcal{Z}[\text{input}]} = \left.\frac{Y(z)}{U(z)}\right|_{\text{Initially relaxed}} \tag{7.29}$$

If we use (7.29) to compute a transfer function, then there is no need to compute its impulse response. Note that if $u[n] = \delta_d[n]$, then $U(z) = 1$, and (7.29) reduces to $H(z) = Y(z)$ which is essentially (7.26). In practical application, we use (7.29) to compute transfer functions. In computing a transfer function, we may assume $u[n] = 0$ and $y[n] = 0$, for all $n < 0$. This will insure the initial relaxedness of the system.

7.7.1   Transfer functions of unit-delay and unit-advance elements

Let $x[n]$ be a positive-time sequence. That is, $x[n] = 0$, for $n < 0$. Note that $x[n-1]$ is the delay of $x[n]$ by one sample and is still positive time. Let us compute the z-transform of $x[n-1]$. By definition, we have

$$\begin{aligned} \mathcal{Z}[x[n-1]] &= \sum_{n=0}^{\infty} x[n-1]z^{-n} = \sum_{n=0}^{\infty} x[n-1]z^{-(n-1)-1} \\ &= z^{-1} \sum_{n=0}^{\infty} x[n-1]z^{-(n-1)} \end{aligned}$$

which can be written as, defining $l := n - 1$ and using $x[-1] = 0$,

$$
\begin{aligned}
\mathcal{Z}[x[n-1]] &= z^{-1} \sum_{l=-1}^{\infty} x[l] z^{-l} = z^{-1} \left[x[-1] z + \sum_{l=0}^{\infty} x[l] z^{-l} \right] \\
&= z^{-1} \left[\sum_{n=0}^{\infty} x[n] z^{-n} \right] = z^{-1} X(z)
\end{aligned}
$$

where we have changed the summing variable from l to n and used (7.24). We see that the time delay of one sampling instant is equivalent to the multiplication of z^{-1} in the z-transform, denoted as

$$
\text{Unit-sample delay} \quad \longleftrightarrow \quad z^{-1}
$$

Thus a system which carries out a unit-sample time delay has transfer function $z^{-1} = 1/z$.

If a positive-time sequence is delayed by two samples, then its z-transform is its original z-transform multiplied by z^{-2}. In general, we have

$$
\mathcal{Z}[x[n-k]] = z^{-k} X(z) \tag{7.30}
$$

for any integer $k \geq 0$ and any positive-time sequence $x[n]$. If k is a negative integer, the formula does not hold as we will show shortly.

Example 7.7.2 Consider a DT system described by (7.5) or

$$
y[n] = 3u[n] - 2u[n-1] + 5u[n-3]
$$

If it is initially relaxed at $n_0 = 0$, we may assume $u[n] = 0$ and $y[n] = 0$ for $n < 0$. Thus $y[n]$ and $u[n]$ are both positive time. Applying the z-transform and using (7.30), we have

$$
Y(z) = 3U(z) - 2z^{-1} U(z) + 5z^{-3} U(z) = (3 - 2z^{-1} + 5z^{-3}) U(z) \tag{7.31}
$$

Thus the transfer function of the system is

$$
H(z) := \frac{Y(z)}{U(z)} = 3 - 2z^{-1} + 5z^{-3} \tag{7.32}
$$

which is the same as (7.27). □

We mention that for TI nonlinear systems, we may use (7.29) to compute transfer functions. However, different input-output pairs will yield different transfer functions. Thus for nonlinear systems, the concept of transfer functions is useless. For TI linear systems, no matter what input-output pair we use in (7.29), the resulting transfer function is always the same. Thus an LTI system can be described by its transfer function.

To conclude this section, we compute the z-transform of $x[n+1]$, the advance or shifting to the left of $x[n]$ by one sample. Be definition, we have

$$
\begin{aligned}
\mathcal{Z}[x[n+1]] &= \sum_{n=0}^{\infty} x[n+1]z^{-n} = \sum_{n=0}^{\infty} x[n+1]z^{-(n+1)+1} \\
&= z\sum_{n=0}^{\infty} x[n+1]z^{-(n+1)}
\end{aligned}
$$

which can be written as, defining $l := n+1$,

$$
\begin{aligned}
\mathcal{Z}[x[n+1]] &= z\sum_{l=1}^{\infty} x[l]z^{-l} = z\left[\sum_{l=0}^{\infty} x[l]z^{-l} - x[0]z^{-0}\right] \\
&= z[X(z) - x[0]]
\end{aligned} \tag{7.33}
$$

Using the same procedure, we can obtain

$$
\mathcal{Z}[x[n+2]] = z^2\left[X(z) - x[0] - x[1]z^{-1}\right] \tag{7.34}
$$

We give an explanation of (7.33). The sequence $x[n+1]$ shifts $x[n]$ one sample to the left. Thus $x[0]$ is shifted from $n=0$ to $n=-1$ and $x[n+1]$ becomes non-positive time even though $x[n]$ is positive time. The z-transform of $x[n+1]$ is defined only for the positive-time part of $x[n+1]$ and, consequently, will not contain $x[0]$. Thus we subtract $x[0]$ from $X(z)$ and then multiply it by z to yield (7.33). Equation (7.34) can be similarly explained. We see that advancing a sequence by one sample is equivalent to the multiplication of z in its z-transform, denoted as

$$\text{Unit-sample advance} \quad \longleftrightarrow \quad z$$

In other words, a system that carries out unit-sample time advance has transfer function z. Note that such a system is not causal and cannot be implemented in real time. Moreover because (7.33) and (7.34) involve $x[0]$ and $x[1]$, their uses are not as convenient as (7.30).

7.8   Composite systems: Transform domain or time domain?

We have introduced four mathematical equations to describe DT LTI systems. They are

1. Convolutions

2. High-order difference equations

3. Sets of first-order difference equations of special form, called state-space (ss) equations

4. DT rational transfer functions

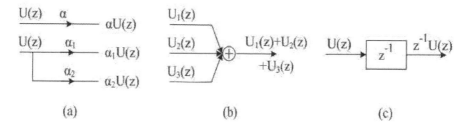

Figure 7.7: Transform domain: (a) Multiplier. (b) Adder. (c) Unit-delay element.

The first three are all expressed in terms of time index and are called *time-domain* descriptions. The last one is based on the z-transform and is called the *transform-domain* description. Note that for DT LTI memoryless systems, the first three descriptions reduce to $y[n] = \alpha u[n]$ and the last description reduces to $Y(z) = \alpha U(z)$, for some constant α.

In the transform domain, we have

$$\mathcal{Z}[\text{output}] = \text{transfer function} \times \mathcal{Z}[\text{input}] \qquad (7.35)$$

This equation is purely algebraic; it does not involve time advance or time delay. For example, consider the unit-delay element shown in Figure 7.2(c) with $y[n] = u[n-1]$. Although the element is denoted by UD, the equation $y[n] = \text{UD}u[n]$ is mathematically meaningless. In the z-transform domain, we have, using (7.30), $Y(z) = z^{-1}U(z)$. Thus the unit-delay element has the transfer function z^{-1} and its input and output are related by an algebraic equation as in (7.35). We plot in Figure 7.7 the three basic elements in Figure 7.2 in the z-transform domain.

A system may be built by interconnecting a number of subsystems. To develop a mathematical equation to describe the system from those of its subsystems, it is simplest to use transfer functions.[5] We demonstrate this for the three basic connections shown in Figure 7.8. Consider two LTI systems described by $Y_i(z) = H_i(z)U_i(z)$, for $i = 1, 2$. Let $Y(z)$ and $U(z)$ be the output and input of the overall system and let $H(z)$ be the overall transfer function. In the *parallel* connection shown in Figure 7.8(a), we have $U_1(z) = U_2(z) = U(z)$ and $Y(z) = Y_1(z) + Y_2(z)$. By direct substitution, we have

$$Y(z) = H_1(z)U_1(z) + H_2(z)U_2(z) = (H_1(z) + H_2(z))U(z) =: H(z)U(z)$$

Thus the overall transfer function of the parallel connection is simply the sum of the two individual transfer functions.

In the *tandem* or *cascade* connection shown in Figure 7.8(b), we have $U(z) = U_1(z)$, $Y_1(z) = U_2(z)$, and $Y_2(z) = Y(z)$. Because

$$\begin{aligned} Y(z) &= Y_2(z) = H_2(z)U_2(z) = H_2(z)Y_1(z) = H_2(z)H_1(z)U_1(z) \\ &= H_2(z)H_1(z)U(z) \end{aligned}$$

we have $H(z) = H_1(z)H_2(z) = H_2(z)H_1(z)$. Thus the transfer function of the tandem connection is simply the product of the two individual transfer functions.

---

[5]In the connection, it is assumed that the original transfer functions will not be affected. See Section 10.4.

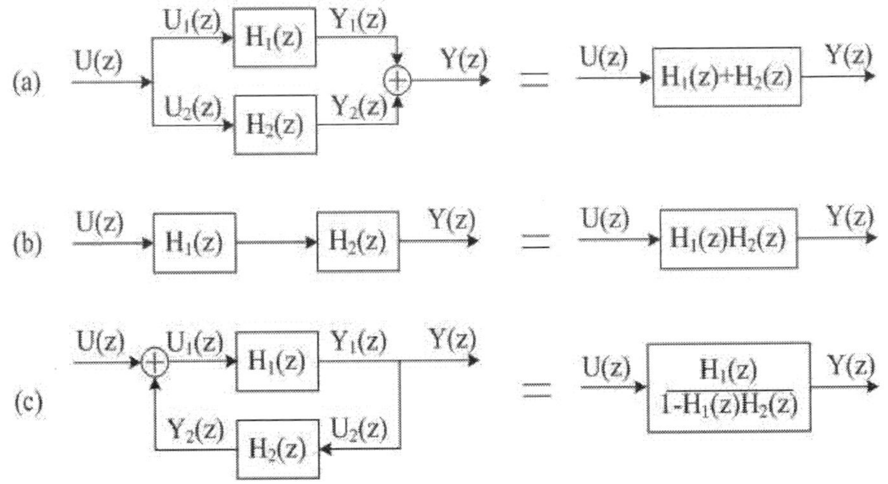

Figure 7.8: (a) Parallel connection. (b) Tandem connection. (c) Positive feedback connection.

In the *positive feedback* connection shown in Figure 7.8(c), we have $U_1(z) = U(z) + Y_2(z)$ and $Y_1(z) = Y(z) = U_2(z)$. By direct substitution, we have

$$Y_2(z) = H_2(z)U_2(z) = H_2(z)Y_1(z) = H_2(z)H_1(z)U_1(z)$$

and

$$U_1(z) = U(z) + Y_2(z) = U(z) + H_2(z)H_1(z)U_1(z)$$

which implies

$$[1 - H_2(z)H_1(z)]U_1(z) = U(z) \quad \text{and} \quad U_1(z) = \frac{U(z)}{1 - H_2(z)H_1(z)}$$

Because $Y(z) = Y_1(z) = H_1(z)U_1(z)$, we have

$$Y(z) = H_1(z)\frac{U(z)}{1 - H_2(z)H_1(z)} = \frac{H_1(z)}{1 - H_1(z)H_2(z)}U(z)$$

Thus the transfer function of the feedback system in Figure 7.8(c) is

$$H(z) := \frac{Y(z)}{U(z)} = \frac{H_1(z)}{1 - H_1(z)H_2(z)} \tag{7.36}$$

We see that the derivation is simple and straightforward.

Before proceeding, we give a simple application of the preceding formulas. Consider the basic block diagram shown in Figure 7.9(a) which is Figure 7.3 in the z-transform domain. Note that every branch is unidirectional. It has three forward paths from $U(z)$ to $Y(z)$ as shown in Figure 7.9(b). The first path is the tandem connection of three unit-delay elements and a multiplier with gain 5. Thus the transfer function of

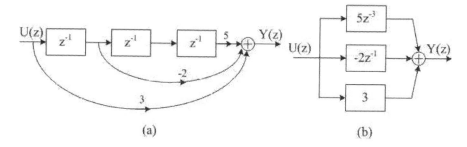

Figure 7.9: Transform domain: (a) Basic block diagram of (7.6). (b) Parallel connection of three paths.

the tandem connection is $z^{-1} \times z^{-1} \times z^{-1} \times 5 = 5z^{-3}$. The second path is the tandem connection of a unit-delay element and a multiplier with gain -2, thus its transfer function is $-2z^{-1}$. The third path from $U(z)$ to $Y(z)$ consists of only one multiplier and its transfer function is 3. The overall transfer function from $U(z)$ to $Y(z)$ is the parallel connection of the three paths and thus equals

$$H(z) = 5z^{-3} - 2z^{-1} + 3$$

This is the same as (7.27).

We now discuss the time-domain description of composite systems. The convolution in (7.10) is often denoted by

$$y[n] = h[n] * u[n]$$

Let $h_i[n]$, for $i = 1, 2$, be the impulse responses of two subsystems and $h[n]$ be the impulse response of an overall system. Then we have

$$h[n] = h_1[n] + h_2[n] \qquad (7.37)$$

in the parallel connection, and

$$h[n] = h_1[n] * h_2[n] = \sum_{k=0}^{\infty} h_1[n-k]h_2[k] \qquad (7.38)$$

in the tandem connection. We see that (7.38) is more complicated than $H_1(z)H_2(z)$ and its derivation is also more complicated. More seriously, there is no simple way of relating $h[n]$ of the feedback connection in Figure 7.8(c) with $h_1[n]$ and $h_2[n]$. Thus convolutions are not used in describing composite systems, especially, feedback systems. Similar remarks apply to ss equations and difference equations. See Reference [C6]. In conclusion, we use exclusively transfer functions in studying composite systems.

7.9 Concluding remarks

In this chapter, we used a DT LTI system with memory $N = 3$ to develop all system concepts and then developed four mathematical equations to describe it. The discussion is applicable to all finite N, no matter how large. Such a system can be described by

1. A discrete convolution.

2. A single high-order difference equation.

3. An ss equation which is a set of first-order difference equations of special form. Its dimension equals the memory length.

4. A rational transfer function.

Among them, ss equations are best for computer computation[6] and real-time processing. Rational transfer functions are best for studying composite systems.

If $N = \infty$ or, equivalently, a system has infinite memory, then the situation changes completely. Even though every DT LTI system with infinite memory can be described by a convolution, it may not be describable by the other three descriptions. This will be studied in the remainder of this text.

Problems

7.1 Consider a DT system whose input $u[n]$ and output $y[n]$ are related by

$$y[n] = 2u[n] + u[n+1]u[n-2]$$

Is it causal? linear? time-invariant?

7.2 Repeat Problem 7.1 for

1. $y[n] = 2u[n] + u[n-2] + 10$
2. $y[n] = 2u[n] + u[n-1] - u[n-2]$
3. $y[n] = u[n-2]$

Note that they all have a memory of two samples.

7.3 Consider a DT system whose input $u[n]$ and output $y[n]$ are related by

$$y[n] = \frac{u^2[n]}{u[n-1]}$$

if $u[n-1] \neq 0$, and $y[n] = 0$ if $u[n-1] = 0$. Show that the system satisfies the homogeneity property but not the additivity property.

7.4* Show that if the additivity property holds, then the homogeneity property holds for all rational number α. Thus if a system has some "continuity" property, then additivity implies homogeneity. Thus in checking linearity, we may check only the additivity property. But we cannot check only the homogeneity property as Problem 7.3 shows.

---

[6] Discrete convolutions can be computed in MATLAB using FFT and inverse FFT. However, ss equations are preferred because they are more efficient, incur less numerical errors, and can be carried out in real time. See Reference [C7].

7.5 Compute the impulse response of a system described by

$$y[n] = 2u[n-1] - 4u[n-3]$$

Is it FIR or IIR? Recall that the impulse response is the output $y[n] = h[n]$ of the system, which is initially relaxed at $n_0 = 0$, excited by the input $u[n] = \delta_d[n]$. If a system is causal, and if $u[n] = 0$ and $y[n] = 0$, for all $n < 0$, then the system is initially relaxed at $n_0 = 0$. In substituting $u[n] = \delta_d[n]$, the condition $u[n] = 0$, for all $n < 0$, is already imbedded in computing the impulse response.

7.6 Consider a DT memoryless system described by $y[n] = 2.5u[n]$. What is its impulse response? Is it FIR or IIR?

7.7 Design a 4-point moving average filter. What is its impulse response? Is it FIR or IIR?

7.8 Compute impulse responses of systems described by

1. $2y[n] - 3y[n-1] = 4u[n]$
2. $y[n] + 2y[n-1] = -2u[n-1] - u[n-2] + 6u[n-3]$

Are they FIR or IIR? Note that in substituting $u[n] = \delta_d[n]$, we have used the condition $u[n] = 0$, for all $n < 0$. In addition, we may assume $y[n] = 0$, for all $n < 0$, in computing impulse responses.

7.9 Verify that if $h[n]$ is given as shown in Figure 7.1(b), then (7.10) reduces to (7.5).

7.10* Verify that the discrete convolution in (7.10) has the additivity and homogeneity properties.

7.11* Verify that (7.10) is time invariant by showing that it has the shifting property in (7.2). That is, if the input is $u[n-n_1]$, for any $n_1 \geq 0$, then (7.10) yields $y[n-n_1]$. Recall that when we apply $u[n-n_1]$, the system is implicitly assumed to be relaxed at n_1 or $u[n] = 0$ and $y[n] = 0$ for all $n < n_1$.

7.12* For the savings account studied in Example 7.3.2, we have $h[n] = (1.00015)^n$, for $n \geq 0$ and $h[n] = 0$, for $n < 0$. If we write its discrete convolution as

$$y[n] = \sum_{k=0}^{\infty} h[n-k]u[k] = \sum_{k=0}^{\infty} (1.00015)^{n-k} u[k]$$

then it is incorrect because it does not use the condition $h[n] = 0$, for $n < 0$ or $h[n-k] = 0$, for $k > n$. The correct expression is

$$y[n] = \sum_{k=0}^{\infty} h[n-k]u[k] = \sum_{k=0}^{n} (1.00015)^{n-k} u[k]$$

Use this equation to develop the recursive difference equation in (7.16). Can you obtain the same difference equation from the incorrect convolution?

7.13 Consider Figure 1.5(c) in which the smoother curve is obtained using a 90-day moving average. What is its convolution or non-recursive difference equation description? Develop a recursive difference equation to describe the moving average. Which description requires less computation?

7.14 Let the output of the adder in Figure 7.2(b) be denoted by $y[n]$. If we write $y[n] = \mathbf{c}\mathbf{u}[n]$, where $\mathbf{u}$ is a 3×1 column vectors with u_i, for $i = 1 : 3$, as its entries, what is $\mathbf{c}$? Is the adder linear and time-invariant?

7.15 The operator shown in Figure 7.10(a) can also be defined as an *adder* in which each entering arrow has a positive or negative sign and its output is $-u_1 + u_2 - u_3$. Verify that Figure 7.10(a) is equivalent to Figure 7.10(b) in which every entering arrow with a negative sign has been changed to a positive sign by inserting a multiplier with gain -1 as shown.

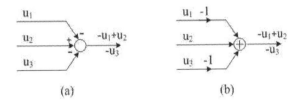

(a)   (b)

Figure 7.10: (a) Adder. (b) Equivalent adder.

7.16 Consider the basic block diagram in Figure 7.4. Develop an ss equation by assigning from left to right the output of each unit-delay element as $x_3[n]$, $x_2[n]$ and $x_1[n]$. Is the ss equation the same as the one in (7.18) and (7.19)? Even though the two ss equations look different, they are equivalent. See Reference [C6].

7.17 Draw a basic block diagram for the five-point moving average defined in Example 7.3.1 and then develop an ss equation to describe it.

7.18 Express the following equations in matrix form

$$
\begin{aligned}
x_1[n+1] &= 2x_1[n] - 2.3x_2[n] + 5u[n] \\
x_2[n+1] &= -1.3x_1[n] + 1.6x_2[n] - 3u[n] \\
y[n] &= -1.5x_1[n] - 3.1x_2[n] + 4u[n]
\end{aligned}
$$

7.19 Consider the savings account described by the first-order difference equation in (7.16). Define $x[n] := y[n-1]$. Can you develop an ss equation of the form shown in (7.22) to describe the account?

7.20 Draw a basic block diagram for (7.16), assign the output of the unit-delay element as a state variable, and then develop an ss equation to describe it. Is it the same as the one developed in Problem 7.19? Note that the block diagram has a *loop*. A block diagram is said to have a loop if it has a closed unidirectional path on which a point can travel along the path and come back to the same point. Note that the block diagram in Figure 7.3 has no loop.

7.21 What are the z-transforms of $\delta_d[n-1]$ and $2\delta_d[n] - 3\delta_d[n-4]$?

7.22 What is the DT transfer function of the savings account described by (7.16)?

7.23 What is the transfer function of the DT system in Problem 7.5? Compute it using (7.26) and using (7.29).

7.24 What is the DT transfer function of the five-point moving average defined in Example 7.3.1?

7.25 Can the order of the tandem connection of two SISO systems with transfer functions $H_i(z)$, with $i = 1, 2$, be interchanged? Can the order of the tandem connection of two MIMO systems with transfer function matrices $H_i(z)$, with $i = 1, 2$, be interchanged?

7.26 Consider the feedback connection shown in Figure 7.11(a). Verify that its transfer function from $U(z)$ to $Y(z)$ is

$$H(z) = \frac{Y(z)}{U(z)} = \frac{H_1(z)}{1 + H_1(z)H_2(z)}$$

as shown in Figure 7.11(b). You can derive it directly or obtain it from Figure 7.8(c) using Figure 7.10.

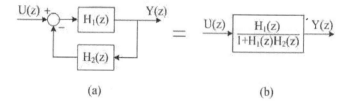

Figure 7.11: (a) Negative feedback system. (b) Overall transfer function.

Chapter 8

CT LTI and lumped systems

8.1 Introduction

We developed in the preceding chapter

1. discrete convolutions,

2. difference equations,

3. state-space equations, and

4. rational transfer functions

to describe DT LTI systems with finite memory. In this chapter, we will develop the four descriptions for the general case. Logically, the next step is to study DT LTI systems with infinite memory. Unfortunately, other than the savings account discussed in Example 7.3.2, no suitable DT systems are available for explaining what will be discussed. Whereas physical examples are available in the CT case. Thus we study in this chapter CT LTI systems with finite and infinite memory.

After introducing the concepts of causality, time-invariance, forced responses, and linearity, we develop an integral convolution to describe a CT LTI system with memory. This part is similar to the DT case in the preceding chapter. We then discuss how to model RLC circuits as LTI systems. Using simple circuit laws, we can develop state-space (ss) equations to describe such systems. We next show that ss equations can be readily used in computer computation and be implemented using op-amp circuits. We also give reasons for downplaying convolutions and high-order differential equations in this text.

In order to develop CT transfer functions, we introduce the Laplace transform. We compute transfer functions for RLC circuits, integrators, and differentiators. Their transfer functions are all rational functions.

We then use transfer functions to classify systems as lumped or distributed. We show that CT LTI systems with finite memory, except for trivial cases, are distributed and their study is complicated. Thus we study in this text only CT LTI and lumped

197

systems which have proper rational transfer functions. We then show that such systems can also be described by ss equations.

Finally we compare ss equations and transfer functions and show that transfer functions may not describe systems fully. However under some conditions, there is no difference between the two descriptions and either one can be used to study systems.

The class of LTI systems is very rich. Some can be described by rational transfer functions, the rest cannot. The latter class is much larger than the former if the analogue between the set of irrational numbers and the set of rational numbers holds. We study in this text only LTI and lumped systems. So does every other text on signals and systems. However the concept of lumpedness is rarely mentioned in those texts.

8.1.1   Forced response and natural response

A CT system is memoryless if its current output $y(t_0)$ depends only on current input $u(t_0)$.[1] It has memory if its current output $y(t_0)$ depends on current input $u(t_0)$, past input $u(t)$ with $t < t_0$, and future input $u(t)$ with $t > t_0$. A system is causal if its current output does not depend on future input. As in the DT case discussed in Section 7.2, the current input of a causal system cannot affect past output; it affects only current and/or future outputs. If a CT system has infinite memory, then the output $y(t_0)$ depends on $u(t)$, for all $t \leq t_0$ or, equivalently, the input $u(t_0)$ affects $y(t)$, for all $t \geq t_0$.

If the characteristics of a system do not change with time, then the system is defined in Section 6.4 as time invariant. For such a system, the output waveform will always be the same no matter at what time instant, called the *initial time*, we start to apply an input. Thus we may assume the initial time to be $t_0 = 0$ in studying time-invariant systems. We study in the remainder of this text only causal and time-invariant systems. We also assume the initial time to be zero or $t_0 = 0$. Note that $t_0 = 0$ is a relative one and is selected by us.

In studying memoryless systems, we used $\{u(t) \to y(t)\}$ in Chapter 6 to denote an input-output pair. This notation implies that $y(t)$ is excited by $u(t)$ alone. Such a notation is not acceptable for a system with memory, because $y(t)$ depends not only on $u(t)$ but also on $u(\tau)$ for $\tau < t$. Thus we first discuss a notation for systems with memory. When we start to study a system at $t_0 = 0$, we will be interested in its input $u(t)$ and output $y(t)$ only for $t \geq t_0 = 0$. Thus we may use

$$u(t), \ t \geq 0 \quad \to \quad y(t), \ t \geq 0$$

to denote an input-output pair. This notation however requires some qualification. If the system has memory, then the output $y(t)$, for $t \geq 0$, also depends on the input applied before $t_0 = 0$. Thus if $u(t) \neq 0$, for some $t < 0$, then the input $u(t)$, for $t \geq 0$, may not excite a unique $y(t)$, for $t \geq 0$, and the notation is not well defined. Thus when we use the notation, we must assume $u(t) = 0$, for all $t < t_0 = 0$. Under this assumption, the notation denotes a unique input-output pair and the output is called a *forced response*, that is,

- Forced response: Output $y(t)$, for $t \geq 0$ excited by $\begin{cases} u(t) = 0 & \text{for all } t < 0 \\ u(t) \neq 0 & \text{for some or all } t \geq 0 \end{cases}$

---

[1]This subsection is the CT counterpart of Subsection 7.2.1.

We defined in Section 6.4 that a CT system is initially relaxed at $t_0 = 0$ if the input $u(t)$, for $t \geq 0$, excites a unique $y(t)$, for $t \geq 0$. Thus a sufficient condition for a system to be initially relaxed at $t_0 = 0$ is that $u(t)$ is identically zero for all $t < 0$. For the class of CT systems to be studied in this text, the condition, as we will discuss in latter sections, can be replaced by a finite number of initial conditions, called an *initial state* denoted by $\mathbf{x}(0)$. If the initial state is zero, then the system is initially relaxed and the excited output is a forced response. Thus a forced response is also called a *zero-state* response, that is,

- Forced response = Zero-state response

$$= \text{Output } y(t), \text{ for } t \geq 0, \text{ excited by } \begin{cases} \text{initial state } \mathbf{x}(0) = \mathbf{0} \text{ (initially relaxed)} \\ u(t) \neq 0 \text{ for some or all } t \geq 0 \end{cases}$$

We next introduce *zero-input* responses defined by

- Zero-input response: Output $y(t)$, for $t \geq 0$ excited by $\begin{cases} u(t) \neq 0 & \text{for some or all } t < 0 \\ u(t) = 0 & \text{for all } t \geq 0 \end{cases}$

It is so named because the input is identically zero for all $t \geq 0$. In this case, the response is determined entirely by the characteristics or the *nature* of the system. Thus it is also called a *natural response*. If a system has a finite number of initial conditions, then we have

- Natural response = Zero-input response

$$= \text{Output } y(t), \text{ for } t > 0, \text{ excited by } \begin{cases} \text{initial state } \mathbf{x}(0) \neq \mathbf{0} \\ u(t) = 0 \text{ for all } t \geq 0 \end{cases}$$

In conclusion, a forced response $y(t)$, for $t \geq 0$, is exited *exclusively* by the input $u(t)$, for $t \geq 0$; whereas, a natural response is excited *exclusively* by the input $u(t)$, for $t < 0$. For a natural response, we require the input to be identically zero for all $t \geq 0$. For a forced response, we require the input to be identically zero for all $t < 0$ which however can be replaced by a set of initial conditions or an initial state for the systems to be studied in this text. If the initial state is zero, then the *net effect* of $u(t)$, for all $t < 0$, on $y(t)$, for all $t \geq 0$, is zero, even though $u(t)$ may not be identically zero for all $t < 0$. See Subsection 7.2.1.

The classification of outputs into natural or forced responses will simplify the development of mathematical equations for the latter. Note that natural responses relate the input applied *before* $t_0 = 0$ and the output appeared *after* $t_0 = 0$. Because their time intervals are different, developing equations to relate such inputs and outputs will be difficult. Whereas, forced responses relate the input and output in the same time interval $[0, \infty)$. Moreover, under the relaxedness condition, every input excites a unique output. Thus developing equations to relate them will be relatively simple. In the remainder of this text, we study mainly forced responses. This is justified because in designing systems, we consider only forced responses.

Before proceeding, we mention that for a memoryless system, there is no initial state and no natural response. A memoryless system is always initially relaxed and the excited output is a forced response. On the other hand, if a system has infinite memory, then the input applied at t_1 affects the output for all t larger than t_1 all the way to infinity. Consequently if $u(t)$ is different from zero for some $t < 0$, the system may not be initially relaxed at any $t_2 > 0$. The only way to make such a system initially relaxed is to turn it off and then to turn it on again. Then the system is initially relaxed.

8.2   Linear time-invariant (LTI) systems

A CT system can be classified dichotomously as causal or noncausal, time-invariant or time varying, linear or nonlinear. As discussed earlier, a system is causal if its current output does not depend on future input. A system is time invariant if its characteristics do not change with time. If so, then the output waveform will always be the same no matter at what time instant the input is applied. We express this as a terminal property. Let

$$u(t),\ t \geq 0 \quad \rightarrow \quad y(t),\ t \geq 0 \tag{8.1}$$

be any input-output pair and let t_1 be any positive number. If the CT system is time invariant, then we have

$$u(t - t_1),\ t \geq t_1 \quad \rightarrow y(t - t_1),\ t \geq t_1 \quad \text{(time shifting)} \tag{8.2}$$

This is called the *time-shifting* property. Recall that the system is implicitly assumed to be initially relaxed at t_1 before applying $u(t - t_1)$.

We next classify a system to be linear or nonlinear. Consider a CT system. Let

$$u_i(t), \quad \text{for } t \geq 0 \rightarrow y_i(t), \quad \text{for } t \geq 0$$

for $i = 1, 2$, be *any* two input-output pairs, and let β be *any* real constant. Then the system is defined to be *linear* if it has the following two properties

$$u_1(t) + u_2(t),\ \ t \geq 0 \rightarrow y_1(t) + y_2(t),\ \ t \geq 0 \ \ \text{(additivity)} \tag{8.3}$$

$$\beta u_1(t),\ \ t \geq 0 \rightarrow \beta y_1(t),\ \ t \geq 0 \ \ \text{(homogeneity)} \tag{8.4}$$

If not, the system is *nonlinear*. These two properties are stated in terms of input and output without referring to the internal structure of a system. Thus it is applicable to any type of system: electrical, mechanical or others.

Consider a CT system. If we can find one or two input-output pairs which do not meet the homogeneity or additivity property, then we can conclude that the system is nonlinear. However to conclude a system to be linear, the two properties must hold for all possible input-output pairs. There are infinitely many of them and there is no way to check them all. Not to mention the requirement that the system be initial relaxed before applying an input. Thus it is not possible to check the linearity of a system by measurement. However we can use the conditions to develop a general equation to describe LTI systems as we show in the next subsection.

8.2.1   Integral convolution

In this subsection, we develop an integral convolution to describe a CT LTI causal system. It is the CT counterpart of the discrete convolution in (7.10) and its development is parallel to the one in Section 7.3.2.

Consider a CT LTI system which is initially relaxed at $t = 0$. Let us apply to the system an impulse with weight 1 at $t = 0$, that is, $u(t) = \delta(t - 0)$. We call the output the *impulse response* of the system. The impulse response will be denoted by $h(t)$. If a system is initially relaxed and causal, then no output will appear before the application of an input. Thus we have $h(t) = 0$, for all $t < 0$. In fact, $h(t) = 0$, for all $t < 0$, is a necessary and sufficient condition for a system to be causal.

Consider a CT LTI system with impulse response $h(t)$. We show that the output $y(t)$ excited by $u(t)$, for $t \geq 0$, can be described by

$$y(t) = \int_{\tau=0}^{\infty} h(t - \tau)u(\tau)d\tau \tag{8.5}$$

for all $t \geq 0$, where the system is assumed to be initially relaxed at $t = 0$. It is an integral equation and is called an *integral convolution* or, simply, a *convolution*. This equation will be used to introduce the concept of transfer functions and a stability condition. However once the concept and the condition are introduced, the equation will not be used again. Thus the reader may glance through the remainder of this subsection.[2]

The derivation of (8.5) is similar to the one of (7.10) but requires a limiting process. Let $h_a(t)$ be the output of the system excited by the input $u(t) = \delta_a(t - 0)$ defined in (2.6) with $t_0 = 0$. It is the pulse plotted in Figure 2.9(a) at $t_0 = 0$. As discussed in Section 2.5.1, an input signal $u(t)$ can be approximated by

$$u(t) \approx \sum_{n=0}^{\infty} u(na)\delta_a(t - na)a$$

This is the same as (2.15) except that T is replaced by a. Now if the system is linear and time invariant, then we have

$$\delta_a(t) \rightarrow h_a(t) \quad \text{(definition)}$$
$$\delta_a(t - na) \rightarrow h_a(t - na) \quad \text{(time shifting)}$$
$$u(na)\delta_a(t - na)a \rightarrow u(na)h_a(t - na)a \quad \text{(homogeneity)}$$
$$\sum_{n=0}^{\infty} u(na)\delta_a(t - na)a \rightarrow \sum_{n=0}^{\infty} u(na)h_a(t - na)a \quad \text{(additivity)} \tag{8.6}$$

The left-hand side of (8.6) equals roughly the input, thus the output of the system is given by

$$y(t) \approx \sum_{n=0}^{\infty} u(na)h_a(t - na)a$$

Let us define $\tau = na$. If $a \rightarrow 0$, the pulse $\delta_a(t)$ becomes the impulse $\delta(t)$, τ becomes a continuous variable, a can be written as $d\tau$, the summation becomes an integration, and the approximation becomes an equality. Thus we have, as $a \rightarrow 0$,

$$y(t) = \int_{\tau=0}^{\infty} h(t - \tau)u(\tau)d\tau$$

---

[2]Correlations are intimately related to convolutions. Correlations can be used to measure the matching between a signal and its time-delayed reflected signal. This is used in radar detection and is outside the scope of this text.

This establishes (8.5).

We mention that the equation reduces to $y(t) = \alpha u(t)$ for an LTI memoryless system. Indeed, for an amplifier with gain α, its impulse response is $h(t) = \alpha\delta(t)$. Thus (8.5) becomes, using (2.14),

$$y(t) = \int_{\tau=0}^{\infty} \alpha\delta(t-\tau)u(\tau)d\tau = \alpha u(\tau)|_{\tau=t} = \alpha u(t)$$

We also mention that convolutions introduced in most texts on signals and systems assume the form

$$y(t) = \int_{\tau=-\infty}^{\infty} h(t-\tau)u(\tau)d\tau \tag{8.7}$$

in which the input is applied from $t = -\infty$. Although the use of (8.7), as we will discuss in Chapter 9, may simplify some derivations, it will suppress some important engineering concepts. Thus we adopt (8.5). Moreover, we stress that convolutions describe only forced responses or, equivalently, are applicable only if systems are initially relaxed.

The derivation of (8.5) is instructional because it uses explicitly the conditions of time invariance and linearity. However, the equation is not used in analysis nor design. See also Section 8.4.3. Thus we will not discuss it further.

8.3   Modeling LTI systems

Most systems are built by interconnecting a number of components. If every component is linear and time-invariant, then so is the overall system.[3] However, no physical component is linear and time invariant in the mathematical sense. For example, a resistor with resistance R is not linear because if the applied voltage is very large, the resistor may burn out. However, within its power limitation, the resistor with its voltage $v(t)$ and current $i(t)$ related by $v(t) = Ri(t)$ is a linear element. Likewise, disregarding saturation, the charge $Q(t)$ stored in a capacitor is related to the applied voltage $v(t)$ by $Q(t) = Cv(t)$, where C is the capacitance, which implies $i(t) = Cdv(t)/dt$; the flux $F(t)$ generated by an inductor is related to its current $i(t)$ by $F(t) = Li(t)$, where L is the inductance, which implies $v(t) = Ldi(t)/dt$. They are linear elements. The resistance R, capacitance C, and inductance L probably will change after 100 years, thus they are not time-invariant. However, their values will remain constant within a number of years. Thus they may be considered as linear and time-invariant. If all resistors, capacitors, and inductors of RLC circuits such as the one in Figures 6.2(a) and 8.1(a) are so modeled, then the circuits are linear and time-invariant.

We give a different example. Consider the mechanical system shown in Figure 8.2(a). It consists of a block with mass m connected to a wall through a spring. The input $u(t)$ is the force applied to the block, and the output $y(t)$ is the displacement measured from the equilibrium position. The spring is a nonlinear element because it will break if it is stretched beyond its elastic limit. However, within some limit, the spring can be described by Hooke's law as

$$\text{Spring force} = ky(t) \tag{8.8}$$

---

[3]Note that the converse is not necessarily true. See Problem 8.34.

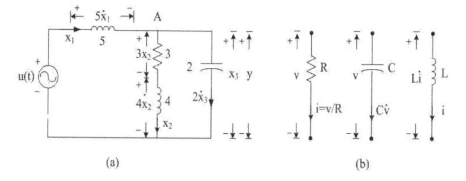

Figure 8.1: (a) RLC circuit. (b) Linear elements.

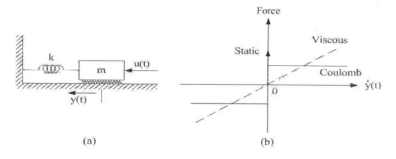

Figure 8.2: (a) Mechanical system. (b) Frictions.

where k is called the *spring constant*. It is a linear element. The friction between the block and the floor is very complex and may consist of three parts: static, Coulomb, and viscous frictions as shown in Figure 8.2(b). Note that the coordinates are friction versus velocity $\dot{y}(t) := dy(t)/dt$. When the block is stationary (its velocity is zero), we need a certain amount of force to overcome its *static friction* to start its movement. Once the block is moving, there is a constant friction, called the *Coulomb friction*, to resist its movement. In addition, there is a friction, called the *viscous friction*, which is proportional to the velocity as shown with a dashed line in Figure 8.2(b) or

$$\text{Viscous friction} = f \times \text{Velocity} = f\dot{y}(t) \qquad (8.9)$$

where f is called the *viscous friction coefficient* or *damping coefficient*. It is a linear relationship. Because of the viscous friction between its body and the air, a sky diver may reach a constant falling speed. If we disregard the static and Coulomb frictions and consider only the viscous friction, then the mechanical system can be modeled as a linear system within the elastic limit of the spring.

In conclusion, most physical systems are nonlinear and time-varying. However, within the time interval of interest and a limited operational range, we can model many physical systems as linear and time-invariant.[4] Thus the systems studied in

---

[4]Physical systems such as heating systems which use on-off control cannot be so modeled.

this text, in fact in most texts, are actually models of physical systems. Modeling is an important problem. If a physical system is properly modeled, we can predict the behavior of the physical system from its model. Otherwise, the behavior of the model may differ appreciably from that of the physical system as the op-amp circuit in Figure 6.9(b) demonstrated.

8.4   State-space (ss) equations

Consider the RLC circuit shown in Figure 8.1(a) with a voltage source as its input and the voltage across the capacitor as its output. It consists of one resistor with resistance 3 ohms (Ω), two inductors with inductances 5 and 4 henrys (H), and one capacitor with capacitance 2 farads (F). Because every element is modeled as linear and time invariant, the resulting circuit is also linear and time invariant. Analogous to the DT case, we can develop a number of mathematical equations to describe the system. In this section we develop its state-space (ss) equation description.

To develop an ss equation, we must first select state variables. For RLC circuits, state variables are associated with energy-storage elements. A resistor whose voltage $v(t)$ and current $i(t)$ are related by $v(t) = Ri(t)$ is a memoryless element; it cannot store energy[5] and its variable cannot be selected as a state variable. A capacitor can store energy in its electric field and its voltage or current can be selected as a state variable. If we select the capacitor voltage $v(t)$ as a state variable, then its current is $C dv(t)/dt =: C\dot{v}(t)$ as shown in Figure 8.1(b). If we select the capacitor current as a state variable, then its voltage is an integration and is not used. An inductor can store energy in its magnetic field and its current or voltage can be selected as a state variable. If we select the inductor current $i(t)$ as a state variable, then its voltage is $L di(t)/dt = L\dot{i}(t)$ as shown in Figure 8.1(b). In the assignment, it is important to assign the polarity of a voltage and the direction of a current, otherwise the assignment is not complete.

Procedure of developing ss equations for RLC circuits

1. Assign all capacitor voltages and all inductor currents as state variables.[6] If the voltage of a capacitor with capacitance C is assigned as $x_i(t)$, then its current is $C\dot{x}_i(t)$. If the current of an inductor with inductance L is assigned as $x_j(t)$, then its voltage is $L\dot{x}_j(t)$.

2. Use Kirchhoff's current or voltage law to express the current or voltage of every resistor in terms of state variables (but not their derivatives) and, if necessary, the input. If the expression is for the current (voltage), then its multiplication (division) by its resistance yields the voltage (current).

3. Use Kirchhoff's current or voltage law to express each $\dot{x}_i(t)$ in terms of state variables and input.

---

[5]Its energy is all dissipated into heat.

[6]There are some exceptions. See Problems 8.26 through 8.28 and Reference [C6]. In essence, all state variables selected must be able to vary independently.

We use an example to illustrate the procedure. Consider the circuit shown in Figure 8.1(a). We assign the 5-H inductor current as $x_1(t)$. Then its voltage is $5\dot{x}_1(t)$ as shown. Next we assign the 4-H inductor current as $x_2(t)$. Then its voltage is $4\dot{x}_2(t)$. Finally, we assign the 2-F capacitor voltage as $x_3(t)$. Then its current is $2\dot{x}_3(t)$. Thus the network has three state variables. Next we express the 3-Ω resistor's voltage or current in terms of state variables. Because the resistor is in series connection with the 4-H inductor, its current is $x_2(t)$. Thus the voltage across the 3-Ω resistor is $3x_2(t)$. Note that finding the voltage across the resistor first will be more complicated. This completes the first two steps of the procedure.

Next applying Kirchhoff's voltage law along the outer loop of the circuit yields

$$5\dot{x}_1(t) + x_3(t) - u(t) = 0$$

which implies

$$\dot{x}_1(t) = -0.2x_3(t) + 0.2u(t) \tag{8.10}$$

Applying Kirchhoff's voltage law along the right-hand-side loop yields

$$x_3(t) - 4\dot{x}_2(t) - 3x_2(t) = 0$$

which implies

$$\dot{x}_2(t) = -0.75x_2(t) + 0.25x_3(t) \tag{8.11}$$

Finally, applying Kirchhoff's current law to the node denoted by A yields

$$x_1(t) - x_2(t) - 2\dot{x}_3(t) = 0$$

which implies

$$\dot{x}_3(t) = 0.5x_1(t) - 0.5x_2(t) \tag{8.12}$$

From Figure 8.1(a), we have

$$y(t) = x_3(t) \tag{8.13}$$

The equations from (8.10) through (8.13) can be arranged in matrix form as

$$\begin{bmatrix} \dot{x}_1(t) \\ \dot{x}_2(t) \\ \dot{x}_3(t) \end{bmatrix} = \begin{bmatrix} 0 & 0 & -0.2 \\ 0 & -0.75 & 0.25 \\ 0.5 & -0.5 & 0 \end{bmatrix} \begin{bmatrix} x_1(t) \\ x_2(t) \\ x_3(t) \end{bmatrix} + \begin{bmatrix} 0.2 \\ 0 \\ 0 \end{bmatrix} u(t) \tag{8.14}$$

$$y(t) = \begin{bmatrix} 0 & 0 & 1 \end{bmatrix} \begin{bmatrix} x_1(t) \\ x_2(t) \\ x_3(t) \end{bmatrix} + 0 \times u(t) \tag{8.15}$$

or, using matrix symbols,

$$\dot{\mathbf{x}}(t) = \mathbf{A}\mathbf{x}(t) + \mathbf{b}u(t) \tag{8.16}$$
$$y(t) = \mathbf{c}\mathbf{x}(t) + du(t) \tag{8.17}$$

with $\mathbf{x} = [x_1 \ x_2 \ x_3]'$ and

$$\mathbf{A} = \begin{bmatrix} 0 & 0 & -0.2 \\ 0 & -0.75 & 0.25 \\ 0.5 & -0.5 & 0 \end{bmatrix}, \quad \mathbf{b} = \begin{bmatrix} 0.2 \\ 0 \\ 0 \end{bmatrix}$$

$$\mathbf{c} = \begin{bmatrix} 0 & 0 & 1 \end{bmatrix}, \quad d = 0$$

The set of two equations in (8.16) and (8.17) is called a (CT) state-space (ss) equation of *dimension* 3 and is the CT counterpart of (7.20) and (7.21). They are identical except that we have the first derivative in the CT case and the first difference or unit-sample advance in the DT case. As in the DT case, we call (8.16) a *state equation* and (8.17) an *output equation*. The vector **x** is called the *state* or *state vector* and its components are called *state variables*. The scalar d is called the *direct transmission* part. Note that the output equation in (8.17) is memoryless. The evolution or dynamic of the system is described by the state equation in (8.16). The preceding procedure is applicable to most simple RLC circuits. For a more systematic procedure, see Reference [C6].

We next consider the mechanical system shown in Figure 8.2(a). It consists of a block with mass m connected to a wall through a spring. We consider the applied force u the input and the displacement y the output. If we disregard the Coulomb and static frictions and consider only the viscous friction, then it is an LTI system within the elastic limit of the spring. The spring force is $ky(t)$, where k is the spring constant, and the friction is $f\,dy(t)/dt = f\dot{y}(t)$, where f is the damping or viscous-friction coefficient. The applied force must overcome the spring force and friction and the remainder is used to accelerate the mass. Thus we have, using Newton's law,

$$u(t) - ky(t) - f\dot{y}(t) = m\ddot{y}(t)$$

or

$$m\ddot{y}(t) + f\dot{y}(t) + ky(t) = u(t) \qquad (8.18)$$

where $\ddot{y}(t) = d^2y(t)/dt^2$ is the second derivative of $y(t)$. It is a second-order LTI differential equation.

For the mechanical system, we can select the position and velocity of the mass as state variables. The energy associated with position is stored in the spring and the kinetic energy is associated with velocity. Let us defined $x_1(t) := y(t)$ and $x_2(t) := \dot{y}(t)$. Then we have

$$\dot{x}_1(t) = \dot{y}(t) = x_2(t)$$

Substituting x_i into (8.18) yields $m\dot{x}_2 + fx_2 + kx_1 = u$ which implies

$$\dot{x}_2(t) = -(k/m)x_1(t) - (f/m)x_2(t) + (1/m)u(t)$$

These two equations and $y(t) = x_1(t)$ can be expressed in matrix form as

$$\begin{bmatrix} \dot{x}_1(t) \\ \dot{x}_2(t) \end{bmatrix} = \begin{bmatrix} 0 & 1 \\ -k/m & -f/m \end{bmatrix} \begin{bmatrix} x_1(t) \\ x_2(t) \end{bmatrix} + \begin{bmatrix} 0 \\ 1/m \end{bmatrix} u(t)$$

$$y(t) = \begin{bmatrix} 1 & 0 \end{bmatrix} \begin{bmatrix} x_1(t) \\ x_2(t) \end{bmatrix} + 0 \times u(t)$$

This ss equation of dimension 2 describes the mechanical system.

8.4.1   Significance of states

We use a simple example to discuss the significance of the state in an ss equation. The example will also be used for comparison in a later subsection. Consider a CT LTI system described by the ss equation

$$\begin{bmatrix} \dot{x}_1(t) \\ \dot{x}_2(t) \end{bmatrix} = \begin{bmatrix} -2 & -10 \\ 1 & 0 \end{bmatrix} \begin{bmatrix} x_1(t) \\ x_2(t) \end{bmatrix} + \begin{bmatrix} 1 \\ 0 \end{bmatrix} u(t)$$

$$y(t) = \begin{bmatrix} 3 & 4 \end{bmatrix} \begin{bmatrix} x_1(t) \\ x_2(t) \end{bmatrix} - 2 \times u(t) \qquad (8.19)$$

It is, as will be discussed in the next chapter, a system with infinite memory. Thus the input applied at any t_1 will affect the output for all $t \geq t_1$. Conversely the output at t_1 will be affected by all input applied before and up to t_1.

Consider the output equation in (8.19) at $t = t_1$ or

$$y(t_1) = \begin{bmatrix} 3 & 4 \end{bmatrix} \mathbf{x}(t_1) - 2u(t_1)$$

where $\mathbf{x}(t_1) = [x_1(t_1) \ x_2(t_1)]'$. As mentioned earlier, $y(t_1)$ depends on $u(t_1)$ and $u(t)$, for all $t < t_1$. The output equation however depends only on $u(t_1)$ and $\mathbf{x}(t_1)$. Thus the state $\mathbf{x}$ at t_1 must summarize the effect of $u(t)$, for all $t < t_1$, on $y(t_1)$. The input $u(t)$, for all $t < t_1$ contains *infinitely many values* of $u(t)$, yet its effect on $y(t_1)$ can be summarized by the *two values* in $\mathbf{x}(t_1)$. Moreover, once $\mathbf{x}(t_1)$ is obtained, the input $u(t)$, for all $t < t_1$, is no longer needed and can be discarded. This is the situation in actual operation of real-world systems. For example, consider the RLC circuit shown in Figure 8.1(a). No matter what input is applied to the circuit, the output will appear instantaneously and in real time. That is, the output at t_1 will appear when $u(t_1)$ is applied, and no external device is needed to store the input applied before t_1. Thus ss equations describe actual operation of real-world systems.

In particular, the initial state $\mathbf{x}(0)$ summarizes the net effect of $u(t)$, for all $t < 0$, on $y(t)$, for $t \geq 0$. Thus if a system can be described by an ss equation, its initial relaxedness can be checked from its initial state. For example, the RLC circuit in Figure 8.1(a) is initially relaxed at $t = 0$ if the initial currents of the two inductors and the initial voltage of the capacitor are zero. There is no need to be concerned with the input applied before $t < 0$. Thus the use of state is extremely convenient.

8.4.2 Computer computation of ss equations

Once an ss equation is developed for a system, the output $y(t)$ of the equation excited by any input $u(t)$, for $t \geq 0$, can be computed using a computer. Even though we can develop an analytical solution for the ss equation, the analytical solution is not needed in its computer computation.

Consider the state equation

$$\dot{\mathbf{x}}(t) = \mathbf{A}\mathbf{x}(t) + \mathbf{b}u(t) \qquad (8.20)$$

The first step in computer computation is to carry out discretization. The simplest way is to use the approximation

$$\dot{\mathbf{x}}(t) \approx \frac{\mathbf{x}(t + \Delta) - \mathbf{x}(t)}{\Delta} \qquad (8.21)$$

where $\Delta > 0$ and is called the *step size*. The smaller Δ, the better the approximation. The approximation becomes exact as Δ approaches 0. Substituting (8.21) into (8.20) yields

$$\mathbf{x}(t + \Delta) = \mathbf{x}(t) + \Delta\left[\mathbf{A}\mathbf{x}(t) + \mathbf{b}u(t)\right] = (\mathbf{I} + \Delta\mathbf{A})\mathbf{x}(t) + \Delta\mathbf{b}u(t) \qquad (8.22)$$

where $\mathbf{I}$ is the unit matrix of the same order as $\mathbf{A}$ (see Problem 3.9). Because $(1+\Delta\mathbf{A})\mathbf{x}$ is not defined, we must use $\mathbf{I}\mathbf{x} = \mathbf{x}$ before summing $\mathbf{x}$ and $\Delta\mathbf{A}\mathbf{x}$ to yield $(\mathbf{I} + \Delta\mathbf{A})\mathbf{x}$. If we compute $\mathbf{x}(t)$ and $y(t)$ at $t = n\Delta$, for $n = 0, 1, 2, \ldots$, then we have

$$
\begin{aligned}
y(n\Delta) &= \mathbf{c}\mathbf{x}(n\Delta) + du(n\Delta) \\
\mathbf{x}((n+1)\Delta) &= (\mathbf{I} + \Delta\mathbf{A})\mathbf{x}(n\Delta) + \Delta\mathbf{b}u(n\Delta)
\end{aligned}
$$

or, by suppressing Δ,

$$
\begin{aligned}
y[n] &= \mathbf{c}\mathbf{x}[n] + du[n] \\
\mathbf{x}[n+1] &= \bar{\mathbf{A}}\bar{\mathbf{x}}[n] + \bar{\mathbf{b}}u[n]
\end{aligned}
\qquad (8.23)
$$

where $\bar{\mathbf{A}} := \mathbf{I} + \Delta\mathbf{A}$ and $\bar{\mathbf{b}} := \Delta\mathbf{b}$. Equation (8.23) is the same as the DT ss equation in (7.22), thus it can be computed recursively using only additions and multiplications as discussed in Section 7.6.1. Moreover it can be computed in real time if the step size is sufficiently large. However (8.23) will generate only the values of $y(t)$ at $t = n\Delta$. Because the output of a CT system is defined for all t, we must carry out interpolation. This can be achieved using MATLAB function `plot` which carries out linear interpolation. See Figure 2.8(c).

The remaining question is how to select a step size. The procedure is simple. We select an arbitrary Δ_1 and carry out the computation. We then select a smaller Δ_2 and repeat the computation. If the result is different from the one using Δ_1, we repeat the process until the computed result is indistinguishable from the preceding one.

The preceding discussion is the basic idea of computer computation of CT ss equations. The discretization procedure used in (8.21) is the simplest and yields the least accurate result for a given Δ. However it can yield a result as accurate as any method if we select Δ to be sufficiently small. The topic of computer computation is a vast one and will not be discussed further. We discuss in the following only the use of some MATLAB functions.

Consider the ss equation

$$
\begin{bmatrix} \dot{x}_1(t) \\ \dot{x}_2(t) \\ \dot{x}_3(t) \end{bmatrix} = \begin{bmatrix} 0 & 0 & -0.2 \\ 0 & -0.75 & 0.25 \\ 0.5 & -0.5 & 0 \end{bmatrix} \begin{bmatrix} x_1(t) \\ x_2(t) \\ x_3(t) \end{bmatrix} + \begin{bmatrix} 0.2 \\ 0 \\ 0 \end{bmatrix} u(t)
$$

$$
y(t) = \begin{bmatrix} 0 & 0 & 1 \end{bmatrix} \begin{bmatrix} x_1(t) \\ x_2(t) \\ x_3(t) \end{bmatrix} + 0 \times u(t) \qquad (8.24)
$$

We use the MATLAB function `lsim`, abbreviation for *linear simulation*, to compute the response of the system excited by the input $u(t) = \sin 2t$, for $t \geq 0$ and the initial state $\mathbf{x}(0) = [0 \quad 0.2 \quad -0.1]'$, where the prime denotes the transpose. We type in an edit window the following:

```
% Program 8.1 (f83.m)
a=[0 0 -0.2;0 -0.75 0.25;0.5 -0.5 0];b=[0.2;0;0];
c=[0 0 1];d=0;
dog=ss(a,b,c,d);
t1=0:1:80;u1=sin(2*t1);x=[0;0.2;-0.1];
y1=lsim(dog,u1,t1,x);

t2=0:0.1:80;u2=sin(2*t2);
y2=lsim(dog,u2,t2,x);

t3=0:0.01:80;u3=sin(2*t3);
y3=lsim(dog,u3,t3,x);
plot(t1,y1,t2,y2,':',t3,y3,'-.',[0 80],[0 0])
xlabel('Time (s)'),ylabel('Amplitude')
```

The first two lines express $\{\mathbf{A}, \mathbf{b}, \mathbf{c}, d\}$ in MATLAB format as in Program 7.1. The third line defines the system. We call the system 'dog', which is defined using the state-space model, denoted by ss. The fourth line t1=0:1:80 indicates the time interval $[0, 80]$ to be computed with increment 1 or, equivalently, with step size $\Delta_1 = 1$ and the corresponding input u1=sin(2*t1). We then use lsim to compute the output y1=lsim(dog,u1,t1,x). The solid line in Figure 8.3 is generated by plot(t1,y1) which also carries out linear interpolation. We then repeat the computation by selecting $\Delta_2 = 0.1$ as its step size and the result y2 is plotted in Figure 8.3 with a dotted line using plot(t2,y2,':'). It is quite different from the solid line. Thus the result obtained using $\Delta_1 = 1$ is not acceptable. At this point we don't know whether the result using $\Delta_2 = 0.1$ will be acceptable. We next select $\Delta_3 = 0.01$ and repeat the computation. We then use plot(t3,y3,'-.') to plot the result in Figure 8.3 using the dash-and-dotted line. It is indistinguishable from the one using $\Delta_2 = 0.1$. Thus we conclude that the response of (8.24) can be computed from its discretized equation if the step size is selected to be 0.1 or smaller. Note that the preceding three plot functions can be combined into one as in the second line from the bottom in Program 8.1. The combined plot function also plots the horizontal axis which is generated by plot([0 80],[0 0]). We save Program 8.1 as an m-file named f83.m. Typing in the command window >> f83 will yield Figure 8.3 in a figure window.

The impulse response of a system was defined in Section 8.2.1 as the output of the system, which is initially relaxed at $t = 0$, excited by an impulse applied at $t = 0$ or $u(t) = \delta(t)$. Likewise the *step response* of a system is defined as the output of the system, which is initially relaxed at $t = 0$, excited by a step function applied at $t = 0$ or $u(t) = 1$, for all $t \geq 0$. MATLAB contains the functions impulse and step which compute the impulse and step responses of systems. The program that follows

```
%Program 8.2 (f84.m)
a=[0 0 -0.2;0 -0.75 0.25;0.5 -0.5 0];b=[0.2;0;0];
c=[0 0 1];d=0;
dog=ss(a,b,c,d);
subplot(2,1,1)
```

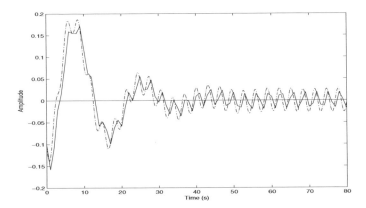

Figure 8.3: Output of (8.23) computed using $\Delta_1 = 1$ (solid line), using $\Delta_2 = 0.1$ (dotted line), and using $\Delta_3 = 0.01$ (dash-and-dotted line).

```
step(dog)
subplot(2,1,2)
impulse(dog)
```

generates the step response in Figure 8.4 (top) and impulse response in Figure 8.4 (bottom). In using `lsim`, we must select a step size and specify the time interval to be computed. The functions `step` and `impulse` however automatically select step sizes and a time interval to be computed. They also automatically generate titles and labels. Thus their use is very simple. We mention that both `step` and `impulse` are based on `lsim`. However they require adaptive selection of step sizes and automatically stop the computation when the responses hardly change. We also mention that `impulse` is applicable only for ss equations with $d = 0$ or, more precisely, simply ignores $d \neq 0$. If $d = 0$, the impulse response can be generated using a nonzero initial state and $u(t) = 0$, for all $t \geq 0$. See Problem 8.6. Thus no impulse is needed in generating impulse responses.

To conclude this section, we mention that an ss equation generates a zero-state response if $\mathbf{x}(0) = \mathbf{0}$ and a zero-input response if $u(t) = 0$, for all $t \geq 0$. Thus ss equations describe both forced or zero-state responses and natural or zero-input responses.

8.4.3   Why we downplay convolutions

In this subsection we compare ss equations and convolutions. Consider the system described by the ss equation in (8.19). The system can also be described by a convolution with impulse response

$$h(t) = -2\delta(t) + 3.02e^{-t}\cos(3t - 0.11)$$

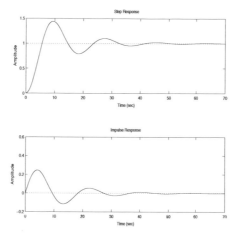

Figure 8.4: (Top) Step response of (8.23). (Bottom) Impulse response of (8.23).

for $t \geq 0$ and $h(t) = 0$, for $t < 0$.[7] Thus the convolution description of the system is

$$
\begin{aligned}
y(t) &= \int_{\tau=0}^{\infty} h(t-\tau)u(\tau)d\tau \\
&= \int_{\tau=0}^{t} \left[-2\delta(t-\tau) + 3.02e^{-(t-\tau)}\cos[3(t-\tau) - 0.11] \right] u(\tau)d\tau \\
&= -2u(t) + \int_{\tau=0}^{t} \left[3.02e^{-(t-\tau)}\cos[3(t-\tau) - 0.11] \right] u(\tau)d\tau \qquad (8.25)
\end{aligned}
$$

Note that the integration upper limit ∞ must be changed to t, otherwise the second integration is incorrect because it does not use the condition $h(t) = 0$, for $t < 0$. We see that the convolution in (8.25) is very complex. More seriously, to compute $y(t_1)$, we need $u(t)$ for all t in $[0, t_1]$. Thus using (8.25) to compute $y(t_1)$ requires the storage of all input applied before t_1. This is not done in practice. Thus even though a convolution describes a system, it does not describe its actual processing.

We next count the numbers of multiplications needed in computer computation of (8.25). Before computing, the convolution must be discretized as

$$
y(N\Delta) = -2u(N\Delta) + \sum_{n=0}^{N} \left[3.02e^{-(N-n)\Delta}\cos[3(N-n)\Delta - 0.11] \right] u(n\Delta)\Delta \qquad (8.26)
$$

Because the summation consists of $(N+1)$ terms and each term requires 6 multiplications, each $y(N\Delta)$ requires $6(N+1) + 1$ multiplications. If we compute $y(N\Delta)$ for

---

[7]It is obtained by computing the transfer function of (8.19) (Problem 8.23) and then taking its inverse Laplace transform (Example 9.3.6).

$N = 0 : \bar{N}$, then the total number of multiplications is[8]

$$\sum_{N=0}^{\bar{N}} (6N + 7) = 6\frac{\bar{N}(\bar{N} + 1)}{2} + 7(\bar{N} + 1) = 3\bar{N}^2 + 10\bar{N} + 7$$

Because its total number of multiplications is proportional to $\bar{N}^2$, it will increase rapidly as $\bar{N}$ increases. For example, computing one thousand points of y requires roughly 3 million multiplications.

We next count the number of multiplications needed in computer computation of the ss equation in (8.19). Its discretized ss equation is, following (8.21) through (8.23),

$$
\begin{aligned}
y(n\Delta) &= \begin{bmatrix} 3 & 4 \end{bmatrix}\mathbf{x}(n\Delta) - 2u(n\Delta) \\
\mathbf{x}([n+1]\Delta) &= \begin{bmatrix} 1 - 2\Delta & -10\Delta \\ \Delta & 1 \end{bmatrix}\mathbf{x}(n\Delta) + \begin{bmatrix} \Delta \\ 0 \end{bmatrix} u(n\Delta)
\end{aligned}
\qquad (8.27)
$$

where Δ is the step size. Computing $\mathbf{x}(n\Delta)$ requires four multiplication. Thus each $y(n\Delta)$ requires a total of 7 multiplications. This number of multiplications is the same for all n. Thus the total number of multiplications to compute $y(n\Delta)$, for $n = 0 : \bar{N}$, is $7(\bar{N} + 1)$ which is much smaller than $3\bar{N}^2$, for $\bar{N}$ large. For example, computing one thousand points of y requires only 7 thousand multiplications in using the ss equation but 3 million multiplications in using the convolution.

We now compare ss equations and convolutions in the following:

1. State-space equations are much easier to develop than convolutions to describe CT LTI systems. To develop a convolution, we must compute first its impulse response. The impulse response can be obtained, in theory, by measurement but cannot be so obtained in practice. Its analytical computation is not simple. Even if $h(t)$ is available, its actual employment is complicated as shown in (8.25). Whereas developing an ss equation to describe a system is generally straightforward as demonstrated in Section 8.4.

2. A convolution relates the input and output of a system, and is called an *external description* or *input-output description*. An ss equation is called an *internal description* because it describes not only the input and output relationship but also the internal structure. A convolution is applicable only if the system is initially relaxed. An ss equation is applicable even if the system is not initially relaxed.

3. A convolution requires an external memory to store the applied input from $t = 0$ onward in order to compute the output. In using an ss equation, the knowledge of $u(t)$, for $t < t_1$, is not needed in determining $y(t_1)$ because the effect of $u(t)$ for all $t < t_1$ on $y(t_1)$ is summarized and stored in the state variables. Thus as soon as the input $u(t_1)$ excites $y(t_1)$ and $\mathbf{x}(t_1)$, it is no longer needed. This is the actual operation of the RLC circuit in Figure 8.1(a). Thus an ss equation describes the actual processing of real-world systems. But a convolution does not.

---

[8]We use the formula $\sum_{N=0}^{\bar{N}} N = \bar{N}(\bar{N} + 1)/2$.

4. A convolution requires a much larger number of operations than an ss equation in their computer computations. The FFT discussed in Section 5.4 can be used to compute convolutions. Even though the number of operations can be reduced, FFT cannot be carried out in real time and may introduce additional numerical errors due to changing real-number computation into complex-number computation. See Reference [C7].

5. An ss equation can be more easily modified to describe time varying or nonlinear systems than a convolution.

In view of the preceding reasons, if a system can be described by a convolution or an ss equation, we prefer the latter. Although every CT LTI system can be described by a convolution, only a small class of CT LTI systems can also be described by ss equations, as we will discuss in a later section. This text studies only this small class of CT LTI systems. Thus we will downplay convolutions in this text.

To conclude this section, we mention that the graphical computation of convolutions is a standard topic in most texts on signals and systems. It involves four steps: shifting, flipping, multiplication and integration. See, for example, Reference [C8]. The procedure is complicated even for very simple $h(t)$ and $u(t)$ and will not be discussed in this text.[9]

8.4.4 Any need to study high-order differential equations?

The differential equation was the first mathematical equation used to study control systems. J. Maxwell developed in the late 1800s a linear differential equation to describe approximately Watt's centrifugal flyball governor for controlling the steam engine and then studied its stability. We also encounter differential equations in earlier physics and mathematics courses. Thus the study of systems often starts with differential equations. However we will downplay differential equations in this text for the reasons to be discussed in this subsection.

A state-space equation is a set of first-order differential equations. The equation in (8.18) is a second-order differential equation which arises from Newton's law. We develop in the next example a third-order differential equation to describe the RLC circuit in Figure 8.1(a).

Example 8.4.1 Consider the circuit in Figure 8.1(a) and replotted in Figure 8.5.[10] The input is a voltage source $u(t)$ and the output is the voltage $y(t)$ across the 2-F capacitor with the polarity shown. Let the current passing through the 2-F capacitor be denoted by $i_1(t)$ and the current passing through the series connection of the 3-Ω resistor and the 4-H inductor be denoted by $i_2(t)$. Then we have

$$i_1(t) = 2\dot{y}(t) \tag{8.28}$$

$$y(t) = 3i_2(t) + 4\dot{i}_2(t) \tag{8.29}$$

---

[9]About fifteen years ago, I saw a questionnaire from a major publisher's editor asking whether convolution can be omitted from a text because it is a difficult topic and has turned off many EE students. I regret to say that I stopped teaching graphical computation of convolution only after 2006.

[10]This example may be skipped without loss of continuity.

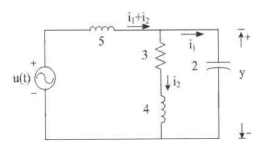

Figure 8.5: RLC circuit

The current passing through the 5-H inductor is $i_1(t) + i_2(t)$. Thus the voltage across the inductor is $5\dot{i}_1(t) + 5\dot{i}_2(t)$. Applying Kirchhoff's voltage law around the outer loop of Figure 8.5 yields

$$5\dot{i}_1(t) + 5\dot{i}_2(t) + y(t) - u(t) = 0 \tag{8.30}$$

In order to develop a differential equation to relate $u(t)$ and $y(t)$, we must eliminate $i_1(t)$ and $i_2(t)$ from (8.28) through (8.30). First we substitute (8.28) into (8.30) to yield

$$10\ddot{y}(t) + 5\dot{i}_2(t) + y(t) = u(t) \tag{8.31}$$

where $\ddot{y}(t) := d^2y(t)/dt^2$, and then differentiate it to yield

$$10y^{(3)}(t) + 5\ddot{i}_2(t) + \dot{y}(t) = \dot{u}(t) \tag{8.32}$$

where $y^{(3)}(t) := d^3y(t)/dt^3$. The summation of (8.31) multiplied by 3 and (8.32) multiplied by 4 yields

$$40y^{(3)}(t) + 5(4\ddot{i}_2(t) + 3\dot{i}_2(t)) + 30\ddot{y}(t) + 4\dot{y}(t) + 3y(t) = 4\dot{u}(t) + 3u(t)$$

which becomes, after substituting the derivative of (8.29),

$$40y^{(3)}(t) + 30\ddot{y}(t) + 9\dot{y}(t) + 3y(t) = 4\dot{u}(t) + 3u(t) \tag{8.33}$$

This is a third-order linear differential equation with constant coefficients. It describes the circuit in Figure 8.5. Note that (8.33) can be more easily developed using a different method. See Section 8.7.1. □

We now compare high-order differential equations and ss equations:

1. For a simple system that has only one or two state variables, there is not much difference in developing a differential equation or an ss equation to describe it. However, for a system with three or more state variables, it is generally simpler to develop an ss equation than a single high-order differential equation because the latter requires to eliminate intermediate variables as shown in the preceding example. Furthermore, the form of ss equations is more compact than the form of differential equations.

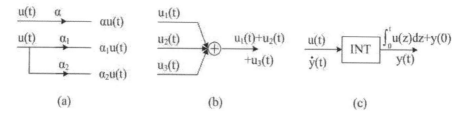

Figure 8.6: Time domain: (a) Multiplier. (b) Adder. (c) Integrator.

2. A high-order differential equation, just as a convolution, is an external description. An ss equation is an internal description. It describes not only the relationship between the input and output but also the internal variables.

3. High-order differential equations are not suitable for computer computation because of the difficulties in discretizing second, third, and higher-order derivatives. State-space equations involve only the discretization of the first derivative, thus they are more suitable for computer computation as demonstrated in Subsection 8.4.1.

4. An ss equation can be easily simulated as a basic block diagram and be implemented using an op-amp circuit as we will discuss in the next section.

5. State-space equations can be more easily extended than high-order differential equations to describe nonlinear systems.

In view of the preceding reasons, there seems no reasons to develop and to study high-order differential equations in this text.

8.5 CT LTI basic elements

State-space equations, as shown in Subsection 8.4.2, can readily be computed using a computer. They can also be easily simulated or implemented using op-amp circuits as we show in this and following subsections.

Consider the CT elements, called *CT basic elements*, shown in Figure 8.6. The element in Figure 8.6(a) denoted by a line with an arrow and a real number α is called a *multiplier* with gain α; its input and output are related by

$$y(t) = \alpha u(t) \quad \text{(multiplier)}$$

Note that α can be positive or negative. If $\alpha = 1$, it is direct transmission and the arrow and α may be omitted. If $\alpha = -1$, it changes only the sign of the input and is called an *inverter*. In addition, a signal may branch out to two or more signals with gain α_i as shown. The element denoted by a small circle with a plus sign as shown in Figure 8.6(b) is called an *adder*. It has two or more inputs denoted by entering arrows and one and only one output denoted by a departing arrow. Its inputs $u_i(t)$, for $i = 1 : 3$ and output $y(t)$ are related by

$$y(t) = u_1(t) + u_2(t) + u_3(t) \quad \text{(adder)}$$

The element in Figure 8.6(c) is an *integrator*. If we assign its input as $u(t)$, then its output $y(t)$ is given by

$$y(t) = \int_{\tau=0}^{t} u(\tau)d\tau + y(0) \quad \text{(Integrator)}$$

This equation is not convenient to use. If we assign its output as $y(t)$, then its input $u(t)$ is given by $u(t) = dy(t)/dt = \dot{y}(t)$. This is simpler. Thus we prefer to assign the output of an integrator as a variable. These elements are the CT counterparts of the DT basic elements in Figure 7.2. They are all linear and time invariant. The multiplier and adder are memoryless. The integrator has memory. Its employment requires the specification of the initial condition $y(0)$. Unless stated otherwise, the initial condition of every integrator will be assumed to be zero.

Integrators are very useful in practice. For example, the width of a table can be measured using a measuring tape. The distance traveled by an automobile however cannot be so measured. But we can measure the speed of the automobile by measuring the rotational speed of the driving shaft of the wheels.[11] The distance can then be obtained by integrating the speed.[12] An airplane cannot measure its own velocity without resorting to outside signals, however it can measure its own acceleration using an accelerometer which will be discussed in Chapter 10. Integrating the acceleration yields the velocity. Integrating once again yields the distance. Indeed, integrators are very useful in practice.

We now discuss op-amp circuit implementations of the three basic elements. Multipliers and adders are implemented as shown in Figures 6.19 and 6.20. Consider the op-amp circuit shown in Figure 8.7(a). It consists of two op-amp circuits. The right-hand-side circuit is an inverter as shown in Figure 6.13(b). If we assign its output as $x(t)$, then its input is $-x(t)$ as shown. Let us assign the input of the left-hand-side op amp as $v_i(t)$. Because $e_- = e_+ = 0$, the current passing through the resistor R and entering the inverting terminal is $v_i(t)/R$ and the current passing through the capacitor and entering the inverting terminal is $-C\dot{x}(t)$. Because $i_-(t) = 0$, we have

$$\frac{v_i(t)}{R} - C\dot{x}(t) = 0 \quad \text{or} \quad v_i(t) = RC\dot{x}(t) \tag{8.34}$$

If we select, for example, $R = 10$ kΩ and $C = 10^{-4}$ Farads, then $RC = 1$ and $v_i(t) = \dot{x}(t)$. In other words, if the input of the op-amp circuit in Figure 8.7(a) is $\dot{x}(t)$, then its output is $x(t)$. Thus the circuit carries out integration and is called an integrator.

8.5.1 Basic block diagram and op-amp circuit implementation of ss equations

We now show that the ss equation in (8.14) and (8.15) can be simulated or built using basic elements. The equation has dimension three and its simulation requires three

---

[11]The rotational speed of a shaft can be measured using centrifugal force, counting pulses, and other methods.

[12]In automobiles, the integration can be carried out mechanically using gear trains.

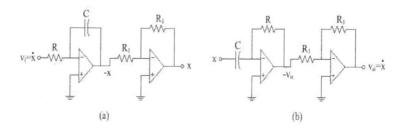

Figure 8.7: (a) Integrator ($RC = 1$). (b) Differentiator ($RC = 1$).

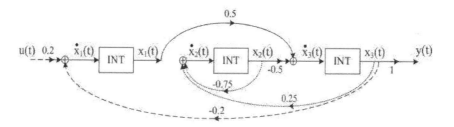

Figure 8.8: Time domain: Basic block diagram of (8.14) and (8.15).

integrators as shown in Figure 8.8. We assign the output of each integrator as a state variable $x_i(t)$. Then its input is $\dot{x}_i(t)$. To carry out simulation, the ss equation must be written out explicitly as in (8.10) through (8.13). We use (8.10) to generate $\dot{x}_1$ as shown in Figure 8.8 using dashed line. We use (8.11) to generate $\dot{x}_2$ as shown using dotted line. Similarly $\dot{x}_3$ and y can be generated using (8.12) and (8.13). The diagram in Figure 8.8 consists of only basic elements and is called a *basic block diagram*. In conclusion, every ss equation can be represented as a basic block diagram. Developing the diagram is simple and straightforward.

Now if every basic element is replaced by an op-amp circuit, then the resulting circuit simulates or implements the ss equation. Note that there are many ways to implement an ss equation using op-amp circuits. The procedure is simple and straightforward. See, for example, References [C5, C6]. This is an important reason of introducing ss equations. In conclusion, ss equations are convenient for computer computation, real-time processing, and op-amp circuit implementation.

8.5.2 Differentiators

The opposite of integration is differentiation. Let us interchange the locations of the capacitor C and resistor R in Figure 8.7(a) to yield the op-amp circuit in Figure 8.7(b). Its right-hand half is an inverter. If we assign its output as $v_o(t)$, then its input is $-v_o(t)$. Let us assign the input of the left-hand-side op-amp circuit as $x(t)$. Because $e_- = e_+ = 0$, the current passing through the resistor R and entering the inverting terminal is $-v_o(t)/R$ and the current passing through the capacitor and entering the inverting terminal is $C\dot{x}(t)$. Because $i_-(t) = 0$, we have

$$-\frac{v_o(t)}{R} + C\dot{x}(t) = 0 \quad \text{or} \quad v_o(t) = RC\dot{x}(t)$$

If $RC = 1$, then the output of Figure 8.7(b) equals the differentiation of its input. Thus the op-amp circuit in Figure 8.7(b) carries out differentiation and is called a *differentiator*.

An integrator is causal. Is a differentiator causal? The answer depends on how we define differentiation. If we define the differentiation as, for $\epsilon > 0$,

$$y(t) = \lim_{\epsilon \to 0} \frac{u(t + \epsilon) - u(t)}{\epsilon}$$

then the output $y(t)$ depends on the future input $u(t + \epsilon)$ and the differentiator is not causal. However if we define the differentiation as, for $\epsilon > 0$,

$$y(t) = \lim_{\epsilon \to 0} \frac{u(t) - u(t - \epsilon)}{\epsilon}$$

then the output $y(t)$ does not depend on any future input and the differentiator is causal. Note that the differentiator is not memoryless; it has a memory of length ϵ which approaches 0. The integrator however has infinite memory as we will discuss in the next chapter.

If a signal contains high-frequency noise, then its differentiation will amplify the noise, whereas its integration will suppress the noise. This is illustrated by an example.

Example 8.5.1 Consider the signal $x(t) = \sin 2t$ corrupted by noise. Suppose the noise can be represented by $n(t) = 0.02 \sin 100t$. Then we have

$$u(t) = x(t) + n(t) = \sin 2t + 0.02 \sin 100t$$

We plot $u(t)$ in Figures 8.9(a) and (c). Because the amplitude of $n(t)$ is small, we have $u(t) \approx x(t)$ and the presence of the noise can hardly be detected from Figure 8.9(a) and (c).

If we apply $u(t)$ to the differentiator in Figure 8.7(b), then its output is

$$\frac{du(t)}{dt} = 2\cos 2t + 0.02 \times 100 \cos 100t = 2\cos 2t + 2\cos 100t \tag{8.35}$$

and is plotted in Figure 8.9(b). We see that the amplitude of the noise is greatly amplified and we can no longer detect $dx(t)/dt$ from Figure 8.9(b). If we apply $u(t)$ to the integrator in Figure 8.8(a), then its output is

$$\begin{aligned}
\int_{\tau=0}^{t} u(\tau)d\tau &= \left.\frac{-\cos 2t}{2}\right|_{\tau=0}^{t} + \left.\frac{-0.02\cos 100t}{100}\right|_{\tau=0}^{t} \\
&= -0.5\cos 2t + 0.5 - 0.0002\cos 100t + 0.0002 \\
&= -0.5\cos 2t + 0.5002 - 0.0002\cos 100t \tag{8.36}
\end{aligned}$$

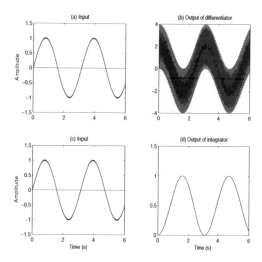

Figure 8.9: (a) Signal corrupted by high-frequency noise. (b) Its differentiation. (c) Signal corrupted by high-frequency noise. (d) Its integration.

and is plotted in Figure 8.9(d). The plot yields essentially the integration of $x(t)$. □

This example shows that differentiators will amplify high-frequency noise, whereas integrators will suppress it. Because electrical systems often contain high-frequency noise, *differentiators should be avoided wherever possible in electrical systems*. Note that a differentiator is not one of the basic elements discussed in Figure 8.6.

Even though differentiators are rarely used in electrical systems, they are widely used in electromechanical systems. A *tachometer* is essentially a generator attached to a shaft which generates a voltage proportional to the angular velocity of the shaft. The rotation of a mechanical shaft is fairly smooth and the issue of high-frequency noise amplification will not arise. Tachometers are widely used in control systems to improve performances.

8.6 Transfer functions – Laplace transform

We introduced integral convolutions, high-order differential equations, and ss equations to describe CT LTI systems. They are all time-domain descriptions. We now introduce a different description.

Consider a CT signal $x(t)$, defined for all t in $(-\infty, \infty)$. Its Laplace transform is defined as

$$X(s) := \mathcal{L}[x(t)] := \int_{t=0}^{\infty} x(t)e^{-st}dt \tag{8.37}$$

where s is a complex variable, called the Laplace-transform variable. The function $X(s)$ is called the Laplace transform of $x(t)$. Note that the Laplace transform is defined only for the positive-time part of $x(t)$. Because the negative-time part is not used, we often assume $x(t) = 0$ for $t < 0$, or $x(t)$ to be positive time. Before proceeding, we mention that the Laplace transform has the following linearity property:

$$\mathcal{L}\left[\beta_1 x_1(t) + \beta_2 x_2(t)\right] = \beta_1 \mathcal{L}[x_1(t)] + \beta_2 \mathcal{L}[x_2(t)]$$

for any constants β_i and any signals $x_i(t)$, for $i = 1, 2$.

Every CT LTI system can be described by (8.5) or

$$y(t) = \int_{\tau=0}^{\infty} h(t - \tau)u(\tau)d\tau \tag{8.38}$$

In this equation, it is implicitly assumed that the system is initially relaxed at $t = 0$ and the input is applied from $t = 0$ onward. If the system is causal, then $h(t) = 0$ for $t < 0$ and no output $y(t)$ will appear before the application of the input. Thus all $u(t)$, $h(t)$, and $y(t)$ in (8.38) are positive time.

Let us apply the Laplace transform to $y(t)$ in (8.38). Then we have

$$
\begin{aligned}
Y(s) &= \mathcal{L}[y(t)] = \int_{t=0}^{\infty} y(t)e^{-st}dt \\
&= \int_{t=0}^{\infty} \left[\int_{\tau=0}^{\infty} h(t - \tau)u(\tau)d\tau\right] e^{-s(t-\tau+\tau)}dt
\end{aligned}
$$

Interchanging the order of integrations yields

$$Y(s) = \int_{\tau=0}^{\infty} u(\tau)e^{-s\tau} \left[\int_{t=0}^{\infty} h(t - \tau)e^{-s(t-\tau)}dt\right] d\tau \tag{8.39}$$

Let us introduce a new variable $\bar{t}$ as $\bar{t} := t - \tau$ where τ is fixed. Then the term inside the brackets becomes

$$\int_{t=0}^{\infty} h(t - \tau)e^{-s(t-\tau)}dt = \int_{\bar{t}=-\tau}^{\infty} h(\bar{t})e^{-s\bar{t}}d\bar{t} = \int_{\bar{t}=0}^{\infty} h(\bar{t})e^{-s\bar{t}}d\bar{t}$$

where we have used the fact that $h(\bar{t}) = 0$, for all $\bar{t} < 0$ because of causality. The last integration is, by definition, the Laplace transform of $h(t)$, that is,

$$H(s) := \mathcal{L}[h(t)] := \int_{t=0}^{\infty} h(t)e^{-st}dt \tag{8.40}$$

Because it is independent of τ, $H(s)$ can be moved outside the integration in (8.39). Thus (8.39) becomes

$$Y(s) = H(s) \int_{\tau=0}^{\infty} u(\tau)e^{-s\tau}d\tau$$

or

$$Y(s) = H(s) \times U(s) = H(s)U(s) \tag{8.41}$$

where $Y(s)$ and $U(s)$ are the Laplace transforms of the output and input, respectively. The function $H(s)$ is the Laplace transform of the impulse response and is called the *CT transfer function*. Because the convolution in (8.38) is applicable only if the system is initially relaxed, so is the transfer function. In other words, *whenever we use a transfer function, the system is assumed to be initially relaxed.*

Using (8.41), we can also define the transfer function $H(s)$ as

$$H(s) = \frac{\mathcal{L}[\text{output}]}{\mathcal{L}[\text{input}]} = \left. \frac{Y(s)}{U(s)} \right|_{\text{Initially relaxed}} \tag{8.42}$$

If we use (8.42), there is no need to compute the impulse response. This is the equation used in practice to compute transfer functions. In computing the transfer function of a system, we may assume $u(t) = 0$ and $y(t) = 0$, for all $t < 0$. This will insure the initial relaxedness of the system.

We mention that for *nonlinear* time-invariant systems, we may use (8.42) to compute transfer functions. However, different input-output pairs will yield different transfer functions for the same system. Thus for nonlinear systems, the concept of transfer functions is useless. For LTI systems, no matter what input-output pair is used in (8.42), the resulting transfer function is always the same. Thus transfer functions can be used to describe LTI systems.

Integral convolutions, ss equations, and high-order differential equations are called *time-domain descriptions*. Transfer functions which are based on the Laplace transform are called the *transform-domain description*. Equations (8.41) and (8.42) are the CT counterparts of (7.25) and (7.29) for DT systems and all discussion in Section 7.8 is directly application. That is, CT transfer functions are most convenient in the study of composite systems including feedback systems. They will also be used in qualitative analysis and design as we will discuss in the remainder of this text.

8.6.1 Transfer functions of differentiators and integrators

Let us compute the Laplace transform of $\dot{x}(t) = dx(t)/dt$. Be definition, we have

$$\mathcal{L}\left[\frac{dx(t)}{dt}\right] = \int_{t=0}^{\infty} \frac{dx(t)}{dt} e^{-st} dt$$

which implies, using integration by parts,

$$\begin{aligned}
\mathcal{L}[\dot{x}(t)] &= x(t)e^{-st}\big|_{t=0}^{\infty} - \int_{t=0}^{\infty} x(t)\left[\frac{de^{-st}}{dt}\right] dt \\
&= 0 - x(0)e^{0} - (-s)\int_{t=0}^{\infty} x(t)e^{-st} dt
\end{aligned}$$

where we have used $e^{-st} = 0$ at $t = \infty$ (this will be discussed in Section 9.3) and $de^{-st}/dt = -se^{-st}$. Thus we have, using (8.37),

$$\mathcal{L}\left[\frac{dx(t)}{dt}\right] = sX(s) - x(0) \tag{8.43}$$

If $x(0) = 0$, then $\mathcal{L}[\dot{x}(t)] = sX(s)$. Thus differentiation in the time domain is equivalent to the multiplication by s in the transform domain, denoted as

$$\frac{d}{dt} \quad \longleftrightarrow \quad s$$

Consequently a differentiator has transfer function s.

Next we compute the Laplace transform of the integration of a signal. Let us define

$$g(t) := \int_{\tau=0}^{t} x(\tau)d\tau \tag{8.44}$$

Clearly we have $g(0) = 0$ and $dg(t)/dt = x(t)$. Thus we have

$$\mathcal{L}[x(t)] = \mathcal{L}[dg(t)/dt]$$

or, using (8.43) and $g(0) = 0$,

$$X(s) = sG(s) - g(0) = sG(s)$$

where $G(s)$ is the Laplace transform of (8.44). In conclusion, we have

$$G(s) := \mathcal{L}[g(t)] = \mathcal{L}\left[\int_{\tau=0}^{t} x(\tau)d\tau\right] = \frac{1}{s}X(s) \tag{8.45}$$

Thus integration in the time domain is equivalent to the multiplication by $1/s = s^{-1}$ in the transform domain, denoted as

$$\int_{t=0}^{t} \quad \longleftrightarrow \quad s^{-1}$$

Consequently an integrator has transfer function $1/s = s^{-1}$.

8.7   Transfer functions of RLC circuits

Consider an RLC circuit such as the one shown in Figure 8.1(a) or 8.5. We developed in Section 8.4 an ss equation and in Subsection 8.4.3 a third-order differential equation to described the circuit. If we apply the Laplace transform to the two equations, we can develop a transfer function to describe the circuit. It is however simpler to develop directly the transfer function using the concept of impedances as we will show in this section.

The voltage $v(t)$ and current $i(t)$ of a resistor with resistance R, a capacitor with capacitance C, and an inductor with inductance L are related, respectively, by

$$v(t) = Ri(t) \quad i(t) = C\frac{dv(t)}{dt} \quad v(t) = L\frac{di(t)}{dt}$$

as shown in Figure 8.1(b). Applying the Laplace transform, using (8.43), and assuming zero initial conditions, we obtain

$$V(s) = RI(s) \quad I(s) = CsV(s) \quad V(s) = LsI(s) \tag{8.46}$$

where $V(s)$ and $I(s)$ are the Laplace transforms of $v(t)$ and $i(t)$. If we consider the current as the input and the excited voltage as the output, then the transfer functions of R, C, and L are R, $1/Cs$, and Ls respectively. They are called *transform impedances*

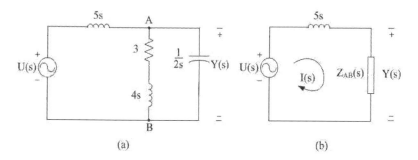

Figure 8.10: (a) RLC circuit in the transform domain. (b) Equivalent circuit.

or, simply, *impedances*.[13] If we consider the voltage as the input and the current as its output, then their transfer functions are called *admittances*.

Using impedances, the voltage and current of every circuit element can be written as $V(s) = Z(s)I(s)$ with $Z(s) = R$ for resistors, $Z(s) = 1/Cs$ for capacitors, and $Z(s) = Ls$ for inductors. They involves only multiplications. In other words, the relationship between the input and output of a circuit element is algebraic in the (Laplace) transform domain; whereas it is calculus (differentiation or integration) in the time domain. Thus the former is much simpler. Consequently, the manipulation of impedances is just like the manipulation of resistances. For example, the resistance of the series connection of R_1 and R_2 is $R_1 + R_2$. The resistance of the parallel connection of R_1 and R_2 is $R_1 R_2/(R_1 + R_2)$. Likewise, the impedance of the series connection of $Z_1(s)$ and $Z_2(s)$ is $Z_1(s) + Z_2(s)$. The impedance of the parallel connection of $Z_1(s)$ and $Z_2(s)$ is

$$\frac{Z_1(s)Z_2(s)}{Z_1(s) + Z_2(s)}$$

Using these two simple rules we can readily obtain the transfer function of the circuit in Figure 8.1(a). Before proceeding, we draw the circuit in Figure 8.10(a) in the Laplace transform domain or, equivalently, using impedances. The impedance of the series connection of the resistor with impedance 3 and the inductor with impedance $4s$ is $3 + 4s$. The impedance $Z_{AB}(s)$ between the two nodes A and B shown in Figure 8.10(a) is the parallel connection of $(4s + 3)$ and the capacitor with impedance $1/2s$ or

$$Z_{AB}(s) = \frac{(4s + 3)(1/2s)}{4s + 3 + 1/2s} = \frac{4s + 3}{8s^2 + 6s + 1}$$

Using $Z_{AB}(s)$, we can simplify the circuit in Figure 8.10(a) as in Figure 8.10(b). The current $I(s)$ around the loop is given by

$$I(s) = \frac{U(s)}{5s + Z_{AB}(s)}$$

---

[13]In some circuit analysis texts, impedances are defined as R, $1/j\omega C$, and $j\omega L$. The definition here is simpler and more general as we will discuss in Section 9.8.2.

Thus the voltage $Y(s)$ across $Z_{AB}(s)$ is

$$Y(s) = Z_{AB}(s)I(s) = \frac{Z_{AB}(s)}{5s + Z_{AB}(s)}U(s)$$

and the transfer function from $u(t)$ to $y(t)$ is

$$H(s) = \frac{Y(s)}{U(s)} = \frac{Z_{AB}(s)}{5s + Z_{AB}(s)}$$

or

$$
\begin{aligned}
H(s) &= \frac{\frac{4s+3}{8s^2+6s+1}}{5s + \frac{4s+3}{8s^2+6s+1}} = \frac{4s+3}{5s(8s^2+6s+1)+4s+3} \\
&= \frac{4s+3}{40s^3+30s^2+9s+3}
\end{aligned}
\tag{8.47}
$$

This transfer function describes the circuit in Figure 8.2(a) or 8.10(a). It is a ratio of two polynomials and is called a *rational transfer function*.

8.7.1   Rational transfer functions and differential equations

This subsection relates rational transfer functions and differential equations. We showed in Section 8.6.1 that differentiation in the time-domain is equivalent to the multiplication by s in the transform domain. Repeating the process, we can show, for $k = 1, 2, 3, \ldots,$

$$s^k \longleftrightarrow \frac{d^k}{dt^k} \tag{8.48}$$

that is, the kth derivative in the time domain is equivalent to the multiplication by s^k in the transform domain. Using (8.48), we can transform a transfer function into a differential equation and vice versa.

Consider the transfer function in (8.47). We write it as

$$(40s^3 + 30s^2 + 9s + 3)Y(s) = (4s + 3)U(s)$$

or

$$40s^3Y(s) + 30s^2Y(s) + 9sY(s) + 3Y(s) = 4sU(s) + 3U(s)$$

which becomes, in the time domain,

$$40y^{(3)}(t) + 30\ddot{y}(t) + 9\dot{y}(t) + 3y(t) = 4\dot{u}(t) + 3u(t)$$

This is the differential equation in (8.33). Thus once a transfer function is obtained, we can readily obtain its differential equation.

Conversely we can readily obtain a transfer function from a differential equation. For example, consider the differential equation in (8.18) or

$$m\ddot{y}(t) + f\dot{y}(t) + ky(t) = u(t)$$

which describes the mechanical system in Figure 8.1(a). Applying the Laplace transform and assuming zero initial conditions, we obtain

$$ms^2Y(s) + fsY(s) + kY(s) = U(s)$$

or

$$(ms^2 + fs + k)Y(s) = U(s)$$

Thus the transfer function of the mechanical system is

$$H(s) := \frac{Y(s)}{U(s)} = \frac{1}{ms^2 + fs + k} \tag{8.49}$$

It is a ratio of a polynomial of degree 0 and a polynomial of degree 2 and is a rational function of s. Indeed, a transfer function can be readily obtained from a differential equation and vice versa.

8.7.2 Proper rational transfer functions

Consider the rational function

$$H(s) = \frac{N(s)}{D(s)} \tag{8.50}$$

where $N(s)$ and $D(s)$ are polynomials of s with real coefficients. The *degree*, denoted by "deg", of a polynomial is defined as the highest power of s with a nonzero coefficient. Depending on the relative degrees of $N(s)$ and $D(s)$, we have the following definitions:

$$
\begin{aligned}
H(s) \text{ improper} &\leftrightarrow \deg N(s) > \deg D(s) &\leftrightarrow |H(\infty)| = \infty \\
H(s) \text{ proper} &\leftrightarrow \deg N(s) \leq \deg D(s) &\leftrightarrow H(\infty) = c \\
H(s) \text{ biproper} &\leftrightarrow \deg N(s) = \deg D(s) &\leftrightarrow H(\infty) = c \neq 0 \\
H(s) \text{ strictly proper} &\leftrightarrow \deg N(s) < \deg D(s) &\leftrightarrow H(\infty) = 0
\end{aligned}
$$

where c is a real number. For example, the rational functions

$$\frac{s^3 + 2}{2s + 3}, \quad s - 2, \quad \frac{s^4}{s^2 + 3s + 1}$$

are improper. The rational functions

$$3, \quad \frac{2s + 5}{s + 1}, \quad \frac{10}{(s^2 + 1.2s)}, \quad \frac{3s^2 + s + 5}{2.5s^3 - 5}$$

are proper. The first two are also biproper, and the last two are strictly proper. Thus proper rational functions include both biproper and strictly proper rational functions. Note that if $H(s) = N(s)/D(s)$ is biproper, so is its inverse $H^{-1}(s) = D(s)/N(s)$.

Properness of $H(s)$ can also be determined from the value of $H(s)$ at $s = \infty$. The rational function $H(s)$ is improper if $|H(\infty)| = \infty$, proper if $H(\infty)$ is a zero or nonzero constant, biproper if $H(\infty)$ is a nonzero constant, and strictly proper if $H(\infty) = 0$.

Consider the improper rational function $H(s) = (s^3 + 2)/(2s + 3)$. We carry out the following direct division

$$
\begin{array}{r}
0.5s^2 \quad -0.75s \quad +1.125 \\
\hline
2s + 3 \;)\; s^3 \quad +0 \cdot s^2 \quad + 0 \cdot s \quad + \quad 2 \\
s^3 \quad +1.5s^2 \\
\hline
-1.5s^2 \quad + 0 \cdot s \quad + \quad 2 \\
-1.5s^2 \quad -2.25s \\
\hline
2.25s \quad + \quad 2 \\
2.25s \quad +3.375 \\
\hline
-1.375
\end{array}
$$

Then we have

$$H(s) = \frac{s^3 + 2}{2s + 3} = \frac{-1.375}{2s + 3} + 0.5s^2 - 0.75s + 1.125 \tag{8.51}$$

If $H(s)$ is the transfer function of a system, then the output $y(t)$ of the system, as will be discussed later in (8.66), will contain the terms $0.5\ddot{u}(t)$ and $-0.75\dot{u}$, where $u(t)$ is the input of the system. As discussed in Subsection 8.5.2, differentiations should be avoided wherever possible in electrical systems. Thus improper rational transfer functions will not be studied. We study in the remainder of this text only proper rational transfer functions. Before proceeding, we mention that *PID* controllers are widely used in control systems. The transfer function of a PID controller is

$$H(s) = k_1 + \frac{k_2}{s} + k_3 s \tag{8.52}$$

where k_1 denotes a *proportional* gain, k_2, an *integration constant*, and k_3, a *differentiation* constant. Clearly the differentiator $k_3 s$ is improper. However, it is only an approximation. In practice, it is implemented as

$$\frac{k_3 s}{1 + s/N} \tag{8.53}$$

where N is a large number. Signals in control systems are generally of low frequency. For such signals, (8.53) can be approximated as $k_3 s$ as we will discuss in Chapter 10. In any case we study in the remainder of this text only proper rational transfer functions.

8.8   Lumped or distributed

The transfer functions in (8.47) and (8.49) are ratios of two polynomials of s. They are rational functions of s. So are the transfer functions of integrators and differentiators which are, respectively, $s/1$ and $1/s$. Are all transfer functions rational functions? The answer is negative as we discuss next.

The transfer function is defined as

$$H(s) = \int_{t=0}^{\infty} h(t)e^{-st}dt$$

It is an infinite integral and the following situations may arise:

1. The infinite integral does not exist. For example, the infinite integrals of e^{t^2} and e^{et} do not exit and their transfer functions are not defined.

2. The infinite integral exists but cannot be expressed in closed form. Most $h(t)$ generated randomly belong to this type.

3. The infinite integral exists and can be expressed in closed form which is not a rational function of s. For example, the Laplace transforms of $1/\sqrt{\pi t}$ and $(\sinh at)/t$ are $1/\sqrt{s}$ and $0.5 \ln[(s+a)/(s-a)]$, respectively. See Reference [Z1].

4. The infinite integral exists and is a rational function of s.

A system whose transfer function is a rational function will be defined as a *lumped* system. Otherwise it is a *distributed* system. In other words, LTI systems can be classified as lumped or distributed.

The RLC circuit in Figure 8.10(a) has the rational function in (8.47) as its transfer function. Thus it is a lumped system. We give a physical interpretation of lumpedness. The circuit consists of one resistor, one capacitor and two inductors. They are called *discrete* or *lumped* components. In their study, we pay no attention to their physical sizes and consider their responses to be instantaneous. This is permitted if (1) the physical size is much less than the shortest wavelength of signals processed by the circuit or (2) the physical size is much less than the distance traveled by the shortest time interval of interest. For example, if the RLC circuit operates in the audio-frequency range (up to 25 kHz), then the shortest wavelength is $3 \times 10^8/(25 \times 10^3)$ or in the order of 10^4 meters. Clearly for audio signals, the RLC circuit can be considered lumped. On the other hand, transmission lines can be hundreds or even thousands of kilometers long and are distributed.

Consider the clock signal shown in Figure 2.5. The signal has a period of 0.5 ns and will travel $0.5 \times 10^{-9} \times 3 \times 10^8 = 0.15$ m or 15 cm (centimeters) in one period. For integrated circuits (IC) of sizes limited to 2 or 3 cm, the responses of the circuits can be considered to be instantaneous and the circuits can be considered lumped. However if the period of the clock signal is reduced to 0.05 ns, then the signal will travel 1.5 cm in one period. In this case, IC circuits cannot be modeled as lumped.

The mechanical system studied in Figure 8.2(a) has the rational function in (8.49) as its transfer function. Thus it is also a lumped system. The physical sizes of the block and spring do not arise in their discussion. If a structure can be modeled as a rigid body, then it can be modeled as lumped. The robotic arm on a space shuttle is very long and is not rigid, thus it is a distributed system. Moving fluid and elastic deformation are also distributed.

We list in the following the differences between lumped and distributed systems;

Lumped	Distributed
Variables are functions of time	Variables are function of time and space
Classical mechanics	Relativistic mechanics
Ohm's and Kirchhoff's laws	Maxwell's wave equations
Rational transfer functions	Irrational transfer functions
Differential equations	Partial differential equations
Finite-dim. ss equations	Infinite-dim or delay form ss equations

If the analogy between the set of rational *numbers* and the set of irrational numbers holds, then the set of rational *functions* is much smaller than the set of irrational functions. In other words, the class of LTI lumped systems is much smaller than the class of LTI distributed systems. The study of LTI distributed systems is difficult. In this text, we study only LTI lumped systems.[14]

8.8.1   Why we do not study CT FIR systems

A DT system is FIR (finite impulse response) or of finite memory if there exits an integer N such that its impulse response $h[n]$ is identically zero for all $n > N$. If no such N exits, the DT system is IIR (infinite impulse response) or of infinite memory. Likewise, we may define a CT system to be FIR or of finite memory if its impulse response $h(t)$ has a finite duration or there exists a finite L such that $h(t) = 0$ for all $t > L$. If no such L exits, it is IIR. In other words, a CT IIR system has infinite memory or has an impulse response of infinite duration.

We defined in the preceding section that a CT system is lumped if its transfer function is a rational function of s and is distributed if its transfer function is not. Likewise, we may define a DT system to be lumped if its transfer function is a rational function of z and distributed if not. As discussed in Chapter 7, the transfer function of every DT FIR system is a rational function. Thus *every DT FIR system is lumped.* DT FIR systems are widely used in practice and many methods are available to design such systems. See, for example, Reference [C7].

Is every CT FIR system lumped as in the DT case? The answer is negative as we show in the following examples.

Example 8.8.1 Consider a CT system whose output $y(t)$ equals the input $u(t)$ delayed by 1.2 seconds, that is, $y(t) = u(t-1.2)$. If the input is an impulse or $u(t) = \delta(t)$, then its impulse response is $h(t) = \delta(t - 1.2)$. It is FIR because $h(t) = 0$, for all $t > 1.2$. Its transfer function is, using the sifting property in (2.14),

$$H(s) = \mathcal{L}[\delta(t - 1.2)] = \int_{t=0}^{\infty} \delta(t - 1.2)e^{-st}dt = e^{-st}\big|_{t=1.2} = e^{-1.2s}$$

It is not a rational function of s. Thus the system is distributed. □

Example 8.8.2 Consider a CT system whose impulse response is $h(t) = 1$, for t in $[0, L]$ and $h(t) = 0$, for $t > L$. It is FIR. Its transfer function is

$$\begin{aligned} H_L(s) &= \int_{t=0}^{\infty} h(t)e^{-st}dt = \int_{t=0}^{L} e^{-st} = \frac{e^{-st}}{-s}\Big|_{t=0}^{L} \\ &= \frac{e^{-sL} - e^0}{-s} = \frac{1 - e^{-sL}}{s} \end{aligned} \tag{8.54}$$

Its numerator is not a polynomial, thus $H_L(s)$ is not a rational function of s and the system is distributed. □

---

[14]Even though all existing texts on signals and systems claim to study LTI systems, they actually study only LTI and lumped systems.

The two CT FIR systems in the preceding examples are distributed. Thus not every CT LTI FIR system is lumped. This is in contrast to the fact that every DT LTI FIR system is lumped.

Are there any CT FIR systems which are lumped? An amplifier with gain α has the impulse response $\alpha\delta(t)$. It is memoryless and FIR because $\alpha\delta(t)$ is zero for all $t \geq a > 0$. Its transfer function is

$$H(s) := \mathcal{L}[\alpha\delta(t)] = \int_{t=0}^{\infty} \alpha\delta(t)e^{-st}dt = \alpha e^{-st}\big|_{t=0} = \alpha = \frac{\alpha}{1}$$

It is a proper rational function, albeit a trivial one. Thus it is lumped. A differentiator has transfer function $s = s/1$ and is, by definition, lumped. The impulse response of a differentiator is $d\delta(t)/dt$. The mathematical meaning of $d\delta(t)/dt$ is not clear. However because we have defined $\delta(t)$ to be identically zero for $t > a > 0$, we also have $d\delta(t)/dt = 0$, for all $t \geq a > 0$. Thus a differentiator is FIR. However it has an improper rational transfer function and is not studied in this text.

We will show in the next chapter that every proper rational transfer function, other than memoryless, is IIR or of infinite memory. In conclusion, excluding improper rational transfer functions and memoryless systems, we have

- All CT LTI FIR systems are distributed and their study is difficult.

- All CT LTI and lumped systems are IIR and we study only this class of systems in this text.

Note that not every system with infinite memory is lumped. Only a very small subset of such systems are lumped.

8.9 Realizations

We defined an LTI system to be lumped if its transfer function is a rational function of s. We show in this section that such a system can also be described by an ss equation.

Consider a proper rational transfer function $H(s)$. The problem of finding an ss equation which has $H(s)$ as its transfer function is called the *realization* problem. The ss equation is called a *realization* of $H(s)$. The name *realization* is justified because the transfer function can be built or implemented through its ss-equation realization using an op-amp circuit. Furthermore, all computation involving the transfer function can be carried out using the ss equation. Thus the realization problem is important in practice.

We use an example to illustrate the realization procedure. Consider the proper rational transfer function

$$H(s) = \frac{Y(s)}{U(s)} = \frac{\bar{b}_1 s^4 + \bar{b}_2 s^3 + \bar{b}_3 s^2 + \bar{b}_4 s + \bar{b}_5}{\bar{a}_1 s^4 + \bar{a}_2 s^3 + \bar{a}_3 s^2 + \bar{a}_4 s + \bar{a}_5} \tag{8.55}$$

with $\bar{a}_1 \neq 0$. The rest of the coefficients can be zero or nonzero. We call $\bar{a}_1$ the denominator's leading coefficient. The first step in realization is to write (8.55) as

$$H(s) = \frac{b_1 s^3 + b_2 s^2 + b_3 s + b_4}{s^4 + a_2 s^3 + a_3 s^2 + a_4 s + a_5} + d =: \frac{N(s)}{D(s)} + d \tag{8.56}$$

where $D(s) := s^4 + a_2 s^3 + a_3 s^2 + a_4 s + a_5$ with $a_1 = 1$. Note that $N(s)/D(s)$ is strictly proper. This can be achieved by dividing the numerator and denominator of (8.55) by $\bar{a}_1$ and then carrying out a direct division as we illustrate in the next example.

Example 8.9.1 Consider the transfer function

$$H(s) = \frac{3s^4 + 5s^3 + 24s^2 + 23s - 5}{2s^4 + 6s^3 + 15s^2 + 12s + 5} \tag{8.57}$$

We first divide the numerator and denominator by 2 to yield

$$H(s) = \frac{1.5s^4 + 2.5s^3 + 12s^2 + 11.5s - 2.5}{s^4 + 3s^3 + 7.5s^2 + 6s + 2.5}$$

and then carry out direct division as follows

$$
\begin{array}{r}
1.5 \\
\hline
s^4 + 3s^3 + 7.5s^2 + 6s + 2.5 \;\overline{)\; 1.5s^4 + 2.5s^3 + 12s^2 + 11.5s - 2.5} \\
1.5s^4 + 4.5s^3 + 11.25s^2 + 9s + 3.75 \\
\hline
-2s^3 + 0.75s^2 + 2.5s - 6.25
\end{array}
$$

Thus (8.57) can be expressed as

$$H(s) = \frac{-2s^3 + 0.75s^2 + 2.5s - 6.25}{s^4 + 3s^3 + 7.5s^2 + 6s + 2.5} + 1.5 \tag{8.58}$$

This is in the form of (8.56). $\lhd$

Now we claim that the following ss equation realizes (8.56) or, equivalently, (8.55):

$$\dot{\mathbf{x}}(t) = \begin{bmatrix} -a_2 & -a_3 & -a_4 & -a_5 \\ 1 & 0 & 0 & 0 \\ 0 & 1 & 0 & 0 \\ 0 & 0 & 1 & 0 \end{bmatrix} \mathbf{x}(t) + \begin{bmatrix} 1 \\ 0 \\ 0 \\ 0 \end{bmatrix} u(t) \tag{8.59}$$

$$y(t) = [b_1 \quad b_2 \quad b_3 \quad b_4] \mathbf{x}(t) + du(t)$$

with $\mathbf{x}(t) = [x_1(t) \quad x_2(t) \quad x_3(t) \quad x_4(t)]'$. The number of state variables equals the degree of the denominator of $H(s)$. This ss equation can be obtained directly from the coefficients in (8.56). We place the denominator's coefficients, except its leading coefficient 1, with sign reversed in the first row of $\mathbf{A}$, and place the numerator's coefficients, without changing sign, directly as $\mathbf{c}$. The constant d in (8.56) is the direct transmission part. The rest of the ss equation have fixed patterns. The second row of $\mathbf{A}$ is $[1 \; 0 \; 0 \cdots]$. The third row of $\mathbf{A}$ is $[0 \; 1 \; 0 \cdots]$ and so forth. The column vector $\mathbf{b}$ is all zero except its first entry which is 1. See Problems 8.19 and 8.20 for a different way of developing (8.59).

To show that (8.59) is a realization of (8.56), we must compute the transfer function of (8.59). We first write the matrix equation explicitly as

$$
\begin{aligned}
\dot{x}_1(t) &= -a_2 x_1(t) - a_3 x_2(t) - a_4 x_3(t) - a_5 x_4(t) + u(t) \\
\dot{x}_2(t) &= x_1(t) \\
\dot{x}_3(t) &= x_2(t) \\
\dot{x}_4(t) &= x_3(t)
\end{aligned}
\tag{8.60}
$$

We see that the four-dimensional state equation in (8.59) actually consists of four first-order differential equations as shown in (8.60). Applying the Laplace transform and assuming zero initial conditions yield

$$
\begin{aligned}
sX_1(s) &= -a_2X_1(s) - a_3X_2(s) - a_4X_3(s) - a_5X_4(s) + U(s) \\
sX_2(s) &= X_1(s) \\
sX_3(s) &= X_2(s) \\
sX_4(s) &= X_3(s)
\end{aligned}
$$

From the second to the last equations, we can obtain

$$
X_2(s) = \frac{X_1(s)}{s}, \quad X_3(s) = \frac{X_2(s)}{s} = \frac{X_1(s)}{s^2}, \quad X_4(s) = \frac{X_1(s)}{s^3} \tag{8.61}
$$

Substituting these into the first equation yields

$$
\left(s + a_2 + \frac{a_3}{s} + \frac{a_4}{s^2} + \frac{a_5}{s^3} \right) X_1(s) = U(s)
$$

or

$$
\left[\frac{s^4 + a_2s^3 + a_3s^2 + a_4s + a_5}{s^3} \right] X_1(s) = U(s)
$$

which implies

$$
X_1(s) = \frac{s^3}{s^4 + a_2s^3 + a_3s^2 + a_4s + a_5} U(s) =: \frac{s^3}{D(s)} U(s) \tag{8.62}
$$

Substituting (8.61) and (8.62) into the following Laplace transform of the output equation in (8.59) yields

$$
\begin{aligned}
Y(s) &= b_1X_1(s) + b_2X_2(s) + b_3X_3(s) + b_4X_4(s) + dU(s) \\
&= \left(\frac{b_1s^3}{D(s)} + \frac{b_2s^3}{D(s)s} + \frac{b_3s^3}{D(s)s^2} + \frac{b_4s^3}{D(s)s^3} \right) U(s) + dU(s) \\
&= \left[\frac{b_1s^3 + b_2s^2 + b_3s + b_4}{D(s)} + d \right] U(s)
\end{aligned}
$$

This shows that the transfer function of (8.59) equals (8.56). Thus (8.59) is a realization of (8.55) or (8.56). The ss equation in (8.59) is said to be in *controllable form*.

Before proceeding, we mention that if $H(s)$ is strictly proper, then no direct division is needed and we have $d = 0$. In other words, there is no direct transmission part from u to y.

Example 8.9.2 Find a realization for the transfer function in Example 8.9.1 or

$$
\begin{aligned}
H(s) &= \frac{3s^4 + 5s^3 + 24s^2 + 23s - 5}{2s^4 + 6s^3 + 15s^2 + 12s + 5} \\
&= \frac{-2s^3 + 0.75s^2 + 2.5s - 6.25}{s^4 + 3s^3 + 7.5s^2 + 6s + 2.5} + 1.5
\end{aligned} \tag{8.63}
$$

Its controllable-form realization is, using (8.59),

$$\dot{\mathbf{x}}(t) = \begin{bmatrix} -3 & -7.5 & -6 & -2.5 \\ 1 & 0 & 0 & 0 \\ 0 & 1 & 0 & 0 \\ 0 & 0 & 1 & 0 \end{bmatrix} \mathbf{x}(t) + \begin{bmatrix} 1 \\ 0 \\ 0 \\ 0 \end{bmatrix} u(t)$$

$$y(t) = [-2 \ 0.75 \ 2.5 \ -6.25]\mathbf{x}(t) + 1.5u(t)$$

We see that the realization can be read out from the coefficients of the transfer function.
□

Example 8.9.3 Find a realization for the transfer function in (8.47). We first normalize its denominator's leading coefficient to 1 to yield

$$H(s) = \frac{4s + 3}{40s^3 + 30s^2 + 9s + 3} = \frac{0.1s + 0.075}{s^3 + 0.75s^2 + 0.225s + 0.075} \tag{8.64}$$

It is strictly proper and $d = 0$.

Using (8.59), we can obtain its realization as

$$\dot{\mathbf{x}}(t) = \begin{bmatrix} -0.75 & -0.225 & -0.075 \\ 1 & 0 & 0 \\ 0 & 1 & 0 \end{bmatrix} \mathbf{x}(t) + \begin{bmatrix} 1 \\ 0 \\ 0 \end{bmatrix} u(t) \tag{8.65}$$

$$y(t) = [0 \ 0.1 \ 0.075]\mathbf{x}(t) + 0 \cdot u(t)$$

This three-dimensional ss equation is a realization of the transfer function of the RLC circuit in Figure 8.2(a) and describes the circuit. □

The ss equation in (8.14) and (8.15) also describes the circuit in Figure 8.2(a); it however looks completely different from (8.65). Even so, they are mathematically equivalent. In fact, it is possible to find infinitely many other realizations for the transfer function. See Reference [C6]. However the controllable-form realization in (8.65) is most convenient to develop and to use.

The MATLAB function tf2ss, an acronym for transfer function to ss equation, carries out realizations. For the transfer function in (8.57), typing in the control window

```
>> n=[3 5 24 23 -5];de=[2 6 15 12 5];
>> [a,b,c,d]=tf2ss(n,de)
```

will yield

$$a = \begin{matrix} -3.0000 & -7.5000 & -6.0000 & -2.5000 \\ 1.0000 & 0 & 0 & 0 \\ 0 & 1.0000 & 0 & 0 \\ 0 & 0 & 1.0000 & 0 \end{matrix}$$

$$b = \begin{matrix} 1 \\ 0 \\ 0 \\ 0 \end{matrix}$$

$$c = \begin{matrix} -2.0000 & 0.7500 & 2.5000 & -6.2500 \end{matrix}$$

$$d = 1.5000$$

This is the controllable-form realization in Example 8.9.2. In using `tf2ss`, there is no need to normalize the leading coefficient and to carry out direct division. Thus its use is simple and straightforward.

We mention that the MATLAB functions `lsim`, `step` and `impulse` discussed in Section 8.4.1 are directly applicable to transfer functions. For example, for the transfer function in (8.63), typing in the command window

```
>> n=[3 5 24 23 -5];d=[2 6 15 12 5];
>> step(n,d)
```

will generate in a figure window its step response. However *the response is not computed directly from the transfer function*. It is computed from its controllable-form realization. Using the realization, we can also readily draw a basic block diagram and then implement it using an op-amp circuit.

Every proper rational transfer function can be realized as an ss equation of the form in (8.16) and (8.17). Such an ss equation can be implemented without using differentiators. If a transfer function is not proper, then the use of differentiators is not avoidable. For example, consider the improper transfer function in (8.51). We write it as

$$H(s) = \frac{s^3 + 2}{2s + 3} = \frac{-0.6875}{s + 1.5} + 1.125 - 0.75s + 0.5s^2$$

Note that the denominator's leading coefficient is normalized to 1. Its controllable-form realization is

$$\begin{aligned} \dot{x}(t) &= -1.5x(t) + u(t) \\ y(t) &= -0.6875x(t) + 1.125u(t) - 0.75\dot{u}(t) + 0.5\ddot{u}(t) \end{aligned} \tag{8.66}$$

The output contains the first and second derivatives of the input. If an input contains high-frequency noise, then the system will amplify the noise and is not used in practice.

8.9.1 From ss equations to transfer functions

Consider the ss equation[15]

$$\begin{aligned} \dot{\mathbf{x}}(t) &= \mathbf{A}\mathbf{x}(t) + \mathbf{b}u(t) \end{aligned} \tag{8.67}$$
$$\begin{aligned} y(t) &= \mathbf{c}\mathbf{x}(t) + du(t) \end{aligned} \tag{8.68}$$

we compute its transfer function from $u(t)$ to $y(t)$. Applying the Laplace transform to (8.67), using the vector version of (8.43), and assuming zero initial state, we have

$$s\mathbf{X}(s) = \mathbf{A}\mathbf{X}(s) + \mathbf{b}U(s)$$

which can be written as, using $s\mathbf{X}(s) = s\mathbf{I}\mathbf{X}(s)$, where $\mathbf{I}$ is a unit matrix,[16]

$$(s\mathbf{I} - \mathbf{A})\mathbf{X}(s) = \mathbf{b}U(s)$$

---

[15]This section may be skipped without loss of continuity.
[16]Note that $s - \mathbf{A}$ is not defined unless $\mathbf{A}$ is scalar.

Premultiplying $(s\mathbf{I} - \mathbf{A})^{-1}$ to the preceding equation yields

$$\mathbf{X}(s) = (s\mathbf{I} - \mathbf{A})^{-1}\mathbf{b}U(s)$$

Substituting this into the Laplace transform of (8.68) yields

$$Y(s) = \mathbf{c}\mathbf{X}(s) + dU(s) = \left[\mathbf{c}(s\mathbf{I} - \mathbf{A})^{-1}\mathbf{b} + d\right]U(s) \qquad (8.69)$$

Thus the transfer function $H(s)$ of the ss equation is

$$H(s) = \mathbf{c}(s\mathbf{I} - \mathbf{A})^{-1}\mathbf{b} + d \qquad (8.70)$$

Although this is a basic equation, it will not be used in this text. For example, for the RLC circuit in Figure 8.1(a), we can readily develop the ss equation in (8.14) and (8.15) to describe the circuit. If we use (8.70) to compute its transfer function, then we must compute the inverse of a matrix. Computing the inverse of a matrix is complicated if the order of the matrix is three or higher. The transfer function can be much more easily computed using impedances as shown in Section 8.7. Even so, we shall mention that the MATLAB function ss2tf, acronym for state-space to transfer function, yields the transfer function of an ss equation. For example, for the ss equation in (8.19), typing in MATLAB

```
>> a=[-2 10;1 0];b=[1;0];c=[3 4];d=-2;
>> [n,de]=ss2tf(a,b,c,d)
```

will yield n=-2 -1 -16 and de=1   2 10 which means that the transfer function of the ss equation is

$$H(s) = \frac{-2s^2 - s - 16}{s^2 + 2s + 10}$$

See Problem 8.22.

8.9.2   Initial conditions

Whenever we use a transfer function, the system is implicitly assumed to be initially relaxed at $t_0 = 0$. If we know $u(t) = 0$, for all t in $(-\infty, 0)$, then the system is initially relaxed at $t_0 = 0$. However for a system with proper rational transfer function, its initial relaxedness can be determined from a finite number of initial conditions; there is no need to track the input $u(t)$, for all $t < 0$. This follows directly from a realization of the transfer function. For example, for the RLC circuit in Figure 8.1(a), if the initial currents in the two inductors and the initial voltage across the capacitor are zero, then the circuit is initially relaxed.

Consider a system with transfer function

$$H(s) = \frac{b_1 s^3 + b_2 s^2 + b_3 s + b_4}{a_1 s^3 + a_2 s^2 + a_3 s + a_4}$$

If we use a realization of $H(s)$ to study the system, then the realization has dimension 3 and the system has three initial conditions. One may then wonder what the three initial conditions are. Depending on the coefficients b_i and a_i, the initial conditions could be

$u(0), \dot{u}(0), \ddot{u}(0)$ or $y(0), \dot{y}(0), \ddot{y}(0)$, or their three independent linear combinations. The relationship is complicated. The interested reader may refer to the second edition of Reference [C6].[17] The relationship is not needed in using the ss equation. Moreover, we study only the case where initial conditions are all zero.

In program 8.1 of Subsection 8.4.2, a system, named dog, is defined as `dog=ss(a,b,c,d)` using the ss-equation model. In MATLAB, we may also define a system using its transfer function. For example, the circuit in Figure 8.1(a) has the transfer function in (8.47). Using its numerator's and denominator's coefficients, we can define

```
nu=[4 3];de=[40 30 9 3];cat=tf(nu,de);
```

If we replace `dog` in Program 8.1 by `cat`, then the program will generate a response as in Figure 8.3 but different. Note that all computation involving transfer functions are carried out using their controllable-form realizations. Moreover when we use a transfer function, all its initial conditions must be assumed to be zero. Thus in using `cat=tf(nu,de)`, MATLAB will automatically ignore the initial conditions or set them to zero. The initial state in Program 8.1 is different from 0, thus the results using `dog` and `cat` will be different. They will yield the same result if the initial state is zero.

8.10 The degree of rational functions – Coprimeness

In this section we introduce the concepts of coprimeness and the degree of rational functions. Consider the polynomial

$$D(s) = a_4 s^4 + a_3 s^3 + a_2 s^2 + a_1 s + a_0 \quad \text{with } a_4 \neq 0 \tag{8.71}$$

where a_i are real numbers, zero or nonzero. The degree of $D(s)$ is defined as the highest power of s with a nonzero coefficient. Thus the polynomial in (8.71) has degree 4 and we call a_4 its *leading coefficient*.

A real or complex number λ is defined as a *root* of $D(s)$ if $D(\lambda) = 0$. For a polynomial with real coefficients, if a complex number $\lambda = \alpha + j\beta$ is a root, so is its complex conjugate $\lambda^* = \alpha - j\beta$. The number of roots equals the degree of its polynomial.

The roots of a polynomial of degree 1 or 2 can be readily computed by hand. Computing the roots of a polynomial of degree 3 or higher is complicated. It is best delegated to a computer. In MATLAB, the function **roots** computes the roots of a polynomial. For example, consider the polynomial of degree 3

$$D(s) = 4s^3 - 3s^2 + 5$$

Typing in MATLAB

```
>> roots([4 -3 0 5])
```

will yields the three roots

$$0.8133 + 0.8743i \quad 0.8133 - 0.8743i \quad -0.8766$$

---

[17]The discussion is omitted in Reference [C6].

The polynomial has one real root and one pair of complex-conjugate roots. Thus the polynomial $D(s)$ can be factored as

$$D(s) = 4(s - 0.8133 - 0.8743i)(s - 0.8133 + 0.8743i)(s + 0.8766)$$

Do not forget the leading coefficient 4 because MATLAB computes actually the roots of $D(s)/4$. Note that $D(s)$, $-2D(s)$, $0.7D(s)$, and $D(s)/4$ all have, by definition, the same set of roots.

A rational function is a ratio of two polynomials. Its degree can be defined as the larger of its denominator's and numerator's degrees. If a rational function is proper, then its degree reduces to the degree of its denominator. This definition however will not yield a unique degree without qualification. For example, consider the proper rational functions

$$H(s) = \frac{3s + 2}{s^2 + 3s - 4} = \frac{3s^2 + 5s + 2}{s^3 + 4s^2 - s - 4} = \frac{3s^3 + 8s^2 + 10s + 4}{s^4 + 5s^3 + 4s^2 - 2s - 8} \tag{8.72}$$

The second rational function is obtained from the first one by multiplying its numerator and denominator by $(s + 1)$ and the last one by multiplying $(s^2 + 2s + 2)$. The degree of $H(s)$ as defined can be 2, 3, 4 or other integers.

Two polynomials are defined to be *coprime* if they have no common roots. For example, the polynomial $(3s + 2)$ has root $-2/3$ and $(s^2 + 3s - 4)$ has roots 1 and -4. The two polynomials have no root in common and thus are coprime. The numerator and denominator of the second rational function in (8.72) have the common factor $(s + 1)$ or common root -1; thus they are not coprime. Nor are the numerator and denominator of the last rational function in (8.72) because they have the common factor $(s^2 + 2s + 2)$ or common roots $-1 \pm j$. Thus in order to define uniquely the degree of a rational function, we require its numerator and denominator to be coprime.

With the preceding discussion, we can now define the *degree* of a proper rational function $H(s) = N(s)/D(s)$ as the degree of its denominator after canceling out all common roots between $N(s)$ and $D(s)$. If $N(s)$ and $D(s)$ are coprime, then the degree of $H(s) = N(s)/D(s)$ equals the degree of $D(s)$. Using this definition, the three rational functions in (8.72) all have degree 2.

One way to check the coprimeness of $N(s)$ and $D(s)$ is to use the MATLAB function roots to compute their roots. Note that if the roots of $N(s)$ are known, there is no need to compute the roots of $D(s)$. For example, consider $N(s) = -2(s+2)(s-1)$ and

$$D(s) = 2s^3 + 4s + 1$$

The polynomial $N(s)$ has roots -2 and 1. We compute $D(-2) = 2(-2)^3 + 4(-2) + 1 = -23 \neq 0$ and $D(1) = 2 + 4 + 1 \neq 0$. Thus $D(s)$ has no roots of $N(s)$ and $N(s)$ and $D(s)$ are coprime. Another way is to use the Euclidean algorithm which is often taught in high schools. Yet another way is to check the non-singularity of a square matrix, called *Sylvester resultant*. See Problem 8.29.

8.10.1   Minimal Realizations

Consider the last rational function in (8.72) or

$$H(s) = \frac{3s^3 + 8s^2 + 10s + 4}{s^4 + 5s^3 + 4s^2 - 2s - 8} \tag{8.73}$$

It is strictly proper and its denominator's leading coefficient is 1. Thus its realization can be obtained as, using (8.59),

$$\dot{\mathbf{x}}(t) = \begin{bmatrix} -5 & -4 & 2 & 8 \\ 1 & 0 & 0 & 0 \\ 0 & 1 & 0 & 0 \\ 0 & 0 & 1 & 0 \end{bmatrix} \mathbf{x}(t) + \begin{bmatrix} 1 \\ 0 \\ 0 \\ 0 \end{bmatrix} u(t) \tag{8.74}$$

$$y(t) = [3 \quad 8 \quad 10 \quad 4]\mathbf{x}(t)$$

This realization has dimension 4.

Now if we cancel the common factor $(s^2 + 2s + 2)$ in (8.73) to yield

$$H(s) = \frac{3s^3 + 8s^2 + 10s + 4}{s^4 + 5s^3 + 4s^2 - 2s - 8} = \frac{3s + 2}{s^2 + 3s - 4} \tag{8.75}$$

then we can obtain the following realization

$$\dot{\mathbf{x}}(t) = \begin{bmatrix} -3 & 4 \\ 1 & 0 \end{bmatrix} \mathbf{x}(t) + \begin{bmatrix} 1 \\ 0 \end{bmatrix} u(t) \tag{8.76}$$

$$y(t) = [3 \quad 2]\mathbf{x}(t)$$

This realization has dimension 2.

A realization of a proper rational function $H(s)$ is called a *minimal realization* if its dimension equals the degree of $H(s)$. It is called minimum because it is not possible to find an ss equation of lesser dimension to realize $H(s)$. Because the degree of $H(s)$ in (8.73) or (8.75) is 2, the realization in (8.74) is not minimal, whereas the realization in (8.76) is minimal. If we use a non-minimal realization to develop a basic block diagram and then built it using an op-amp circuit, then the number of components used will be unnecessarily large. Thus there is no reason to use a non-minimal realization. Whenever we are given a transfer function, we should first cancel out all common factors, if there are any, between its numerator and denominator. We then use (8.59) to find a realization. The realization is automatically minimal.

8.11 Do transfer functions describe systems fully?

An ss equation describes not only the relationship between its input and output but also its internal variables, thus it is called an *internal description*. Just like a convolution and high-order differential equation, a transfer function is an *external description* because it does not reveal anything inside the system. Moreover, a transfer function describes only an initially relaxed system, whereas an ss equation is still applicable even if the system is not initially relaxed. In other words, a transfer function describes only forced responses, whereas an ss equation describes forced and natural responses. Thus it is natural to ask: Does the transfer function describe fully a system or is there any information of the system missing from the transfer function? This is discussed next.

Consider the network shown in Figure 8.11(a). It has an LC loop connected in series with the current source. Even though the input will excite some response in the

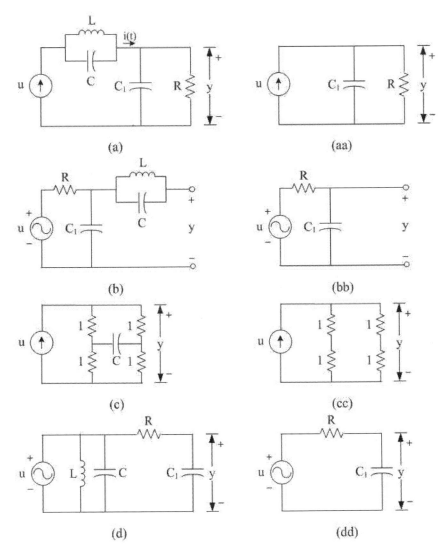

Figure 8.11: (a) Unobservable circuit. (aa) Reduced circuit. (b) Uncontrollable circuit. (bb) Reduced circuit. (c) Uncontrollable and unobservable circuit. (cc) Reduced circuit. (d) Unobservable circuit. (dd) Reduced circuit.

LC loop, because the current $i(t)$ shown always equals the input $u(t)$, the circuit, as far as the output $y(t)$ is concerned, can be reduced to the one in Figure 8.11(aa). In other words, because the response in the LC loop will not appear at the output terminal, the two circuits in Figures 8.11(a) and 8.11(aa) have the same transfer function. Thus the transfer function of Figure 8.11(a) contains no information of the LC loop and cannot describe the circuit fully.

For the circuit in Figure 8.11(b), if the initial conditions of L and C are zero, the current in the LC loop will remain zero no matter what input is applied. Thus the LC loop will not be excited by any input and, consequently, will not contribute any voltage to the output, and the circuit can be reduced to the one in Figure 8.11(bb) in computing its transfer function. For the circuit in Figure 8.11(c), because the four resistors are the same, if the initial voltage across the capacitor is zero, the voltage will remain zero no matter what input is applied. Thus the transfer functions of the two circuits in Figures 8.11(c) and 8.11(cc) are the same. The circuit in Figure 8.11(d) is the dual of Figure 8.11(a). The capacitor with capacitance C and the inductor with inductance L connected in parallel with the voltage source $u(t)$ can be deleted in computing the transfer function of the network. In conclusion, all circuits in Figures 8.11(a) through (d) cannot be described fully by their transfer functions.

8.11.1 Complete characterization

Consider a system with transfer function $H(s)$. The system is defined to be *completely characterized* by its transfer function if the number of energy storage elements in the system equals the degree of the transfer function. We first give some examples.

For the circuit in Figure 8.11(aa), we have $Y(s) = H_a(s)U(s)$, where $H_a(s)$ is the impedance of the parallel connection of R and $1/C_1s$ or

$$H_a(s) = \frac{R(1/C_1s)}{R + 1/C_1s} = \frac{R}{RC_1s + 1} \tag{8.77}$$

It has degree 1 which equals the number of energy storage element in the RC circuit in Figure 8.11(aa). Thus the circuit is completely characterized by the transfer function. The circuit in Figure 8.11(a) also has (8.77) as its transfer function but has three energy storage elements (two capacitors and one inductor). Thus the circuit is not completely characterized by its transfer function.

For the circuit in Figure 8.11(bb), we have $Y(s) = H_b(s)U(s)$ with

$$H_b(s) = \frac{1/C_1s}{R + 1/C_1s} = \frac{1}{RC_1s + 1} \tag{8.78}$$

It has degree one. Thus the transfer function characterizes completely the circuit in Figure 8.11(bb). To find the transfer function of the circuit in Figure 8.11(b), we first assume that the circuit is initially relaxed, or the current in the LC loop is zero. Now no matter what input is applied, the current in the LC loop will remain zero. Thus the circuit in Figure 8.11(b) also has (8.78) as its transfer function but is not completely characterized by it. Likewise, the circuits in Figures 8.11(d) and (dd) have (8.78) as their transfer function. But the former is completely characterized by (8.78) but not the latter.

For the circuit in Figure 8.11(cc), we have $Y(s) = H_c(s)U(s)$ with

$$H_c(s) = \frac{2 \times 2}{2 + 2} = \frac{4}{4} = 1 \tag{8.79}$$

It has degree zero and characterizes completely the circuit in Figure 8.11(cc) because it has no energy storage element. On the other hand, the circuit in Figure 8.11(c) has one energy storage element and is not completely characterized by its transfer function given in (8.79).

What is the physical significance of complete characterization? In practice, we try to design a simplest possible system to achieve a given task. As far as the input and output are concerned, all circuits on the left-hand side of Figure 8.11 can be replaced by the simpler circuits on the right-hand side. In other words, the circuits on the left-hand side have some redundancies or some unnecessary components, and we should not design such systems. It turns out that such redundancies can be detected from their transfer functions. If the number of energy storage elements in a system is larger than the degree of its transfer function, then the system has some redundancy. Most practical systems, unless inadvertently designed, have no redundant components and are completely characterized by their transfer functions.[18]

To conclude this subsection, we mention that complete characterization deals only with energy-storage elements; it pays no attention to non-energy storage elements such as resistors. For example, the circuit in Figure 8.11(cc) is completely characterized by its transfer function $H(s) = 1$. But the four resistors with resistance 1 can be replaced by a single resistor. Complete characterization does not deal with this problem.

8.11.2   Equivalence of ss equations and transfer functions

All the circuits on the left hand side of Figure 8.11 are not completely characterized by their transfer functions. They however can all be described fully by their ss equations. For example, consider the RLC circuit in Figure 8.11(a) with $L = 2$H, $C = 4$F, $C_1 = 1$F, and $R = 5\Omega$. If we assign the inductor's current (flowing from left to right) as $x_1(t)$, the 4F-capacitor's voltage (with positive polarity at left) as $x_2(t)$, and the 1F-capacitors' voltage as $x_3(t)$, then we can develop the following ss equation

$$\begin{bmatrix} \dot{x}_1(t) \\ \dot{x}_2(t) \\ \dot{x}_3(t) \end{bmatrix} = \begin{bmatrix} 0 & 0.5 & 0 \\ -0.25 & 0 & 0 \\ 0 & 0 & -0.2 \end{bmatrix} \begin{bmatrix} x_1(t) \\ x_2(t) \\ x_3(t) \end{bmatrix} + \begin{bmatrix} 0 \\ 0.25 \\ 1 \end{bmatrix} u(t)$$

$$y(t) = \begin{bmatrix} 0 & 0 & 1 \end{bmatrix} \begin{bmatrix} x_1(t) \\ x_2(t) \\ x_3(t) \end{bmatrix} + 0 \times u(t) \tag{8.80}$$

to describe the circuit (Problem 8.30). This is a 3-dimensional ss equation and is more descriptive than the transfer function $H_a(s) = 5/(5s + 1)$. For example, an input may excite a sinusoidal oscillation with frequency $1/\sqrt{LC}$ in the LC loop. This can be discerned from the ss equation in (8.80) but not from the transfer function $H_a(s)$ in

---

[18]The redundancy in this section is different from the redundancy purposely introduced for safety reason such as to use two or three identical systems to achieve the same control in the space station.

(8.77). Likewise, we can develop ss equations to describe fully the circuits in Figures 8.11(b) through (d). In conclusion, *if a system is not completely characterized by its transfer function, its ss equation is more descriptive than its transfer function.*

Although the ss equations describing the circuits in Figures 8.11(a) through (d) are more descriptive, they all have some deficiency. They cannot be both controllable and observable[19] Thus the redundancy of a system can also be checked from its ss equation. The checking however is much more complicated than the one based on transfer functions. In any case, if a system is not completely characterized by its transfer function, then the system has some redundant components. This type of system should be discarded. There is no need to develop an ss equation to describe it.

Most systems designed in practice do not have unnecessary components. For such a system, will its ss equation be more descriptive than its transfer function? The answer is negative as we discuss next. If a system has no redundant components and if we develop an ss equation and a transfer function to describe the system, then we have

$$\text{No. of energy-storage elements} \; = \; \text{Degree of its transfer function}$$
$$= \; \text{Dimension of its ss equation}$$

In this case, there is no difference in using its transfer function or ss equation to describe the system. To be more specific, all internal characteristics of the system will appear in the transfer function. The response excited by any set of initial conditions can be generated from its transfer function by applying some input. In other words, every natural or zero-input response of the system can be generated as a forced or zero-state response. Thus nothing is lost in using transfer functions. In conclusion, *if a system is completely characterized by its transfer function, no information is lost in using the transfer function in analysis and design* even though the transfer function is an external description and describes only an initially relaxed system. See Reference [C6]. In fact most design methods discussed in texts are based on transfer functions. Once a satisfactory transfer function is designed, we then use its minimal realization to carry out op-amp circuit implementation. Then the resulting system has no redundant components.

8.12 Concluding remarks

To conclude this chapter, we mention that systems can be classified dichotomously as shown in Figure 8.12. This text studies only linear, time-invariant, and lumped systems. In addition, we study only causal systems in order to be physically realizable. This class of systems can be described by

1. Integral convolutions

2. State-space (ss) equations

3. High-order differential equations

---

[19]They can be either controllable or observable but not both. All design methods require ss equations to be both controllable and observable.

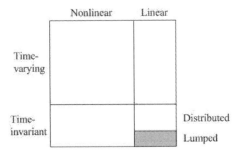

Figure 8.12: Classification of systems.

4. Proper rational transfer functions

The first three are in the time domain. The last one is in the transform domain. The ss equation is an internal description; the other three are external descriptions. If a system is completely characterized by its transfer function, then the transfer function and state-space descriptions of the system are equivalent.

State-space (ss) equations can be

1. computed in real time,

2. computed without incurring numerical sensitivity issue to be discussed in Subsection 9.3.2,

3. computed very efficiently, and

4. implemented easily using op-amp circuits.

The preceding four merits were developed without using any analytical property of ss equations. Thus the discussion was self contained. All other three descriptions lack at least one of the four merits. Thus ss equations are most important in computation and implementation. Are ss equations important in discussing general properties of systems and in design? This will be answered in the next chapter.

A system is modeled as a black box as shown in Figure 6.1. If the system is LTI and lumped, then it can be represented as shown in Figure 8.13(a) in the transform domain. In the time domain, it can be represented as in Figure 8.13(b) using an ss equation. In the diagram, a single line denotes a single variable and double lines denote two or more variables. The output $y(t_1)$ in Figure 8.13(b) depends on $u(t_1)$ and $\mathbf{x}(t_1)$ and will appear at the same time instant as $u(t_1)$. If we use a convolution, then the output $y(t_1)$ depends on $u(t)$, for $t \leq t_1$ and the system requires an external memory to memorize past input. If we use a high-order differential equation, then the representation will involve feedback from the output and its derivatives. This is not the case in using an ss equation. Thus the best representation of a system in the time domain is the one shown in Figure 8.13(b).

Problems

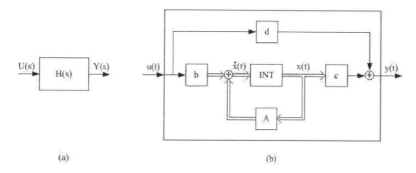

Figure 8.13: (a) Transform-domain representation of CT system. (b) Time-domain representation.

8.1 Consider the RLC circuit in Figure 8.1(a). What is its ss equation if we consider the voltage across the 5-H inductor as the output?

8.2 Consider the RLC circuit in Figure 8.1(a). What is its ss equation if we consider the current passing through the 2-F capacitor as the output?

8.3 Find ss equations to describe the circuits shown in Figure 8.14. For the circuit in Figure 8.14(a), choose the state variables x_1 and x_2 as shown.

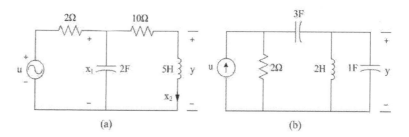

Figure 8.14: (a) Circuit with two state variables. (b) Circuit with three state variables.

8.4 Use the MATLAB function lsim to compute the response of the circuit in Figure 8.14(a) excited by the input $u(t) = 1 + e^{-0.2t} \sin 10t$ for t in $[0, 30]$ and with the initial inductor's current and initial capacitor's voltage zero. Select at least two step sizes.

8.5 Use the MATLAB functions step and impulse to compute the step and impulse responses of the circuit shown in Figure 8.14(b).

8.6* The impulse response of the ss equation in (8.16) and (8.17) with $d = 0$ is, by definition, the output excited by $u(t) = \delta(t)$ and $\mathbf{x}(0) = \mathbf{0}$. Verify that it equals the output excited by $u(t) = 0$, for all $t \geq 0$ and $\mathbf{x}(0) = \mathbf{b}$. Note that if $d \neq 0$, then the impulse response of the ss equation contains $d\delta(t)$. A computer however cannot generate an impulse. Thus in computer computation of impulse responses, we must assume $d = 0$. See also Problem 9.3.

8.7 Develop basic block diagrams to simulate the circuits in Figure 8.14.

8.8 If we replace all basic elements in the basic block diagram for Figure 8.14(a) obtained in Problem 8.7 by the op-amp circuits shown in Figures 6.19, 6.20, and 8.7(a), how many op amps will be used in the overall circuit?

8.9 Consider the basic block diagram shown in Figure 8.15. Assign the output of each integrator as a state variable and then develop an ss equation to describe the diagram.

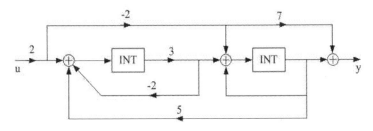

Figure 8.15: Basic block diagram.

8.10 Consider the op-amp circuit shown in Figure 8.16 with $RC = 1$ and a, b, and c are positive real constants. Let the inputs be denoted by $v_i(t)$, for $i = 1 : 3$. Verify that if the output is assigned as $x(t)$, then we have

$$\dot{x}(t) = -[av_1(t) + bv_2(t) + cv_3(t)]$$

Thus the circuit acts as amplifiers, an adder, and an integrator.

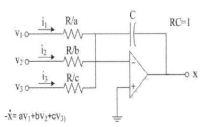

Figure 8.16: Op-amp circuit that acts as three inverting amplifiers, one adder and one integrator.

8.11 In Problem 8.10, if the output of the op-amp circuit in Figure 8.16 is assigned as $-x(t)$, what is its relationship with $v_i(t)$, for $i = 1 : 3$?

8.12 Verify that the op-amp circuit in Figure 8.17 with $RC = 1$ simulates the RLC circuit shown in Figure 8.14(a). Note that x_1 and x_2 in Figure 8.17 correspond to those in Figure 8.14(a). Compare the number of op amps used in Figure 8.17 with the one used in Problem 8.8? This shows that op-amp implementations of a system are not unique.

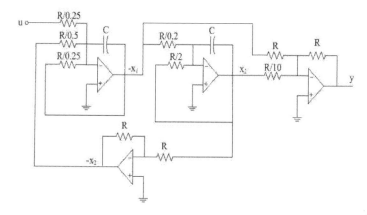

Figure 8.17: Op-amp circuit implementation of the RLC circuit in Figure 8.14(a).

8.13 Use impedances to compute the transfer functions of the circuits in Figure 8.14.

8.14 Classify the following rational functions:

$$-10, \quad \frac{2s^2 + 1}{3s - 1} \quad \frac{2s^2 + 1}{3s^2 - 1} \quad \frac{2s^2 + 1}{s^{10}}$$

8.15 Find a realization of the transfer function of the circuit in Figure 8.14(a) computed in Problem 8.13. Compare this ss equation with the one in Problem 8.3. Note that although they look different, they describe the same RLC circuit and are equivalent. See Reference [C6]. Implement the ss equation using the structure in Figure 8.17. Which implementation uses a smaller total number of components? Note that the use of controllable-form realization for implementation will generally use a smaller number of components. However the physical meaning of its state variables may not be transparent.

8.16 Use the transfer function obtained in Problem 8.13 to find a differential equation to describe the circuit in Figure 8.14(b).

8.17 Find realizations for the transfer functions

$$H_1(s) = \frac{3s^2 + 1}{2s^2 + 4s + 5} \quad \text{and} \quad H_2(s) = \frac{1}{2s^3}$$

8.18 Verify that the one-dimensional ss equation

$$\begin{aligned} \dot{x}(t) &= -ax(t) + u(t) \\ y(t) &= kx(t) + du(t) \end{aligned}$$

realizes the transfer function $H(s) = k/(s+a) + d$. Note that the ss equation is a special case of (8.59) with dimension 1.

8.19 Consider the transfer function

$$H(s) = \frac{b}{s^3 + a_2 s^2 + a_3 s + a_4} = \frac{Y(s)}{U(s)}$$

or, equivalently, the differential equation

$$y^{(3)}(t) + a_2 \ddot{y}(t) + a_3 \dot{y}(t) + a_4 y(t) = bu(t) \tag{8.81}$$

where $y^{(3)}(t) = d^3 y(t)/dt^3$ and $\ddot{y}(t) = d^2 y(t)/dt^2$. Verify that by defining $x_1(t) := \ddot{y}(t)$, $x_2(t) := \dot{y}(t)$, and $x_3(t) := y(t)$, we can transform the differential equation into the following ss equation

$$\begin{bmatrix} \dot{x}_1(t) \\ \dot{x}_2(t) \\ \dot{x}_3(t) \end{bmatrix} = \begin{bmatrix} -a_2 & -a_3 & -a_4 \\ 1 & 0 & 0 \\ 0 & 1 & 0 \end{bmatrix} \begin{bmatrix} x_1(t) \\ x_2(t) \\ x_3(t) \end{bmatrix} + \begin{bmatrix} b \\ 0 \\ 0 \end{bmatrix} u(t)$$

$$y(t) = [0 \quad 0 \quad 1] \begin{bmatrix} x_1(t) \\ x_2(t) \\ x_3(t) \end{bmatrix}$$

What will be the form of the ss equation if we define $x_1(t) = y(t)$, $x_2(t) = \dot{y}(t)$, and $x_3(t) = \ddot{y}(t)$.

8.20* Consider the transfer function

$$H(s) = \frac{b_1 s^2 + b_2 s + b_3}{s^3 + a_2 s^2 + a_3 s + a_4} = \frac{Y(s)}{U(s)} \tag{8.82}$$

or, equivalently, the differential equation

$$y^{(3)}(t) + a_2 \ddot{y}(t) + a_3 \dot{y}(t) + a_4 y(t) = b_1 \ddot{u}(t) + b_2 \dot{u}(t) + b_3 u(t) \tag{8.83}$$

If we define state variables as in Problem 8.19, then the resulting ss equation will involve $\dot{u}(t)$ and $\ddot{u}(t)$ and is not acceptable. Let us introduce a new variable $v(t)$ defined by

$$V(s) := \frac{1}{s^3 + a_2 s^2 + a_3 s + a_4} U(s)$$

or, equivalently, by the differential equation

$$v^{(3)}(t) + a_2 \ddot{v}(t) + a_3 \dot{v}(t) + a_4 v(t) = u(t) \tag{8.84}$$

Verify

$$\frac{Y(s)}{V(s)} = b_1 s^2 + b_2 s + b_3$$

and

$$y(t) = b_1 \ddot{v}(t) + b_2 \dot{v}(t) + b_3 v(t) \tag{8.85}$$

Note that (8.85) can be directly verified from (8.83) and (8.84) without using transfer functions but will be more complex. Verify that by defining $x_1(t) = \ddot{v}(t)$, $x_2(t) = \dot{v}(t)$, and $x_3(t) = v(t)$, we can express (8.84) and (8.85) as

$$
\begin{bmatrix} \dot{x}_1(t) \\ \dot{x}_2(t) \\ \dot{x}_3(t) \end{bmatrix} = \begin{bmatrix} -a_2 & -a_3 & -a_4 \\ 1 & 0 & 0 \\ 0 & 1 & 0 \end{bmatrix} \begin{bmatrix} x_1(t) \\ x_2(t) \\ x_3(t) \end{bmatrix} + \begin{bmatrix} 1 \\ 0 \\ 0 \end{bmatrix} u(t)
$$

$$
y(t) = [b_1 \quad b_2 \quad b_3] \begin{bmatrix} x_1(t) \\ x_2(t) \\ x_3(t) \end{bmatrix}
$$

This is the ss equation in (8.59) with $d = 0$. Note that the relationships between $x_i(t)$ and $\{u(t), y(t)\}$ are complicated. Fortunately, we never need the relationships in using the ss equation.

8.21 Find realizations of

$$
H_1(s) - \frac{2(s-1)(s+3)}{s^3 + 5s^2 + 8s + 6} \quad \text{and} \quad H_2(s) = \frac{s^3}{s^3 + 2s - 1}
$$

Are they minimal? If not, find one.

8.22* The transfer function of (8.19) was computed using MATLAB in Subsection 8.9.1 as

$$
H(s) = \frac{-2s^2 - s - 16}{s^2 + 2s + 10}
$$

Now verify it analytically using (8.70) and the following inversion formula

$$
\begin{bmatrix} a & b \\ c & d \end{bmatrix}^{-1} = \frac{1}{ad - bc} \begin{bmatrix} d & -b \\ -c & a \end{bmatrix}
$$

8.23 What are the transfer functions from u to y of the three circuits in Figure 8.18? Is every circuit completely characterized by its transfer function? Which circuit is the best to be used as a voltage divider?

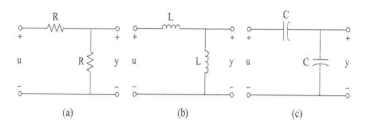

Figure 8.18: Voltage dividers.

8.24 Consider the circuit shown in Figure 8.19(a) where the input $u(t)$ is a current source and the output can be the voltage across or the current passing through the impedance Z_4. Verify that either transfer function is independent of the impedance Z_1. Thus in circuit design, one should not connect any impedance in series with a current source.

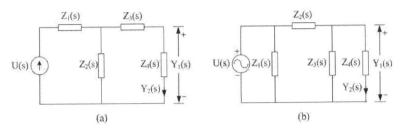

(a)   (b)

Figure 8.19: (a) Circuit with redundant Z_1. (b) Circuit with redundant Z_1.

8.25 Consider the circuit shown in Figure 8.19(b) where the input $u(t)$ is a voltage source and the output can be the voltage across or the current passing through the impedance Z_4. Verify that either transfer function is independent of the impedance Z_1. Thus in circuit design, one should not connect any impedance in parallel with a voltage source.

8.26 Consider the circuit shown in Figure 8.20(a). (a) Because the two capacitor voltages are identical, assign only one capacitor voltage as a state variable and then develop a one-dimensional ss equation to describe the circuit. Note that if we assign both capacitor voltages as state variables $x_i(t)$, for $i = 1, 2$, then $x_1(t) = x_2(t)$, for all t, and they cannot act independently. (b) Compute the transfer function of the circuit. Is the circuit completely characterized by its transfer function?

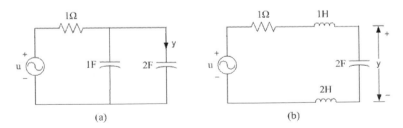

(a)   (b)

Figure 8.20: (a) Circuit in which not all capacitor voltages are assigned as state variables. (b) Circuit in which not all inductor currents are assigned as state variables..

8.27 Consider the circuit shown in Figure 8.20(b). (a) Because the two inductor currents are identical, assign only one inductor current as a state variable and then develop a two-dimensional ss equation to describe the circuit. (b) Compute the transfer function of the circuit. Is the circuit completely characterized by its transfer function?

8.28* Consider the circuit shown in Figure 8.21. (a) Verify that if we assign the 1-F capacitor voltage as $x_1(t)$, the inductor current as $x_2(t)$, and the 2-F capacitor voltage as $x_3(t)$, then the circuit can be described by

$$\dot{\mathbf{x}}(t) = \begin{bmatrix} 0 & 1/3 & 0 \\ 0 & -2 & 1 \\ 0 & -1/3 & 0 \end{bmatrix} \mathbf{x}(t) + \begin{bmatrix} 2/3 \\ 0 \\ 1/2 \end{bmatrix} \dot{u}(t)$$

$$y(t) = \begin{bmatrix} 0 & -2 & 1 \end{bmatrix} \mathbf{x}(t)$$

where $\mathbf{x} = [x_1 \ x_2 \ x_3]'$. Is it in the standard ss equation form? (b) Verify that if we assign only $x_1(t)$ and $x_2(t)$ (without assign $x_3(t)$), then the circuit can be described by

$$\dot{\mathbf{x}}(t) = \begin{bmatrix} 0 & 1/3 \\ -1 & -2 \end{bmatrix} \mathbf{x}(t) + \begin{bmatrix} 0 \\ 1 \end{bmatrix} u(t) + \begin{bmatrix} 2/3 \\ 0 \end{bmatrix} \dot{u}(t)$$

$$y(t) = \begin{bmatrix} -1 & -2 \end{bmatrix} \mathbf{x}(t)$$

where $\mathbf{x} = [x_1 \ x_2]'$. Is it in the standard ss equation form? (c) Compute the transfer function of the circuit. Is the circuit completely characterized by its transfer function? Note that $x_1(t) + x_3(t) = u(t)$, for all t, thus the 2-F capacitor is, in some sense, redundant.

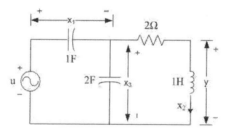

Figure 8.21: Circuit which cannot be described by a standard ss equation.

8.29* Consider the rational function

$$H(s) = \frac{N(s)}{D(s)} = \frac{N_2 s^2 + N_1 s + N_0}{D_2 s^2 + D_1 s + D_0}$$

One way to check the coprimeness of its numerator and denominator is to solve the equation

$$\frac{N_2 s^2 + N_1 s + N_0}{D_2 s^2 + D_1 s + D_0} = \frac{B_1 s + B_0}{A_1 s + A_0}$$

or the polynomial equation

$$(D_2 s^2 + D_1 s + D_0)(B_1 s + B_0) = (N_2 s^2 + N_1 s + N_0)(A_1 s + A_0)$$

Verify that equating the coefficients of the same power of s^i, for $i = 0 : 3$, will yield

$$
\begin{bmatrix}
D_0 & N_0 & 0 & 0 \\
D_1 & N_1 & D_0 & N_0 \\
D_2 & N_2 & D_1 & N_1 \\
0 & 0 & D_2 & N_2
\end{bmatrix}
\begin{bmatrix}
B_0 \\
-A_0 \\
B_1 \\
-A_1
\end{bmatrix}
=
\begin{bmatrix}
0 \\
0 \\
0 \\
0
\end{bmatrix}
$$

The square matrix of order 4 is called the *Sylvester resultant*. It is a basic result that if the square matrix is nonsingular, then no nonzero solutions A_i and B_i exist in the matrix equation. In this case, $N(s)/D(s)$ cannot be reduced to $B(s)/A(s)$. thus $N(s)$ and $D(s)$ are coprime. If the square matrix is singular, then nonzero solutions A_i and B_i exist. In this case, $N(s)/D(s)$ can be reduced to $B(s)/A(s)$. Thus $N(s)$ and $D(s)$ are not coprime. All pole-placement and model-matching designs using transfer functions can be transformed into solving linear algebraic equations that contain Sylvester resultants. See References [C5] and [C6].

8.30 Verify that the RLC circuit in Figure 8.11(a) with $L = 2$H, $C = 4$F, $C_1 = 1$F, and $R = 5\Omega$ can be described by the ss equation in (8.80).

8.31 To find the transfer function of the ss equation in (8.80), typing in MATLAB

```
>> a=[0 0.5 0;-0.25 0 0;0 0 -0.2];b=[0;0.25;1];c=[0 0 1];d=0;
>> [n,d]=ss3tf(a,b,c,d)
```

will yield n=0 1 0 0.125 and d=1 0.2 0.125 0.025. What is the transfer function? What is its degree?

8.32 The convolution in (8.5) can be reduced as

$$
y(t) = \int_{\tau=0}^{t} h(t - \tau)u(\tau)d\tau
$$

by using the causality condition $h(t - \tau) = 0$, for $\tau > t$. Verify the following commutative property

$$
y(t) = \int_{\tau=0}^{t} h(t - \tau)u(\tau)d\tau = \int_{\tau=0}^{t} u(t - \tau)h(\tau)d\tau
$$

8.33 Let $y_q(t)$ be the output of a system excited by a step input $u(t) = 1$ for all $t \geq 0$. Use the second integration in Problem 8.32 to verify

$$
h(t) = \frac{dy_q(t)}{dt}
$$

where $h(t)$ is the impulse response of the system.

8.34 Consider the circuit shown in Figure 8.22(a) which consists of two ideal diodes connected in parallel but in opposite directions. The voltage $v(t)$ and the current $i(t)$ of an ideal diode is shown in Figure 8.22(b). If the diode is forward biased or $v(t) > 0$, the diode acts as short circuit and the current $i(t)$ will flow. If the diode is reverse biased or $v(t) < 0$, the diode acts as open circuit and no current can flow. Thus the current of a diode can flow only in one direction. Is an ideal diode a linear element? Is the system from u to y linear? What is its transfer function?

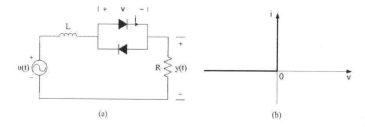

Figure 8.22: (a) Linear system (from u to y) that contains nonlinear elements. (b) Characteristic of ideal diode.

8.35 Consider $N(s) = 2(s-1)(s+2)$ and $D(s) = s^3 + 3s^2 - 3$. Verify their coprimeness, without computing all roots of $D(s)$, using the following two methods:

1. Check whether or not that $D(s)$ contains the roots of $N(s)$.
2. Use the Euclidian algorithm.

Note that it can also be checked using Problem 8.20.

8.36 Verify that the three circuits in Figures 8.20 and 8.21 are not completely characterized by their transfer functions.

8.37 Compute the transfer function of the circuit in Figure 8.23(a). What is its degree? Is the circuit completely characterized by its transfer function? Verify that the circuit in Figure 8.23(b) has the same transfer function as the one in Figure 8.23(b).

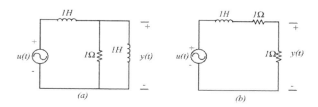

Figure 8.23: (a) Circuit with two energy-storage elements. (b) Circuit with one energy-storage element.

Chapter 9

Qualitative analysis of CT LTI lumped systems

9.1 Introduction

The study of systems consists four parts: modeling, developing mathematical equations, analysis, and design. As an introductory course, we study only LTI lumped systems. Such systems can be described by, as discussed in the preceding chapter, convolutions, high-order differential equations, ss equations, and rational transfer functions.

The next step in the study is to carry out analysis. There are two types of analyses: quantitative and qualitative. In quantitative analysis, we are interested in the outputs excited by some inputs. This can be easily carried out using ss equations and MATLAB as discussed in the preceding chapter. In qualitative analysis, we are interested in general properties of systems. We use transfer functions to carry out this study because it is simplest and most transparent among the four descriptions. We will introduce the concepts of poles, zeros, stability, transient and steady-state responses and frequency responses. These concepts will be introduced using two types of input signals: step functions and sinusoids. There is no need to consider general signals. Even so, once a system is designed using these concepts, the system is applicable to any signal.

The transfer functions studied in this chapter are limited to proper rational functions with real coefficients. When we use a transfer function to study a system, the system is implicitly assumed to be initially relaxed (all its initial conditions are zero) at $t = 0$ and the input is applied from $t = 0$ onward. Recall that $t = 0$ is the instant we start to study the system and is selected by us.

9.1.1 Design criteria – time domain

Before plunging into qualitative analysis, we discuss some basic criteria in designing systems. This will justify what will be discussed in this chapter.

We use mercury-in-glass thermometers and spring weighing scales to discuss the issues. A conventional thermometer uses the expansion of mercury to measure a body's temperature. It usually takes 60 seconds or more for the thermometer to give a final

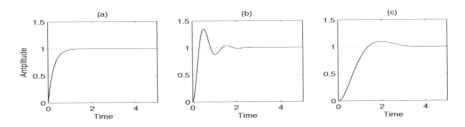

Figure 9.1: Three possible step responses.

reading. If we consider the thermometer as a system, then the body's temperature will be the input, and the thermometer's reading will be the output.

A weighing scale built with a spring can be used to measure a body's weight. When a person steps on the scale, the reading will start to oscillate and then settle down in a couple of seconds to a reading. The reading will give the weight of the person.

Let us use $u(t)$ to denote the input and $y(t)$ the output. For the two examples, $u(t)$ is just a constant, for $t \geq 0$, or a step function with an unknown amplitude. The output $y(t)$ of the thermometer will increase monotonically to a constant as shown in Figure 9.1(a); the output $y(t)$ of the weighing scale will approach oscillatorily a constant as shown in Figure 9.1(b). An important question about the two systems is the correctness of the final reading. Does the final reading gives the actual body's temperature or weight? This depends entirely on the calibration. If a calibration is correct, then the final reading will be accurate.

Another question regarding the two systems is the time for the output to reach the final reading. We call the time the *response time*. The response time of the thermometer is roughly 60 seconds and that of the spring scale is about 2 seconds. If a weighing scale takes 60 seconds to reach a final reading, no person will buy such a scale. The response time of mercury-in-glass thermometers is very long. If we use other technology such as infrared sensing, then the response time can be shorten significantly. For example, thermometers now used in hospitals take only 4 seconds or less to give a final digital reading.

Yet another question regarding the two systems is the waveform of $y(t)$ before reaching the final reading. Generally, $y(t)$ may assume one of the waveforms shown in Figure 9.1. The reading of the thermometer takes the form in Figure 9.1(a). The spring scale takes the form in Figure 9.1(b); it oscillates and then settles down to the final value. For some systems, $y(t)$ may go over and then come back to the final value, without appreciable oscillation, as shown in Figure 9.1(c). If $y(t)$ goes over the final reading as shown in Figures 9.1(b) and (c), then the response is said to have *overshoot*.

The accuracy of the final reading, response time, and overshoot are in fact three standard specifications in designing control systems. For example, in designing an elevator control, the elevator floor should line up with the floor of the intended story (accuracy). The speed of response should be fast but not too fast to cause uncomfortable for the passengers. The overshoot of an elevator will cause the feeling of nausea.

The preceding three specifications are in the time domain. In designing filters, the specifications are given mainly in the frequency domain as we will discuss later.

Keeping the preceding discussion in mind, we will start our qualitative analysis of LTI lumped systems.

9.2 Poles and zeros

Let a be any *nonzero* real or complex number. Then we have $0/a = 0$ and $|a|/0 = \infty$. Note that $0/0$ is not defined. We use these to define poles and zeros. Consider a proper rational transfer function $H(s)$. A real or complex number λ is called a *zero* of $H(s)$ if $H(\lambda) = 0$. It is called a *pole* if $|H(\lambda)| = \infty$. For example, consider

$$H(s) = \frac{2s^2 - 2s - 4}{s^3 + 5s^2 + 17s + 13} \tag{9.1}$$

We compute its value at $s = 1$:

$$H(1) = \frac{2 - 2 - 4}{1 + 5 + 17 + 13} = \frac{-4}{36}$$

which is finite and different from 0. Thus $s = 1$ is not a zero nor a pole. Next we compute $H(s)$ at $s = -1$ as

$$H(-1) = \frac{2(-1)^2 - 2(-1) - 4}{(-1)^3 + 5(-1)^2 + 17(-1) + 13} = \frac{0}{0}$$

It is not defined. In this case, we must use l'Hôpital's rule to find its value as

$$H(-1) = \frac{4s - 2}{3s^2 + 10s + 17}\bigg|_{s=-1} = \frac{4(-1) - 2}{3(-1)^2 + 10(-1) + 17} = \frac{-6}{10}$$

which is finite and different from 0. Thus $s = -1$ is not a pole nor a zero. We compute

$$H(2) = \frac{2 \cdot 4 - 2 \cdot 2 - 4}{8 + 5 \cdot 4 + 17 \cdot 2 + 13} = \frac{0}{75} = 0$$

Thus 2 is a zero of $H(s)$.

Now we claim that for a proper rational function $H(s) = N(s)/D(s)$, if $N(s)$ and $D(s)$ are coprime, then every root of $D(s)$ is a pole of $H(s)$ and every root of $N(s)$ is a zero of $H(s)$. Indeed, If $N(s)$ and $D(s)$ are coprime and if p is a root of $D(s)$ or $D(p) = 0$, then $N(p) \neq 0$. Thus we have $|H(p)| = |N(p)|/0 = \infty$ and p is a pole of $H(s)$. If q is a zero of $N(s)$ or $N(q) = 0$, then $D(q) \neq 0$. Thus we have $H(q) = 0/D(q) = 0$ and q is a zero of $H(s)$. This establishes the assertion. Using the MATLAB function roots, we factor (9.1) as

$$H(s) = \frac{2(s - 2)(s + 1)}{(s + 1)(s + 2 + j3)(s + 2 - j3)} = \frac{2(s - 2)}{(s + 2 + j3)(s + 2 - j3)} \tag{9.2}$$

See Section 8.10. Thus $H(s)$ has zero at 2 and poles at $-2 - j3$ and $-2 + j3$. Note that if $H(s)$ has only real coefficients, then complex conjugate poles or zeros must appear in pairs. Note also that if a transfer function is strictly proper as in (9.2), then we have

$H(\infty) = 0$, and $\lambda = \infty$ could be a zero. However, ∞ is not a number, thus ∞ is not a zero by our definition. In other words, we consider only finite poles and finite zeros.

A transfer function expressed as a ratio of two polynomials can be transformed into a ratio of zeros and poles in MATLAB by calling the function `tf2zp`, an acronym for transfer function to zero/pole. For example, consider

$$H(s) = \frac{4s^4 - 16s^3 + 8s^2 - 48s + 180}{s^5 + 6s^4 + 29.25s^3 + 93.25s^2 + 134s + 65} \tag{9.3}$$

It is a ratio of two polynomials. The transfer function can be expressed in MATLAB as `n=[4 -16 8 -48 180]` for the numerator and `d=[1 6 29.25 93.25 134 65]` for the denominator. Typing in the command window of MATLAB

```
>> n=[4 -16 8 -48 180];d=[1 6 29.25 93.25 134 65];
>> [z,p,k]=tf2zp(n,d)
```

will yield `z=[-1+2i -1-2i 3 3]`; `p=[-0.5+4i -0.5-4i -2+0i -2-0i -1]`; `k=4`. It means that the transfer function can be expressed as

$$H(s) = \frac{4(s+1-2j)(s+1+2j)(s-3)^2}{(s+0.5-4j)(s+0.5+4j)(s+2)^2(s+1)} \tag{9.4}$$

It has zeros and poles as shown and gain 4. This is said to be in the *zero/pole/gain* form. Note that zeros and poles do not specify uniquely a transfer function. We must also specify the gain k.

A pole or zero is called *simple* if it appears only once; it is called *repeated* if it appears twice or more. For example, the transfer function in (9.3) or (9.4) has two simple zeros at $-1 \pm 2j$ and three simple poles at $-0.5 \pm 4j$ and -1. It has a repeated zero at 3 with multiplicity 2 and a repeated pole at -2 also with multiplicity 2.

Poles and zeros are often plotted on a complex plane, called a complex s-plane, with the horizontal and vertical axes denote respectively the real and imaginary parts. For example, the poles and zeros of (9.4) are plotted in Figure 9.2 where poles are denoted by crosses and zeros by small circles. If a transfer function has only real coefficients as in (9.3), then complex-conjugate poles or zeros must appear in pairs as in (9.4). Thus the plot of poles and zeros will be symmetric with respect to the horizontal or real axis. Note that the multiplicity of a repeated pole or zero is denoted by an integer enclosed by brackets. Note also that the gain $k = 4$ does not appear in the figure. Thus strictly speaking, a pole-zero plot does not specify a transfer function fully.

9.3   Some Laplace transform pairs

Using the definition of the Laplace transform alone, we developed in the preceding chapter the concepts of transfer functions and impedances. Now we will discuss more about the Laplace transform.

Consider a CT signal $x(t)$ defined for all t in $(-\infty, \infty)$. Its Laplace transform is defined as

$$X(s) := \mathcal{L}[x(t)] = \int_{t=0}^{\infty} x(t)e^{-st}dt \tag{9.5}$$

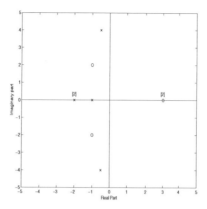

Figure 9.2: Poles (x) and zeros (o).

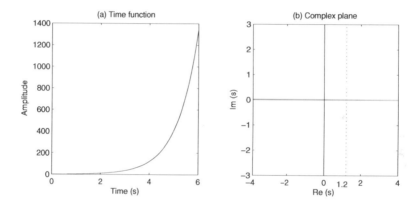

Figure 9.3: (a) The function $e^{1.2t}$, for $t \geq 0$. (b) The region of convergence of its Laplace transform (Re $s > 1.2$).

where s is a complex variable. Note that the Laplace transform is defined only for the positive-time part of $x(t)$. Because its negative-time part is not used, we often assume $x(t) = 0$, for $t < 0$. The inverse Laplace transform of $X(s)$ is defined as

$$x(t) = \mathcal{L}^{-1}[X(s)] = \frac{1}{2\pi j} \int_{c-j\infty}^{c+j\infty} X(s)e^{st}ds \tag{9.6}$$

where c is a real constant and will be discussed shortly. We first use an example to discuss the issue involved in using (9.5).

Example 9.3.1 Consider the signal $x(t) = e^{1.2t}$, for $t \geq 0$. This function grows exponentially to ∞ as $t \to \infty$ as shown in Figure 9.3(a). Its Laplace transform is

$$
\begin{aligned}
X(s) &= \int_0^\infty e^{1.2t}e^{-st}dt = \int_0^\infty e^{-(s-1.2)t}dt = \frac{-1}{s-1.2}\, e^{-(s-1.2)t}\Big|_{t=0}^{\infty} \\
&= \frac{-1}{s-1.2}\left[e^{-(s-1.2)t}\Big|_{t=\infty} - e^{-(s-1.2)t}\Big|_{t=0}\right]
\end{aligned} \tag{9.7}
$$

The value of $e^{-(s-1.2)t}$ at $t=0$ is 1 for all s. However, the value of $e^{-(s-1.2)t}$ at $t=\infty$ can be, depending on s, zero, infinity, or undefined. To see this, we express the complex variable as $s=\sigma+j\omega$, where $\sigma=\text{Re } s$ and $\omega=\text{Im } s$ are, respectively, the real and imaginary part of s. Then we have

$$e^{-(s-1.2)t}=e^{-(\sigma-1.2)t}e^{-j\omega t}=e^{-(\sigma-1.2)t}(\cos\omega t-j\sin\omega t)$$

If $\sigma=1.2$, the function reduces to $\cos\omega t-j\sin\omega t$ whose value is not defined as $t\to\infty$. If $\sigma<1.2$, then $e^{-(\sigma-1.2)t}$ approaches infinity as $t\to\infty$. If $\sigma>1.2$, then $e^{-(\sigma-1.2)t}$ approaches 0 as $t\to\infty$. In conclusion, we have

$$e^{-(s-1.2)t}\Big|_{t=\infty}=\begin{cases}\infty \text{ or } -\infty & \text{for Re } s<1.2\\ \text{undefined} & \text{for Re } s=1.2\\ 0 & \text{for Re } s>1.2\end{cases}$$

Thus (9.7) is not defined for Re $s\le 1.2$. However, if Re $s>1.2$, then (9.7) reduces to

$$X(s)=\mathcal{L}[e^{1.2t}]=\int_0^\infty e^{1.2t}e^{-st}dt=\frac{-1}{s-1.2}[0-1]=\frac{1}{s-1.2} \qquad (9.8)$$

This is the Laplace transform of $e^{1.2t}$. □

The Laplace transform in (9.8) is, strictly speaking, defined only for Re $s>1.2$. The region Re $s>1.2$ or the region on the right-hand side of the dotted vertical line shown in Figure 9.3(b) is called the *region of convergence*. The region of convergence is important if we use the integration formula in (9.6) to compute the inverse Laplace transform of $X(s)=1/(s-1.2)$. For the example, if c is selected to be larger than 1.2, then the formula will yield $x(t)=e^{1.2}$, for $t\ge 0$ and $x(t)=0$, for $t<0$, a positive-time function. If c is selected incorrectly, the formula will yield a signal $x(t)=0$, for $t>0$ and $x(t)\ne 0$, for $t\le 0$, a negative-time signal. In conclusion, without specifying the region of convergence, (9.6) may not yield the original $x(t)$, for $t\ge 0$. In other words, the relationship between $x(t)$ and $X(s)$ is *not* one-to-one without specifying the region of convergence. Fortunately in practical application, we study only positive-time signals and consider the relationship between $x(t)$ and $X(s)$ to be *one-to-one*. Once we obtain a Laplace transform $X(s)$ from a positive-time signal, we automatically assume the inverse Laplace transform of $X(s)$ to be the original positive-time signal. Thus there is no need to consider the region of convergence and to use the inversion formula in (9.6). Indeed when we developed transfer functions in the preceding chapter, the region of convergence was not mentioned. Nor will it appear again in the remainder of this text. Moreover we will automatically set the value of $e^{(a-s)t}$, for any real or complex a, to zero at $t=\infty$.

Before proceeding, we mention that the two-sided Laplace transform of $x(t)$ is defined as

$$X_{II}(s):=\mathcal{L}_{II}[x(t)]:=\int_{t=-\infty}^\infty x(t)e^{-st}dt$$

For such transform, the region of convergence is essential because the same $X_{II}(s)$ may have many different two-sided $x(t)$. Thus its study is much more complicated than the (one-sided) Laplace transform introduced in (9.5). Fortunately, the two-sided Laplace transform is rarely, if not never, used in practice. Thus its discussion is omitted.

We compute in the following Laplace transforms of some positive-time functions. Their inverse Laplace transforms are the original positive-time functions. Thus they form one-to-one pairs.

Example 9.3.2 Find the Laplace transform of the impulse $\delta(t)$. By definition, we have

$$\Delta(s) = \int_0^\infty \delta(t)e^{-st}dt$$

Using the sifting property in (2.14), we have[1]

$$\Delta(s) = e^{-st}\big|_{t=0} = e^0 = 1$$

Thus we have $\mathcal{L}[\delta(t)] = 1$. $\square$

Example 9.3.3 Find the Laplace transform of $x(t) = e^{at}$, for $t \geq 0$, where a is a real or complex number. By definition, we have

$$
\begin{aligned}
X(s) &= \int_0^\infty e^{at}e^{-st}dt = \frac{1}{a-s}e^{(a-s)t}\Big|_{t=0}^\infty \\
&= \frac{1}{a-s}\left[e^{(a-s)t}\Big|_{t=\infty} - e^{(a-s)\cdot 0}\right] = \frac{1}{a-s}[0-1] = \frac{1}{s-a}
\end{aligned}
$$

where we have used $e^{(a-s)t} = 0$ at $t = \infty$. Thus we have

$$\mathcal{L}[e^{at}] = \frac{1}{s-a}$$

If $a = 0$, the function e^{at} reduces to 1, for all $t \geq 0$ and is a step function. Thus we have

$$\mathcal{L}[\text{step function}] = \mathcal{L}[1] = \frac{1}{s}$$

If $a = \sigma + j\omega_0$, then we have

$$\mathcal{L}[e^{(\sigma+j\omega_0)t}] = \frac{1}{s-\sigma-j\omega_0}$$

$\square$

Example 9.3.4 Using the result in the preceding example, we have

$$
\begin{aligned}
\mathcal{L}[e^{\sigma t}\sin\omega_0 t] &= \mathcal{L}\left[e^{\sigma t} \cdot \frac{e^{j\omega_0 t} - e^{-j\omega_0 t}}{2j}\right] = \frac{1}{2j}\left[\mathcal{L}[e^{(\sigma+j\omega_0)t}] - \mathcal{L}[e^{(\sigma-j\omega_0)t}]\right] \\
&= \frac{1}{2j}\left[\frac{1}{s-\sigma-j\omega_0} - \frac{1}{s-\sigma+j\omega_0}\right] \\
&= \frac{1}{2j} \cdot \frac{s-\sigma+j\omega_0 - (s-\sigma-j\omega_0)}{(s-\sigma)^2 - (j\omega_0)^2} \\
&= \frac{1}{2j} \cdot \frac{2j\omega_0}{(s-\sigma)^2 + (\omega_0)^2} = \frac{\omega_0}{(s-\sigma)^2 + \omega_0^2} \quad (9.9)
\end{aligned}
$$

---

[1]Recall from Section 2.5 that when an integration interval includes or touches an impulse, the impulse is assumed to be included entirely inside the integration interval.

which reduces to, for $\sigma = 0$,

$$\mathcal{L}[\sin \omega_0 t] = \frac{\omega_0}{s^2 + \omega_0^2}$$

Likewise, we have

$$\mathcal{L}[e^{\sigma t} \cos \omega_0 t] = \frac{s - \sigma}{(s - \sigma)^2 + \omega_0^2}$$

and, for $\sigma = 0$,

$$\mathcal{L}[\cos \omega_0 t] = \frac{s}{s^2 + \omega_0^2}$$

□

Before proceeding, we develop the formula

$$\mathcal{L}[tx(t)] = -\frac{dX(s)}{ds}$$

Indeed, by definition, we have

$$\begin{aligned}
\frac{d}{ds}X(s) &= \frac{d}{ds}\int_0^\infty x(t)e^{-st}dt = \int_0^\infty x(t)\left[\frac{d}{ds}e^{-st}\right]dt \\
&= \int_0^\infty (-t)x(t)e^{-st}dt
\end{aligned}$$

which is, by definition, the Laplace transform of $-tx(t)$. This establishes the formula.

Using the formula and $\mathcal{L}[e^{-at}] = 1/(s + a)$, we can establish

$$\begin{aligned}
\mathcal{L}[te^{-at}] &= -\frac{d}{ds}\left[\frac{1}{s + a}\right] = \frac{1}{(s + a)^2} \\
\mathcal{L}[t^2 e^{-at}] &= -\frac{d}{ds}\left[\frac{1}{(s + a)^2}\right] = \frac{2!}{(s + a)^3} \\
&\vdots \\
\mathcal{L}[t^k e^{-at}] &= \frac{k!}{(s + a)^{k+1}}
\end{aligned}$$

where $k! = 1 \cdot 2 \cdot 3 \cdots k$. We list the preceding Laplace transform pairs in Table 9.1. Note that all Laplace transforms in the table are strictly proper rational functions except the first one which is biproper.

9.3.1   Inverse Laplace transform

In this subsection, we discuss how to compute the time function of a Laplace transform. We use examples to discuss the procedure.

Example 9.3.5 Consider a system with transfer function

$$H(s) = \frac{s^2 - 10}{2s^2 - 4s - 6}$$

$x(t),\ t \geq 0$	$X(s)$
$\delta(t)$	1
1 or $q(t)$	$\frac{1}{s}$
t	$\frac{1}{s^2}$
t^k (k : positive integer)	$\frac{k!}{s^{k+1}}$
e^{-at} (a : real or complex)	$\frac{1}{s+a}$
$t^k e^{-at}$	$\frac{k!}{(s+a)^{k+1}}$
$\sin \omega_0 t$	$\frac{\omega_0}{s^2 + \omega_0^2}$
$\cos \omega_0 t$	$\frac{s}{s^2 + \omega_0^2}$
$t \sin \omega_0 t$	$\frac{2\omega_0 s}{(s^2 + \omega_0^2)^2}$
$t \cos \omega_0 t$	$\frac{s^2 - \omega_0^2}{(s^2 + \omega_0^2)^2}$
$e^{-at} \sin \omega_0 t$	$\frac{\omega_0}{(s+a)^2 + \omega_0^2}$
$e^{-at} \cos \omega_0 t$	$\frac{s+a}{(s+a)^2 + \omega_0^2}$

Table 9.1: Laplace transform pairs.

We compute its step response, that is, the output excited by the input $u(t) = 1$, for $t \geq 0$. The Laplace transform of the input is $U(s) = 1/s$. Thus the output in the transform domain is

$$Y(s) = H(s)U(s) = \frac{s^2 - 10}{2(s^2 - 2s - 3)} \cdot \frac{1}{s} =: \frac{N_y(s)}{D_y(s)}$$

To find the output in the time domain, we must compute the inverse Laplace transform of $Y(s)$. The procedure consists of two steps: expanding $Y(s)$ as a sum of terms whose inverse Laplace transforms are available in a table such as Table 9.1, and the table look-up. Before proceeding, we must compute the roots of the denominator $D_y(s)$ of $Y(s)$. We then expand $Y(s)$ as

$$Y(s) = \frac{s^2 - 10}{2(s+1)(s-3)s} = k + r_1 \frac{1}{s+1} + r_2 \frac{1}{s-3} + r_3 \frac{1}{s} \qquad (9.10)$$

This is called *partial fraction expansion*. We call k the *direct term*, and r_i, for $i = 1 : 3$, the *residues*. In other words, every residue is associated with a root of $D_y(s)$ and the direct term is not. Once all r_i and k are computed, then the inverse Laplace transform of $Y(s)$ is, using Table 9.1,

$$y(t) = k\delta(t) + r_1 e^{-t} + r_2 e^{3t} + r_3$$

for $t \geq 0$. Note that the inverse Laplace transform of $r_3/s = r_3/(s-0)$ is $r_3 e^{0 \times t} = r_3$, for all $t \geq 0$, which is a step function with amplitude r_3.

We next discuss the computation of k and r_i. They can be computed in many ways. We discuss in the following only the simplest. Equation (9.10) is an identity and holds for any s. If we select $s = \infty$, then the equation reduces to

$$Y(\infty) = 0 = k + r_1 \times 0 + r_2 \times 0 + r_3 \times 0$$

which implies

$$k = Y(\infty) = 0$$

Thus the direct term is simply the value of $Y(\infty)$. It is zero if $Y(z)$ is strictly proper.

Next we select $s = -1$, then we have

$$Y(-1) = -\infty = k + r_1 \times \infty + r_2 \times \frac{1}{-1-3} + r_3 \times \frac{1}{-1}$$

This equation contains ∞ on both sides of the equality and cannot be used to solve any r_i. However, multiplying (9.10) by $s + 1$ to yield

$$Y(s)(s+1) = \frac{s^2 - 10}{2(s-3)s} = k(s+1) + r_1 + r_2\frac{s+1}{s-3} + r_3\frac{s+1}{s}$$

and then substituting s by -1, we obtain

$$\begin{aligned} r_1 &= Y(s)(s+1)|_{s+1=0} = \left.\frac{s^2 - 10}{2(s-3)s}\right|_{s=-1} \\ &= \frac{(-1)^2 - 10}{2(-4)(-1)} = \frac{-9}{8} = -1.125 \end{aligned}$$

Using the same procedure, we can obtain

$$\begin{aligned} r_2 &= Y(s)(s-3)|_{s-3=0} = \left.\frac{s^2 - 10}{2(s+1)s}\right|_{s=3} \\ &= \frac{9 - 10}{2(4)(3)} = \frac{-1}{24} = -0.0417 \end{aligned}$$

and

$$\begin{aligned} r_3 &= X(s)s|_{s=0} = \left.\frac{s^2 - 10}{2(s+1)(s-3)}\right|_{s=0} \\ &= \frac{-10}{2(1)(-3)} = \frac{-10}{-6} = \frac{5}{3} = 1.6667 \end{aligned}$$

This completes solving k and r_i. Note the MATLAB function `residue` carries out partial fraction expansions. The numerator's coefficients of $Y(s)$ can be expressed in MATLAB as `ny=[1 0 -10]`; its denominator's coefficients can be expressed as `dy=[2 -4 -6 0]`. Typing in the control window

```
>> ny=[1 0 -10];dy=[2 -4 -6 0];
>> [r,p,k]=residue(ny,dy)
```

will yield `r=-0.0417  -1.125 1.6667`, `p=3  -1  0` and `k=[  ]`. They are the same as computed above. In conclusion, (9.10) can be expanded as

$$Y(s) = -1.125\frac{1}{s+1} - 0.0417\frac{1}{s-3} + 1.6667\frac{1}{s}$$

and its inverse Laplace transform is

$$y(t) = -1.125e^{-t} - 0.0417e^{-3t} + 1.6667$$

for $t \geq 0$. This is the step response of the system.□

Example 9.3.6 Consider a system with transfer function[2]

$$H(s) = \frac{-2s^2 - s - 16}{s^2 + 2s + 10} = \frac{-2s^2 - s - 16}{(s + 1 - j3)(s + 1 + j3)} \tag{9.11}$$

We compute its impulse response or the output excited by the input $u(t) = \delta(t)$. The Laplace transform of $\delta(t)$ is 1. Thus the impulse response in the transform domain is

$$Y(s) = H(s)U(s) = H(s) \times 1 = H(s)$$

which implies that the impulse response in the time domain is simply the inverse Laplace transform of the transfer function. This also follows directly from the definition that the transfer function is the Laplace transform of the impulse response.

The denominator of (9.11) has the pair of complex conjugate roots $-1 \pm j3$. Thus $H(s)$ can be factored as

$$H(s) = \frac{-2s^2 - s - 16}{(s + 1 - j3)(s + 1 + j3)} = k + \frac{r_1}{s + 1 - j3} + \frac{r_2}{s + 1 + j3}$$

Using the procedure in the preceding example, we can compute

$$k = H(\infty) = -2$$

and

$$r_1 = \frac{-2s^2 - s - 16}{s + 1 + j3}\bigg|_{s = -1 + j3} = 1.5 - j0.1667 = 1.51e^{-j0.11}$$

The computation involves complex numbers and is complicated. It is actually obtained using the MATLAB function `residue`. Because all coefficients of (9.11) are real numbers, the residue r_2 must equal the complex conjugate of r_1, that is, $r_2 = r_1^* = 1.51e^{j0.11}$. Thus $H(s)$ in (9.11) can be expanded in partial fraction expansion as

$$H(s) = -2 + \frac{1.51e^{-j0.11}}{s + 1 - j3} + \frac{1.51e^{j0.11}}{s + 1 + j3}$$

and its inverse Laplace transform is, using Table 9.1,

$$h(t) = -2\delta(t) + 1.51e^{-j0.11}e^{-(1-j3)t} + 1.51e^{j0.11}e^{-(1+j3)t}$$

for $t \geq 0$. It consists of two complex-valued functions. They can be combined to become a real-valued function as

$$\begin{aligned} h(t) &= -2\delta(t) + 1.51e^{-t}\left[e^{j(3t-0.11)} + e^{-j(3t-0.11)}\right] \\ &= -2\delta(t) + 1.51e^{-t} \times 2\cos(3t - 0.11) = -2\delta(t) + 3.02e^{-t}\cos(3t - 0.11) \end{aligned}$$

---

[2]The reader may glance through this example.

for $t \geq 0$. This is the impulse response used in (8.25). Because $h(t)$ has infinite duration, the system has infinite memory. $\square$

A great deal more can be said regarding the procedure. We will not discuss it further for the reasons to be given in the next subsection. Before proceeding, we mention that a system with a proper rational transfer function of degree 1 or higher has infinite memory. Indeed, the rational transfer function can be expressed as a sum of terms in Table 9.1. The inverse Laplace transform of each term is of infinite duration. For example, the transfer function of an integrator is $H(s) = 1/s$; its impulse response is $h(t) = 1$, for all $t \geq 0$. Thus the integrator has infinite memory.

9.3.2   Reasons for not using transfer functions in computing responses

Consider a system with transfer function $H(s)$. If we use $H(s)$ to compute the response of the system excited by some input, we must first compute the Laplace transform of the input and then multiply it by $H(s)$ to yield $Y(s) = H(s)U(s)$. The inverse Laplace transform of $Y(s)$ then yields the output $y(t)$ in the time domain. For some simple inputs, this approach will yield closed-form solutions of $y(t)$ and will be used in the next section to discuss general properties of systems. For computer computation, this approach is not used for the following reasons:

1. The Laplace transform of a signal $u(t)$ is defined for all t in $[0, \infty)$, thus we cannot compute $U(s)$ until the entire signal has been received. Thus the computation cannot be carried out in real time.

2. From Table 9.1 we see that the Laplace transform of a signal is a rational function of s only if the signal is a step, pure sinusoid, or more generally, a complex exponential and must last until $t = \infty$. For a real-world signal $u(t)$, its Laplace transform $U(s)$ is rarely a rational function of s. In this case, there is no simple analytical way of computing its time response.

3. Even if $Y(s)$ is a rational function of s, the procedure is not desirable from the computational point of view. The procedure requires computing the roots of the denominator of $Y(s)$. Such a computation is sensitive to parameter variations. For example, the roots of

$$D(s) = s^4 + 7s^3 + 18s^2 + 20s + 8$$

are $-1, -2, -2$, and -2. However, the roots of

$$\bar{D}(s) = s^4 + 7.001s^3 + 17.999s^2 + 20s + 8$$

are $-0.998, -2.2357$, and $-1.8837 \pm 0.1931j$. We see that the two coefficients of $D(s)$ change less than 0.1%, but all roots of $D(s)$ change their values greatly. Thus the roots of a polynomial is very sensitive to its coefficients. Any procedure that requires computing roots is not desirable in computer computation.

In conclusion, transfer functions are not used directly in computer computation. In contrast, computation using ss equations can be carried out in real time, does not involve any transformation, and involves only additions and multiplications. Thus ss equations are most suited for computer computation. Consequently computation involving transfer functions should be transformed into computation involving ss equations. This is the realization problem discussed in Section 8.9.

9.4 Step responses – Roles of poles and zeros

In this section we use step responses to discuss the roles of poles and zeros of transfer functions. In the transform domain, the output and input of a system with transfer function $H(s)$ are related by

$$Y(s) = H(s)U(s) \tag{9.12}$$

Although the equation is not used, as discussed in the preceding subsection, in computer computation of responses, we will use it to discuss general properties of systems and to develop concepts used in design. These can be achieved by considering only two types of inputs: step functions and pure sinusoids. There is no need to consider other inputs. The overshoot and response time discussed in Section 9.1.1 are in fact defined on step responses.

The step response, in the transform domain, of a system with transfer function $H(s)$ is $Y(s) = H(s)/s$. If $Y(s)$ has a simple pole at p_i, then its partial fraction expansion will contain the term

$$\frac{r_i}{s - p_i} = \mathcal{L}\left[r_i e^{p_i t}\right] \tag{9.13}$$

with

$$r_i = H(s)(s - p_i)\big|_{s=p_i} \tag{9.14}$$

as discussed in the preceding section. If $H(s)$ has a simple complex pole $\alpha + j\beta$, then it also has its complex conjugate $\alpha - j\beta$ as a pole because we study $H(s)$ only with real coefficients. Even though (9.13) still holds for a simple complex pole, it is simpler to expand $Y(s)$ to contain the term

$$\frac{r_1 s + r_2}{(s + \alpha)^2 + \beta^2} = \mathcal{L}\left[k_1 e^{-\alpha t} \sin(\beta t + k_2)\right] \tag{9.15}$$

where all six parameters are real numbers. Computing k_1 and k_2 from the other four numbers are generally complicated as shown in Example 9.3.6. Fortunately, we need only the form in our discussion. Next we use an example to discuss repeated poles.

Example 9.4.1 Consider a system with transfer function

$$H(s) = \frac{3s^2 - 10}{s^3 + 6s^2 + 12s^2 + 8} = \frac{3s^2 - 10}{(s + 2)^3} \tag{9.16}$$

The proper transfer function has a repeated pole at -2 with multiplicity 3. Its step response in the transform domain is given by

$$Y(s) = H(s)U(s) = \frac{3s^2 - 10}{(s + 2)^3}\frac{1}{s} \tag{9.17}$$

To find its time response, we expand it into terms whose inverse Laplace transforms are available in Table 9.1. The table contains

$$\mathcal{L}\left[e^{-at}\right] = \frac{1}{s+a} \qquad \mathcal{L}\left[te^{-at}\right] = \frac{1}{(s+a)^2} \qquad \mathcal{L}\left[t^2e^{-at}\right] = \frac{2}{(s+a)^3}$$

Thus if we expand (9.17) as

$$Y(s) = k_0 + r_1\frac{1}{s+2} + r_2\frac{1}{(s+2)^2} + r_3\frac{2}{(s+2)^3} + k_u\frac{1}{s} \qquad (9.18)$$

then its inverse Laplace transform is

$$y(t) = k_0\delta(t) + r_1e^{-2t} + r_2te^{-2t} + r_3t^2e^{-2t} + k_u \qquad (9.19)$$

for $t \geq 0$. Note that $k_0 = Y(\infty) = 0$ and k_u can be computed as, using (9.14),

$$k_u = Y(s)s|_{s=0} = H(s)\frac{1}{s} \times s\Big|_{s=0} = H(s)|_{s=0} = H(0) = \frac{-10}{2^3} = -1.25 \qquad (9.20)$$

But we cannot use (9.14) to compute r_1 because -2 is not a simple pole. There are other formulas for computing r_i, for $i = 1 : 3$. They will not be discussed because we are interested only in the form of the response.□

In conclusion, if $H(s)$ contains a repeated pole at p_i with multiplicity 3, then its step response $Y(s)$ will contain the terms

$$\frac{r_1s^2 + r_2s + r_3}{(s-p_i)^3} = \mathcal{L}\left[k_1e^{p_it} + k_2te^{p_it} + k_3t^2e^{p_it}\right] \qquad (9.21)$$

for some real r_i and k_i.

With the preceding discussion, we can now discuss the role of poles and zeros. Consider the following transfer functions

$$H_1(s) = \frac{4(s-3)^2(s+1+j2)(s+1-j2)}{(s+1)(s+2)^2(s+0.5+j4)(s+0.5-j4)} \qquad (9.22)$$

$$H_2(s) = \frac{-12(s-3)(s^2+2s+5)}{(s+1)(s+2)^2(s^2+s+16.25)} \qquad (9.23)$$

$$H_3(s) = \frac{-20(s^2-9)}{(s+1)(s+2)^2(s^2+s+16.25)} \qquad (9.24)$$

$$H_4(s) = \frac{60(s+3)}{(s+1)(s+2)^2(s^2+s+16.25)} \qquad (9.25)$$

$$H_5(s) = \frac{180}{(s+1)(s+2)^2(s^2+s+16.25)} \qquad (9.26)$$

They all have the same set of poles. In addition, they have $H_i(0) = 180/65 = 2.77$, for $i = 1 : 5$. Even though they have different sets of zeros, their step responses, in the transform domain, are all of the form

$$Y_i(s) = H_i(s)\frac{1}{s} = k_0 + \frac{k_1}{s+1} + \frac{r_2s + r_3}{(s+2)^2} + \frac{r_4s + r_5}{(s+0.5)^2 + 4^2} + \frac{k_u}{s}$$

with $k_0 = Y_i(\infty) = 0$ and

$$k_u = H_i(s)\frac{1}{s} \cdot s\bigg|_{s=0} = H_i(0) = 2.77$$

Thus their step responses in the time domain are all of the form

$$y_i(t) = k_1 e^{-t} + k_2 e^{-2t} + k_3 t e^{-2t} + k_4 e^{-0.5t} \sin(4t + k_5) + 2.77 \qquad (9.27)$$

for $t \geq 0$. This form is determined entirely by the poles of $H_i(s)$ and $U(s)$. The pole -1 generates the term $k_1 e^{-t}$. The repeated pole -2 generates the terms $k_2 e^{-2t}$ and $k_3 t e^{-2t}$. The pair of complex-conjugate poles at $-0.5 \pm j4$ generates the term $k_4 e^{-0.5t} \sin(4t + k_5)$. The input is a step function with Laplace transform $U(s) = 1/s$. Its pole at $s = 0$ generates a step function with amplitude $H_i(0) = 2.77$. We see that the zeros of $H_i(s)$ do not play any role in determining the form in (9.27). They affect only the coefficients k_i.

We plot the step responses of $H_i(s)$, for $i = 1 : 5$ in Figure 9.4. They are obtained in MATLAB by typing in an edit window the following

```
%Program 9.1 (f94.m)
d=[1 6 29.25 93.25 134 65];
n1=[4 -16 8 -48 180];n2=[-12 12 12 180];
n3=[-20 0 180];n4=[60 180];n5=180;
t=0:0.001:10;
y1=step(n1,d,t);y2=step(n2,d,t);y3=step(n3,d,t);
y4=step(n4,d,t);y5=step(n5,d,t);
plot(t,y1,t,y2,':',t,y3,'--',t,y4,'-.',t,y5)
```

It is then saved with file name f94.m. Typing in the command window >> f94 will yield Figure 9.4 in a figure window. Note that the responses are not computed directly from the transfer functions. They are computed in the time domain using their controllable-form realizations discussed in Section 8.9. We see that all responses approach 2.77 as $t \to \infty$, but the responses right after the application of the input are all different even though their step responses are all of the form in (9.27). This is due to different sets of k_i. In conclusion, poles dictate the general form of responses; zeros affect only the coefficients k_i. Thus we conclude that zeros play lesser role than poles in determining responses of systems.

9.4.1 Responses of poles as $t \to \infty$

In this subsection we discuss responses of poles. Before proceeding, it is convenient to divide the complex s-plane shown in Figure 9.2 into three parts: right-half plane (RHP), left-half plane (LHP), and imaginary- or $j\omega$-axis. In this definition, RHP and LHP do not include the $j\omega$-axis.

Let $\alpha + j\beta$ be a pole. The pole is inside the RHP if its real part is positive or $\alpha > 0$. It is inside the LHP if $\alpha < 0$, and on the $j\omega$-axis if $\alpha = 0$. Note that the imaginary part β does not play any role in this classification. Note also that the pole $\alpha + j\beta$

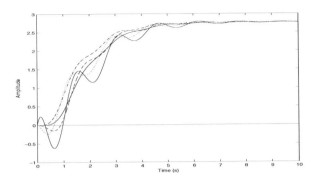

Figure 9.4: Step responses of the transfer functions in (9.22) through (9.26).

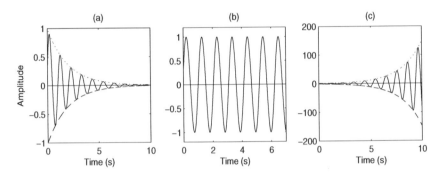

Figure 9.5: (a) $e^{-0.5t}\sin(6t+0.3)$. (b) $\sin(6t+0.3)$. (c) $e^{0.5t}\sin(6t+0.3)$

contributes the factor $(s-\alpha-j\beta)$, not $(s+\alpha+j\beta)$, into the denominator of a transfer function.

The time response of a real pole at α or the factor $1/(s-\alpha)$ is $e^{\alpha t}$. The response approaches 0 if $\alpha < 0$, a constant if $\alpha = 0$, and ∞ if $\alpha > 0$, as $t \to \infty$. If a transfer function has a complex pole $\alpha + j\beta$, then it also has $\alpha - j\beta$ as its pole. The pair of complex-conjugate poles $\alpha \pm j\beta$ will generate the time response

$$k_1 e^{\alpha t}\sin(\beta t + k_2)$$

for some real constants k_1 and k_2. Note that the real part α governs the *envelope* of the time function because

$$\left|k_1 e^{\alpha t}\sin(\beta t + k_2)\right| \le \left|k_1 e^{\alpha t}\right|$$

and its imaginary part β governs the *frequency* of oscillation as shown in Figure 9.5. If $\alpha < 0$ or the pair of complex-conjugate poles lies inside the LHP, then the time function vanishes oscillatorily as shown in Figure 9.5(a). If $\alpha = 0$ or the pair lies on the $j\omega$-axis, then the time response is a pure sinusoid, which will not vanish nor grow unbounded as shown in Figure 9.5(b). If $\alpha > 0$ or the pair lies inside the RHP, then the time function grows to $\pm\infty$ oscillatorily as shown in Figure 9.5(c).

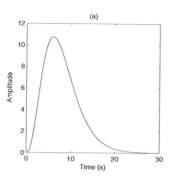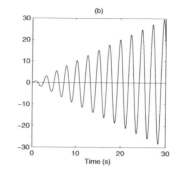

Figure 9.6: (a) $t^2 e^{-0.5t}$. (b) $t\sin(2.6t)$

We next discuss repeated poles. For repeated poles, we have

$$\mathcal{L}^{-1}\left[\frac{b_1 s + b_0}{(s-\alpha)^2}\right] = k_1 e^{\alpha t} + k_2 t e^{\alpha t}$$

$$\mathcal{L}^{-1}\left[\frac{b_3 s^3 + b_2 s^2 + b_1 s + b_0}{[(s-\alpha)^2 + \omega_0^2]^2}\right]$$

$$= k_1 c^{\alpha t}\sin(\omega_0 t + k_2) + k_3 t e^{\alpha t}\sin(\omega_0 t + k_4)$$

If $\alpha > 0$, both $e^{\alpha t}$ and $t e^{\alpha t}$ approach ∞ as $t \to \infty$. Thus if a pole, simple or repeated, real or complex, lying inside the RHP, then its time function approaches infinity as $t \to \infty$.

If $\alpha < 0$ such as $\alpha = -0.5$, then $\mathcal{L}^{-1}[1/(s+0.5)^2] = te^{-0.5t}$ is a product of ∞ and zero as $t \to \infty$. To find its value, we must use l'Hôpital's rule as

$$\lim_{t\to\infty} te^{-0.5t} = \lim_{t\to\infty}\frac{t}{e^{0.5t}} = \lim_{t\to\infty}\frac{1}{0.5e^{0.5t}} = 0$$

Using the same procedure, we can shown $\mathcal{L}^{-1}[1/(s+0.5)^3] = 0.5t^2 e^{-0.5t} \to 0$ as $t \to \infty$. Thus we conclude that the time function of a pole lying inside the LHP, simple or repeated, real or complex, will approach zero as $t \to \infty$. This is indeed the case as shown in Figure 9.6(a) for $t^2 e^{-0.5t}$.[3]

The situation for poles with $\alpha = 0$, or pure imaginary poles, is more complex. The time function of a simple pole at $s = 0$ is a constant for all $t \geq 0$; it will not grow unbounded nor approach zero as $t \to \infty$. The time response of a simple pair of pure-imaginary poles is $k_2 \sin(\beta t + k_2)$ which is a sustained oscillation for all $t \geq 0$ as shown in Figure 9.5(b). However the time function of a repeated pole at $s = 0$ is t which approaches ∞ as $t \to \infty$. The time response of a repeated pair of pure-imaginary poles contains $k_3 t \sin(\beta t + k_4)$ which approaches ∞ or $-\infty$ oscillatorily as $t \to \infty$ as shown in Figure 9.6(b). In conclusion, we have

---

[3]Note that the exponential function $e^{\alpha t}$, for any negative real α, approaches 0 much faster than that t^k, for any negative integer k, approaches 0 as $t \to \infty$. Likewise, the exponential function $e^{\alpha t}$, for any positive real α, approaches ∞ much faster than that t^k, for any positive integer k, approaches ∞ as $t \to \infty$. Thus we have $t^{10}e^{-0.0001t} \to 0$ as $t \to \infty$ and $t^{-10}e^{0.0001t} \to \infty$ as $t \to \infty$. In other words, real exponential functions dominate over polynomial functions as $t \to \infty$.

Pole Location	Response as $t \to \infty$
LHP, simple or repeated	0
RHP, simple or repeated	∞ or $-\infty$
$j\omega$-axis, simple	constant or sustained oscillation
$j\omega$-axis, repeated	∞ or $-\infty$

We see that the response of a pole approaches 0 as $t \to \infty$ *if and only if* the pole lies inside the LHP.

9.5   Stability

This section introduces the concept of stability for systems. In general, if a system is not stable, it may burn out or saturate (for an electrical system), disintegrate (for a mechanical system), or overflow (for a computer program). Thus every system designed to process signals is required to be stable. Let us give a formal definition.

Definition 9.1 A system is BIBO (bounded-input bounded-output) stable or, simply, stable if *every* bounded input excites a bounded output. Otherwise, the system is said to be unstable.[4] $\square$

A signal is bounded if it does not grow to ∞ or $-\infty$. In other words, a signal $u(t)$ is bounded if there exists a constant M_1 such that $|u(t)| \leq M_1 < \infty$ for all $t \geq 0$. We first give an example to illustrate the concept.[5]

Example 9.5.1 Consider the circuit shown in Figure 9.7(a). The input $u(t)$ is a current source; the output $y(t)$ is the voltage across the capacitor. The impedances of the inductor and capacitor are respectively s and $1/4s$. The impedance of their parallel connection is $s(1/4s)/(s + 1/4s) = s/(4s^2 + 1) = 0.25s/(s^2 + 0.25)$. Thus the input and output of the circuit are related by

$$Y(s) = \frac{0.25s}{s^2 + 0.25}U(s) = \frac{0.25s}{s^2 + (0.5)^2}U(s)$$

If we apply a step input $(u(t) = 1$, for $t \geq 0)$, then its output is

$$Y(s) = \frac{0.25s}{s^2 + (0.5)^2}\frac{1}{s} = \frac{0.25}{s^2 + (0.5)^2}$$

---

[4]Some texts define a system to be stable if every pole of its transfer function has a negative real part. That definition is applicable only to LTI lumped systems. Definition 9.1 is applicable to LTI lumped and distributed systems and is more widely adopted.

[5]Before the application of an input, the system is required to be initially relaxed. Thus the BIBO stability is defined for forced or zero-state responses. Some texts introduce the concept of stability using a pendulum and an inverted pendulum or a bowl and an inverted bowl. Strictly speaking, such stability is defined for natural or zero-input responses and should be defined for an equilibrium state. Such type of stability is called *asymptotic stability* or *marginal stability*. See Reference [C6]. A pendulum is marginally stable if there is no air friction, and is asymptotically stable if there is friction. An inverted pendulum is neither marginally nor asymptotically stable.

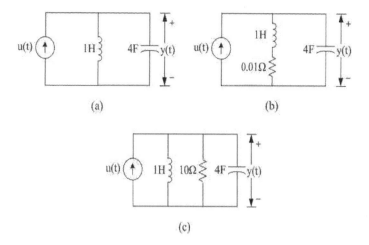

Figure 9.7: (a) Unstable circuit. (b) and (c) Stable circuits.

which implies $y(t) = 0.5 \sin 0.5t$. This output is bounded. If we apply $u(t) = \sin 3t$, then the output is

$$Y(s) = \frac{0.25s}{s^2 + (0.5)^2} \frac{3}{s^2 + 9}$$

which implies

$$y(t) = k_1 \sin(0.5t + k_2) + k_3 \sin(3t + k_4)$$

for some constants k_i. This output is bounded. Thus the outputs excited by the bounded inputs $u(t) = 1$ and $u(t) = \sin 3t$ are bounded. Even so, we cannot conclude that the circuit is BIBO stable because we have not yet checked all possible bounded inputs.

Now let us apply the bounded input $u(t) = \sin 0.5t$. Then the output is

$$Y(s) = \frac{0.25s}{s^2 + 0.25} \frac{0.5}{s^2 + 0.25} = \frac{0.125s}{(s^2 + 0.25)^2}$$

Now because $Y(s)$ has a repeated imaginary poles at $\pm 0.5j$, its time response is of the form

$$y(t) = k_1 \sin(0.5t + k_2) + k_3 t \sin(0.5t + k_4)$$

for some constants k_i and with $k_3 \neq 0$. The response grows unbounded. In other words, the bounded input $u(t) = \sin 0.5t$ excites an unbounded output. Thus the circuit is not BIBO stable.[6] □

Example 9.5.2 Consider the op-amp circuits in Figures 6.9(a) and (b). Because $v_o(t) = v_i(t)$, if an input $v_i(t)$ is bounded, so is the output. Thus the two circuits are BIBO stable. Note that they are stable based on the memoryless model of the op amp. If the op amp is modeled to have memory, then the former is stable and the latter is not as we will discuss in the next chapter. □

---

[6]The circuit is marginally stable. See the previous footnote.

Every LTI memoryless system can be described by $y(t) = \alpha u(t)$, for some finite α. Clearly if $u(t)$ is bounded, so is $y(t)$. Thus such a system is always stable. Other than memoryless systems, we cannot use Definition 9.1 to check the stability of a system because there are infinitely many bounded inputs to be checked. Fortunately, stability is a property of a system and can be determined from its mathematical descriptions.

Theorem 9.1 An LTI system with impulse response $h(t)$ is BIBO stable if and only if $h(t)$ is absolutely integrable in $[0, \infty)$; that is,

$$\int_0^\infty |h(t)|dt \leq M < \infty \tag{9.28}$$

for some constant M. $\square$

Proof Recall that we study only causal systems and all impulse responses have the property $h(t) = 0$, for $t < 0$. The input $u(t)$ and output $y(t)$ of any LTI system which is initially relaxed at $t = 0$ can be related by (8.5) or

$$y(t) = \int_{\tau=0}^\infty h(t - \tau)u(\tau)d\tau \tag{9.29}$$

We use this equation to prove the theorem.

We first show that the system is BIBO stable under the condition in (9.28). Indeed, if $u(t)$ is bounded or $|u(t)| \leq M_1$ for all $t \geq 0$, then we have

$$
\begin{aligned}
|y(t)| &= \left| \int_{\tau=0}^\infty h(t - \tau)u(\tau)d\tau \right| \leq \int_{\tau=0}^\infty |h(t - \tau)||u(\tau)|d\tau \\
&\leq M_1 \int_{\tau=0}^\infty |h(t - \tau)|d\tau
\end{aligned}
\tag{9.30}
$$

The integrand $h(t - \tau)u(\tau)$ in the first integration in (9.30) may be positive or negative and its integration may cancel out each other. This will not happen in the second integration because its integrand $|h(t - \tau)||u(\tau)|$ is positive for all t. Thus we have the first inequality in (9.30). The second inequality follows from $|u(t)| \leq M_1$. Let us introduce a new variable $\bar{\tau} := t - \tau$, where t is fixed. Then we have $d\bar{\tau} = -d\tau$ and (9.30) becomes

$$
\begin{aligned}
|y(t)| &\leq -M_1 \int_{\bar{\tau}=t}^{-\infty} |h(\bar{\tau})|d\bar{\tau} = M_1 \int_{\bar{\tau}=-\infty}^t |h(\bar{\tau})|d\bar{\tau} \\
&= M_1 \int_{\bar{\tau}=0}^t |h(\bar{\tau})|d\bar{\tau} \leq M_1 \int_{\bar{\tau}=0}^\infty |h(\bar{\tau})|d\bar{\tau} = M_1 M
\end{aligned}
\tag{9.31}
$$

where we have used the causality condition $h(t) = 0$, for all $t < 0$. Because (9.31) holds for all $t \geq 0$, the output is bounded. This show that under the condition in (9.28), every bounded input will excite a bounded output. Thus the system is BIBO stable.

Next we show that if $h(t)$ is not absolutely integrable, then there exists a bounded input that will excite an output which will approach ∞ or $-\infty$ as $t \to \infty$. Because ∞ is not a number, the way to show $|y(t)| \to \infty$ is to show that no matter how large M_2

is, there exists a t_1 with $y(t_1) > M_2$. If $h(t)$ is not absolutely integrable, then for any arbitrarily large M_2, there exists a t_1 such that

$$\int_{t-0}^{t_1} |h(t)|dt = \int_{\bar{\tau}=0}^{t_1} |h(\bar{\tau})|d\bar{\tau} \geq M_2$$

Let us select an input as, for all t in $[0, t_1]$,

$$u(t) = \left\{ \begin{array}{ll} 1 & \text{if } h(t_1 - t) \geq 0 \\ -1 & \text{if } h(t_1 - t) < 0 \end{array} \right.$$

For this bounded input, the output $y(t)$ at $t = t_1$ is

$$\begin{aligned} y(t_1) &= \int_{\tau=0}^{\infty} h(t_1 - \tau)u(\tau)d\tau = \int_{\tau=0}^{\infty} |h(t_1 - \tau)|d\tau = -\int_{\bar{\tau}=t_1}^{-\infty} |h(\bar{\tau})|d\bar{\tau} \\ &= \int_{\bar{\tau}=-\infty}^{t_1} |h(\bar{\tau})|d\bar{\tau} = \int_{\bar{\tau}=0}^{t_1} |h(\bar{\tau})|d\bar{\tau} \geq M_2 \end{aligned}$$

where we have introduced $\bar{\tau} = t_1 - \tau$, $d\bar{\tau} = -d\tau$, and used the causality condition. This shows that if $h(t)$ is not absolutely integrable, then there exists a bounded input that will excite an output with an arbitrarily large magnitude. Thus the system is not stable. This establishes the theorem.□

Example 9.5.3 Consider the circuit shown in Figure 9.7(a). Its transfer function was computed in Example 9.5.1 as $H(s) = 0.25s/(s^2 + 0.5^2)$. Its inverse Laplace transform or the impulse response of the circuit is, using Table 9.1, $h(t) = 0.25\cos 0.5t$. Although

$$\int_{t=0}^{\infty} 0.25\cos 0.5t dt$$

is finite because the positive and negative areas will cancel out, we have

$$\int_{t=0}^{\infty} |0.25\cos 0.5t| \, dt = \infty$$

See the discussion regarding Figure 3.3. Thus the circuit is not stable according to Theorem 9.1.□

Theorem 9.1 is applicable to every LTI lumped or distributed system. However it is not used in practice because impulse responses of systems are generally not available. We will use it to develop the next theorem.

Theorem 9.2 An LTI lumped system with proper rational transfer function $H(s)$ is stable if and only if every pole of $H(s)$ has a negative real part or, equivalently, all poles of $H(s)$ lie inside the left half s-plane.□

We discuss only the basic idea in developing Theorem 9.2 from Theorem 9.1. The impulse response of a system is the inverse Laplace transform of its transfer function. If the system is lumped, its transfer function $H(s)$ is a rational function.

If $H(s)$ has a pole inside the right half s-plane (RHP), then its impulse response grows to infinity and cannot be absolutely integrable. If $H(s)$ has simple poles on the $j\omega$-axis, then its impulse response is a step function or a sinusoid. Both are not absolutely integrable. In conclusion, if $H(s)$ has one or more poles on the $j\omega$-axis or lying inside the RHP, then the system is not stable.

Next we argue that if all poles lying inside the left half s-plane (LHP), then the system is stable. To simplify the discussion, we assume that $H(s)$ has only simple poles and can be expanded as

$$H(s) = \sum_i \frac{r_i}{s - p_i} + k_0$$

where the poles p_i can be real or complex. Then the impulse response of the system is

$$h(t) = \sum_i r_i e^{p_i t} + k_0 \delta(t) \qquad (9.32)$$

for $t \geq 0$. We compute

$$\int_0^\infty |h(t)| dt = \int_0^\infty \left| \sum_i r_i e^{p_i t} + k_0 \delta(t) \right| dt \leq \sum_i \int_0^\infty |r_i e^{p_i t}| dt + k_0 \qquad (9.33)$$

If $p_i = \alpha_i + j\beta_i$ and if $\alpha_i < 0$, then

$$\int_0^\infty \left| r_i e^{(\alpha_i + j\beta_i)t} \right| dt = \int_0^\infty \left| r_1 e^{\alpha_i t} \right| dt = \frac{|r_i|}{\alpha_i} \left. e^{\alpha_i t} \right|_{t=0}^\infty = \frac{|r_i|}{\alpha_i}(0 - 1) = \frac{|r_i|}{-\alpha_i} =: M_i < \infty$$

where we have used $\left| e^{j\beta_i t} \right| = 1$, for every β_i and for all t and $e^{\alpha_i t} = 0$ at $t = \infty$ for $\alpha_i < 0$. Thus if every pole of $H(s)$ has a negative real part, then every term in (9.33) is absolutely integrable. Consequently $h(t)$ is absolutely integrable and the system is BIBO stable. This establishes Theorem 9.2. Note that the stability of a system depends only on the poles. A system can be stable with its zeros anywhere (inside the RHP or LHP, or on the $j\omega$-axis).

We study in this text only LTI lumped systems, thus we use mostly Theorem 9.2 to check stability. For example, the circuit in Figure 9.7(a) has, as discussed in Example 9.5.1, transfer function $0.25s/(s^2 + 0.25)$ which has poles at $\pm j0.5$. The poles have zero real part, thus the system is not BIBO stable. An amplifier with a finite gain α is always stable following from Definition 9.1. It also follows from Theorem 9.2 because its transfer function is α which has no pole and no condition to be met.

Corollary 9.2 An LTI lumped system with impulse response $h(t)$ is BIBO stable if and only if $h(t) \to 0$ as $t \to \infty$. $\square$

The impulse response is the inverse Laplace transform of $H(s)$. As discussed in Subsection 9.4.1, the response of a pole approaches zero as $t \to \infty$ if and only if the pole lies inside the LHP. Now if all poles of $H(s)$ lie inside the LHP, then all its inverse Laplace transforms approach zero as $t \to \infty$. Thus we have the corollary. This corollary will be used to check stability by measurement.

9.5.1 What holds for lumped systems may not hold for distributed systems

We mention that the results for LTI lumped systems may not be applicable to LTI distributed systems. For example, consider an LTI system with impulse response

$$h(t) = \frac{1}{t+1} \tag{9.34}$$

for $t \geq 0$ and $h(t) = 0$ for $t < 0$. Note that its Laplace transform can be shown to be an irrational function of s. Thus the system is distributed. Even though $h(t)$ approaches zero as $t \to \infty$, it is not absolutely integrable because

$$\int_0^\infty |h(t)|dt = \int_0^\infty \frac{1}{t+1}dt = \left. \log(t+1) \right|_0^\infty = \infty$$

Thus the system is not BIBO stable and Corollary 9.2 cannot be used.

We give a different example. Consider an LTI system with impulse response

$$h_1(t) = \sin(t^2/2)$$

for $t \geq 0$ and $h_1(t) = 0$ for $t < 0$. Its Laplace transform is not a rational function of s and the system is distributed. Computing its Laplace transform is difficult, but it is shown in Reference [K5] that its Laplace transform meets some differential equation and is analytic (has no singular point or pole) inside the RHP and on the imaginary axis. Thus the system is BIBO stable if Theorem 9.2 *were* applicable. The system is actually not BIBO stable according to Theorem 9.1 because $h_1(t)$ is not absolutely integrable. In conclusion, Theorem 9.2 and its corollary are not applicable to LTI distributed systems. Thus care must be exercised in applying theorems; we must check every condition or word in a theorem.

9.5.2 Stability check by one measurement

The stability of an LTI lumped system can be checked from its rational transfer function. It turns out that it can also be easily checked by one measurement. This can be deduced from Corollary 9.2 and the following theorem.

Theorem 9.3 An LTI lumped system with proper rational transfer function $H(s)$ is stable if and only if its step response approaches a zero or nonzero constant as $t \to \infty$.

If $u(t) = 1$, for $t \geq 0$, then we have $U(s) = 1/s$ and

$$Y(s) = H(s)U(s) = H(s)\frac{1}{s} = \frac{H(0)}{s} + \text{terms due to all poles of } H(s) \tag{9.35}$$

Note that the residue associated with every pole of $H(s)$ will be nonzero because $U(s) = 1/s$ has no zero to cancel any pole of $H(s)$. In other words, the step input will excite every pole of $H(s)$. See Problem 9.7. The inverse Laplace transform of (9.35) is of the form

$$y(t) = H(0) + \text{linear combination of time responses of all poles of } H(s)$$

for $t \geq 0$. If $H(s)$ is stable, then the time responses of all its poles approach zero as $t \to \infty$. Thus we have $y(t) \to H(0)$ as $t \to \infty$.

Next we consider the case where $H(s)$ is not stable. If $H(s)$ has a pole at $s = 0$, then $Y(s) = H(s)U(s) = H(s)/s$ has a repeated pole at $s = 0$ and its step response will approach ∞ as $t \to \infty$. If $H(s)$ contains a pair of pure imaginary poles at $\pm j\omega_0$, then its step response contains $k_1 \sin(\omega_0 t + k_2)$ which does not approach a constant as $t \to \infty$. If $H(s)$ has one or more poles in the RHP, then its step response will approach ∞ or $-\infty$. In conclusion, if $H(s)$ is not stable, then its step response will not approach a constant as $t \to \infty$; it will grow unbounded or remain in oscillation. This establishes the theorem.$\square$

If a system is known to be linear, time-invariant and lumped, then its stability can be easily checked by measurement. If its step response grows unbounded, saturates, or remains in oscillation, then the system is not stable. In fact, we can apply any input for a short period of time to excite all poles of the system. We then remove the input. If the response does not approach zero as $t \to \infty$, then the system is not stable. If a system is not stable, it must be redesigned.

Resistors with positive resistances, inductors with positive inductances, and capacitors with positive capacitances are called *passive elements*. Any circuit built with these three types of elements is an LTI lumped system. Resistors dissipate energy. Although inductors and capacitors can store energy, they cannot generate energy. Thus when an input is removed from any RLC network, the energy stored in the inductors and capacitors will eventually dissipate in the resistors and consequently the response eventually approaches zero. Note that the LC circuit in Figure 9.7(a) is a model. In reality, every physical inductor has a small series resistance as shown in Figure 9.7(b), and every physical capacitor has a large parallel resistance as shown in Figure 9.7(c). Thus all practical RLC circuits are stable, and their stability study is unnecessary. For this reason, before the advent of active elements such as op amps, stability is not studied in RLC circuits. It is studied only in designing feedback systems. This practice appears to remain to this date. See, for example, References [H3, I1, S2].

As a final remark, we mention that systems are models of physical systems. If a model is properly chosen, then a physical system is stable if its model (system) is stable. On the other hand, if a model is not selected correctly, then a physical system may not be stable even if its model is stable as discussed in Section 6.7.2. Fortunately the stability of a physical system can be easily detected. If an actual op-amp circuit jumps to a saturation region when a very small input is applied, the circuit is not stable and must be redesigned. If an actual mechanical system starts to vibrate, we must stop immediately its operation and redesign the physical system.

9.5.3 The Routh test

Consider a system with proper rational transfer function $H(s) = N(s)/D(s)$. We assume that $N(s)$ and $D(s)$ are coprime or have no common roots. Then the poles of $H(s)$ are the roots of $D(s)$. If $D(s)$ has degree three or higher, computing its roots by hand is not simple. This, however, can be carried out using a computer. For example, consider

$$D(s) = 2s^5 + s^4 + 7s^3 + 3s^2 + 4s + 2 \qquad (9.36)$$

$$\begin{array}{c|cccc}
 & s^6 & a_0 & a_2 & a_4 & a_6 \\
k_1 = a_0/a_1 & s^5 & a_1 & a_3 & a_5 & \\
k_2 = a_1/b_1 & s^4 & b_1 & b_2 & b_3 & \\
k_3 = b_1/c_1 & s^3 & c_1 & c_2 & & \\
k_4 = c_1/d_1 & s^2 & d_1 & d_2 & & \\
k_5 = d_1/e_1 & s & e_1 & & & \\
 & s^0 & f_1 & & &
\end{array}$$

$[b_0\ b_1\ b_2\ b_3] = $ (1st row)-k_1(2nd row)
$[c_0\ c_1\ c_2] = $ (2nd row)-k_2(3rd row)
$[d_0\ d_1\ d_2] = $ (3rd row)-k_3(4th row)
$[e_0\ e_1] = $ (4th row)-k_4(5th row)
$[f_0\ f_1] = $ (5th row)-k_5(6th row)

Table 9.2: The Routh Table

Typing in the command window of MATLAB

```
>> d=[2 1 7 3 4 2];roots(d)
```

will yield $-0.0581 \pm j1.6547, 0.0489 \pm j0.8706, -0.4799$. It has one real root and two pairs of complex conjugate roots. One pair has a positive real part 0.0489. Thus any transfer function with (9.36) as its denominator is not BIBO stable.

A polynomial is defined to be *CT stable* if all its roots have negative real parts. We discuss a method of checking whether or not a polynomial is CT stable without computing its roots. We use the following $D(s)$ to illustrate the procedure:

$$D(s) = a_0 s^6 + a_1 s^5 + a_2 s^4 + a_3 s^3 + a_4 s^2 + a_5 s + a_6 \quad \text{with } a_0 > 0 \qquad (9.37)$$

We require the leading coefficient a_0 to be positive. If $a_0 < 0$, we apply the procedure to $-D(s)$. Because $D(s)$ and $-D(s)$ have the same set of roots, if $-D(s)$ is CT stable, so is $D(s)$. The polynomial in (9.37) has degree 6 and seven coefficients a_i, $i = 0 : 6$. We use the coefficients to form the first two rows of the table, called the Routh table, in Table 9.2. They are placed, starting from a_0, alternatively in the first and second rows. Next we compute $k_1 = a_0/a_1$, the ratio of the first entries of the first two rows. We then subtract from the first row the product of the second row and k_1. The result $[b_0\ b_1\ b_2\ b_3]$ is placed at the right-hand side of the second row, where

$$b_0 = a_0 - k_1 a_1 = 0 \quad b_1 = a_2 - k_1 a_3 \quad b_2 = a_4 - k_1 a_5 \quad b_3 = a_6 - k_1 \cdot 0 = a_6$$

Note that b_0 is always zero because of $k_1 = a_0/a_1$. We discard b_0 and place $[b_1\ b_2\ b_3]$ in the third row as shown in the table. The fourth row is obtained using the same procedure from its previous two rows. That is, we compute $k_2 = a_1/b_1$, the ratio of the first entries of the second and third rows. We subtract from the second row the product of the third row and k_2. The result $[c_0\ c_1\ c_2]$ is placed at the right-hand side of the third row. We discard $c_0 = 0$ and place $[c_1\ c_2]$ in the fourth row as shown. We repeat the process until the row corresponding to $s^0 = 1$ is computed. If the degree of $D(s)$ is n, the table contains $n + 1$ rows.[7]

We discuss the size of the Routh table. If the degree n of $D(s)$ is even, the first row has one more entry than the second row. If n is odd, the first two rows have the

---

[7]This formulation of the Routh table is different from the conventional cross-product formulation. This formulation might be easier to be grasped and is more in line with its DT counterpart.

same number of entries. In both cases, the number of entries decreases by one at odd powers of s. For example, the numbers of entries in the rows of s^5, s^3, s decrease by one from their previous rows. The last entries of all rows corresponding to even powers of s are the same. For example, we have $a_6 = b_3 = d_2 = f_1$ in Table 9.2.

Theorem 9.4 A polynomial with a positive leading coefficient is CT stable if and only if every entry in the Routh table is positive. If a zero or a negative number appears in the table, then $D(s)$ is not a CT stable polynomial.□

A proof of this theorem can be found in Reference [2nd ed. of C6]. We discuss here only its employment. A necessary condition for $D(s)$ to be CT stable is that all its coefficients are positive. If $D(s)$ has missing terms (coefficients are zero) or negative coefficients, then the first two rows of its Routh table will contain 0's or negative numbers. Thus it is not a CT stable polynomial. For example, the polynomials

$$D(s) = s^4 + 2s^2 + 3s + 10$$

and

$$D(s) = s^4 + 3s^3 - s^2 + 2s + 1$$

are not CT stable. On the other hand, a polynomial with all positive coefficients may not be CT stable. For example, the polynomial in (9.36) with all positive coefficients is not CT stable as discussed earlier. We can also verify this by applying the Routh test. We use the coefficients of

$$D(s) = 2s^5 + s^4 + 7s^3 + 3s^2 + 4s + 2$$

to form

$$
\begin{array}{r|ccc}
 & s^5 & 2 & 7 & 4 \\
k_1 = 2/1 & s^4 & 1 & 3 & 2 \qquad [0 \ \ 1 \ \ 0] \\
 & s^3 & 1 & 0 \\
\end{array}
$$

A zero appears in the table. Thus we conclude that the polynomial is not CT stable. There is no need to complete the table.

Example 9.5.3 Consider the polynomial

$$D(s) = 4s^5 + 2s^4 + 14s^3 + 6s^2 + 8s + 3$$

We form

$$
\begin{array}{r|ccc}
 & s^5 & 4 & 14 & 8 \\
k_1 = 2/1 & s^4 & 2 & 6 & 3 \qquad [0 \ \ 2 \ \ 2] \\
k_2 = 2/2 & s^3 & 2 & 2 & \qquad [0 \ \ 4 \ \ 3] \\
k_3 = 2/4 & s^2 & 4 & 3 & \qquad [0 \ \ 0.5] \\
k_4 = 4/0.5 & s^1 & 0.5 & & \qquad [0 \ \ 3] \\
 & s^0 & 3 \\
\end{array}
$$

Every entry in the table is positive, thus the polynomial is CT stable.□

9.6 Steady-state and transient responses

In this section, we introduce the concepts of steady-state and transient responses. Consider a system with transfer function $H(s)$. Let $y(t)$ be the output excited by some input $u(t)$. We call

$$y_{ss}(t) = \lim_{t \to \infty} y(t) \tag{9.38}$$

the *stead-state (ss)* response and

$$y_{tr}(t) = y(t) - y_{ss}(t) \tag{9.39}$$

the *transient (tr)* response of the system excited by the input $u(t)$. We first give examples.

Example 9.6.1 Consider

$$H(s) = \frac{6}{(s+3)(s-1)} \tag{9.40}$$

Its step response in the transform domain is

$$Y(s) = H(s)U(s) - \frac{6}{(s+3)(s-1)s} = \frac{6/12}{s+3} + \frac{6/4}{s-1} - \frac{6/3}{s}$$

Thus its step response is

$$y(t) = 0.5e^{-3t} + 1.5e^{t} - 2 \tag{9.41}$$

for $t \geq 0$.

Because e^{-3t} approaches zero as $t \to \infty$, the steady-state response of (9.41) is

$$y_{ss}(t) = 1.5e^{t} - 2$$

and the transient response is

$$y_{tr}(t) = 0.5e^{-3t}$$

for $t \geq 0$. We see that the steady-state response consists of the time functions due to the pole of the input and a pole of the system. Because the pole $s = 1$ is in the RHP, we have $y_{ss}(t) \to \infty$ as $t \to \infty$. Note that $H(s)$ in (9.40) is not stable, and its steady-state response generally approaches infinite no matter what input is applied. $\square$

Example 9.6.2 Consider a system with transfer function

$$H(s) = \frac{24}{s^3 + 3s^2 + 8s + 12} = \frac{24}{(s+2)(s+0.5+j2.4)(s+0.5-j2.4)} \tag{9.42}$$

Its step response in the transform domain is

$$Y(s) = H(s)U(s) = \frac{24}{(s+2)(s+0.5+j2.4)(s+0.5-j2.4)} \cdot \frac{1}{s}$$

To find its time response, we expand it as

$$Y(s) = \frac{k_1}{s+2} + \frac{r_1 s + r_2}{(s+0.5)^2 + 2.4^2} + \frac{k_4}{s}$$

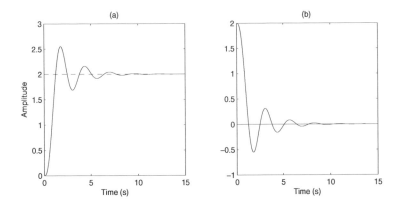

Figure 9.8: (a) Step response of (9.42). (b) Its transient response.

with

$$k_4 = Y(s)s|_{s=0} = H(0) = 2$$

Thus the step response of the system is of the form

$$y(t) = H(0) + k_1 e^{-2t} + k_2 e^{-0.5t} \sin(2.4t + k_3) \tag{9.43}$$

for $t \geq 0$ and is plotted in Figure 9.8(a).

Because e^{-2t} and $e^{-0.5t}$ approach zero as $t \to \infty$, the steady-state response of (9.43) is

$$y_{ss}(t) = H(0) \tag{9.44}$$

and the transient response is

$$y_{tr}(t) = k_1 e^{-2t} + \bar{k}_2 e^{-0.5t} \sin(2.4t + \bar{k}_3) \tag{9.45}$$

We see that the steady-state response is determined by the pole of the input alone and the transient response is due to all poles of the transfer function in (9.42). We plot in Figure 9.8(b) $-y_{tr}(t)$ for easier viewing. □

Systems are designed to process signals. The most common processions are amplification and filtering. The processed signals or the outputs of the systems clearly should be dictated by the signals to be processed. If a system is not stable, its output will generally grow unbounded. Clearly such a system cannot be used. On the other hand, if a system is stable, after its transient response dies out, the processed signal is dictated by the applied signal. Thus the transient response is an important criterion in designing a system. The faster the transient response vanishes, the faster the response reaches steady state.

9.6.1   Time constant and Response time of stable systems

Consider the transfer function in (9.42). Its step response approaches the steady state $y_{ss}(t) = H(0) = 2$ as shown in Figure 9.8(a) and its transient response approaches zero

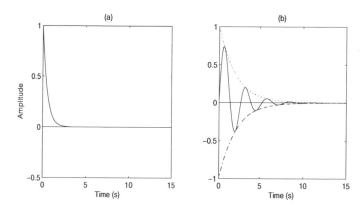

Figure 9.9: (a) The function e^{-2t}. (b) The function $e^{-0.5t}\sin(2.4t)$.

as shown in Figure 9.8(b). The steady state response corresponds to the final reading discussed in Section 9.1.1. Thus an important issue in designing a system is the time for the response to reach stead state or, equivalently, for the transient to become zero. This is discussed in this subsection.

Recall from Section 2.7 that the real exponential function e^{-2t} is defined to have a time constant of $1/2 = 0.5$ second and takes five time constants or 2.5 seconds to have its magnitude decreases to less than 1% of its peak value as shown in Figure 9.9(a). Thus we may consider e^{-2t} to have reached zero in 2.5 seconds in engineering. Because $e^{-2t} = \mathcal{L}^{-1}[1/(s+2)]$, we may define the time constant of a real pole α, with $\alpha < 0$, as $1/|\alpha|$. Note that if $\alpha \geq 0$, its response will not approach zero and its time constant is not defined.

For a pair of complex-conjugate poles $\alpha \pm j\beta$ with $\alpha < 0$, we will define its time constant as $1/|\alpha|$. This definition is justified because

$$\left| e^{\alpha t}\sin(\beta t + \theta) \right| \leq \left| e^{\alpha t} \right|$$

for all t. Thus the response will decrease to less than one percentage of its peak value in five time constants. For example, the time constant of $-0.5 \pm j2.4$ is $1/|-0.5| = 2$, and the response due to the pair of poles vanishes in five time constant or 10 seconds as shown in Figure 9.9(b).

With the preceding discussion, we may define the *time constant* of a stable $H(s)$ as

$$
\begin{aligned}
t_c &= \frac{1}{\text{Smallest real part in magnitude of all poles}} \\
&= \frac{1}{\text{Smallest distance from all poles to the } j\omega\text{-axis}}
\end{aligned}
\tag{9.46}
$$

If $H(s)$ has many poles, the larger the real part in magnitude, the faster its time response approaches zero. Thus the pole with the smallest real part in magnitude dictates the time for the transient response to approach zero. For example, the transfer function in (9.42) has three poles -2 and $-0.5 \pm j2.4$. The smallest real part in magnitude is 0.5. Thus the time constant of (9.42) is $1/0.5 = 2$.

The response time of a stable system will be defined as the time for its step response to reach its steady-state response or, correspondingly, for its transient response to become zero. Mathematically speaking, the transient response becomes zero only at $t = \infty$. In engineering, the transient response however will be considered to have reach zero when its magnitude becomes less than 1% of its peak value. Thus we will consider the transient response to have reach zero or the step response to have reach steady-state response in five time constant, that is,

$$\text{Response time} = 5 \times \text{time constant}$$

For example, the time constants of the transfer functions in (9.42) and (9.22) through (9.26) all equal 2 seconds and it takes 10 seconds for their step responses to reach steady-state as shown in Figures 9.8(a) and 9.4. Thus the response times of the transfer functions all equal 10 seconds. Note the rule of five time constants should be considered only as a guide, not a strict rule. For example, the transfer function $2/(s + 0.5)^3$ has time constant 2 and its time response $t^2 e^{-0.5t}$ approaches zero in roughly 25 s as shown in Figure 9.6(a), instead of 10 s. See also Problem 9.16. In general, if the poles of a stable transfer function are not clustered in a small region, the rule is applicable.

The time constant of a stable transfer function is defined from its poles. Its zeros do not play any role. For example, the transfer functions from (9.22) through (9.26) have the same set of poles but different zeros. Although their transient responses are different, their step responses all approach steady state in roughly 10 seconds. Thus the time constant and response time depend mainly on the poles of a transfer function.

The final reading discussed in Subsection 9.1.1 corresponds to the steady-state response and the response time is the time for the response to reach steady state. These are time-domain specifications. In the design of filters, the specifications are mainly in the frequency domain as we discuss next.

9.7   Frequency responses

In this section, we introduce the concept of frequency responses. Consider a transfer function $H(s)$. Its values along the $j\omega$-axis — that is, $H(j\omega)$ — is called the *frequency response*. In general, $H(j\omega)$ is complex-valued and can be expressed in polar form as[8]

$$H(j\omega) = A(\omega)e^{j\theta(\omega)} \tag{9.47}$$

where $A(\omega)$ and $\theta(\omega)$ are real-valued functions of ω and $A(\omega) \geq 0$. We call $A(\omega)$ the *magnitude response* and $\theta(\omega)$ the *phase response*. If $H(s)$ contains only real coefficients, then

$$A(\omega) = A(-\omega) \quad \text{(even)} \tag{9.48}$$

and

$$\theta(\omega) = -\theta(-\omega) \quad \text{(odd)} \tag{9.49}$$

This follows from $[H(j\omega)]^* = H(-j\omega)$, which implies

$$[A(\omega)e^{j\theta(\omega)}]^* = A(\omega)e^{-j\theta(\omega)} = A(-\omega)e^{j\theta(-\omega)}$$

---

[8]This part is identical to frequency spectra of signals discussed in the second half of Section 4.3, especially from (4.26) onward, thus its discussion will be brief.

where we have used $A^* = A$ and $\theta^* = \theta$. Thus if $H(s)$ has no complex coefficients, then its magnitude response is even and its phase response is odd. Because of these properties, we often plot magnitude and phase responses only for $\omega \geq 0$ instead of for all ω in $(-\infty, \infty)$. We give examples.

Before proceeding, we mention that if $H(s)$ has only real coefficients, then $H(0)$ is a real number. If $H(0) \geq 0$, then its phase is 0 and meets (9.49). If $H(0) < 0$, then its phase is π (rad) or 180°. It still meets (9.49) because $180 = -180$ (mod 360) as discussed in Subsection 3.2.1.

Example 9.7.1 Consider $H(s) = 2/(s+2)$. Its frequency response is

$$H(j\omega) = \frac{2}{j\omega + 2}$$

We compute

$$\omega = 0: \quad H(0) = \frac{2}{2} = 1 = 1 \cdot e^{j0^o}$$

$$\omega = 1: \quad H(j1) = \frac{2}{j1 + 2} = \frac{2e^{j0^o}}{2.24e^{j26.6^o}} = 0.9e^{-j26.6^o}$$

$$\omega = 2: \quad H(j2) = \frac{2}{j2 + 2} = \frac{1}{1 + j} = \frac{1}{1.414e^{j45^o}} = 0.707e^{-j45^o}$$

$$\omega \to \infty: \quad H(j\omega) \approx \frac{2}{j\omega} = re^{-j90^o}$$

Note that as $\omega \to \infty$, the frequency response can be approximated by $2/j\omega$. Thus its magnitude $r = 2/\omega$ approaches zero but its phase approaches -90°. From the preceding computation, we can plot the magnitude and phase responses as shown in Figures 9.10(a) and (b) with solid lines. They are actually obtained in MATLAB by typing in an edit window the following

```
%Program 9.2 (f910.m)
w=0:0.01:40;s=j*w;
H=2./(s+2);
subplot(1,2,1)
plot(w,abs(H)),title('(a)')
ylabel('Magnitude response'),xlabel('Frequency (rad/s)'),axis square
subplot(1,2,2)
plot(w,angle(H)*180/pi),title('(b)')
ylabel('Phase response (deg)'),xlabel('Frequency (rad/s)'),axis square
axus([0 40 -180 0])
```

The first two lines compute the values of $H(s)$ at $s = j\omega$, with ω ranging from 0 to 40 rad/s with increment 0.1. The function abs computes the absolute value or magnitude and angle computes the angle or phase. We save Program 9.1 as an m-file named f910.m. Typing in the command window >> f910 will yield Figure 9.10 in a figure window.

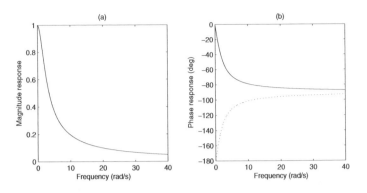

Figure 9.10: (a) Magnitude responses of $H(s)$ (solid line) and $H_1(s)$ (dotted line). (b) Phase responses of $H(s)$ (solid line) and $H_1(s)$ (dotted line).

Example 9.7.2 We repeat Example 9.7.1 for $H_1(s) = 2/(s-2)$. The result is shown in Figures 9.10 with dotted lines. Note that the dotted line in Figure 9.10(a) overlaps with the solid line. See Problem 9.18.□

We now discuss the implication of stability and physical meaning of frequency responses. Consider a system with stable transfer function $H(s)$ with real coefficients. Let us apply the input $u(t) = ae^{j\omega_0 t}$, where a is a real constant. Note that this input is not real-valued. However, if we use the real-valued input $u(t) = \sin \omega_0 t$ or $\cos \omega_0 t$, then the derivation will be more complex. The Laplace transform of $ae^{j\omega_0 t}$ is $a/(s - j\omega_0)$. Thus the output of the system is given by

$$Y(s) = H(s)U(s) = H(s)\frac{a}{s - j\omega_0} = k_u \frac{1}{s - j\omega_0} + \text{terms due to poles of } H(s)$$

Because $H(s)$ is stable, it has no poles on the imaginary axis. Thus $j\omega_0$ is a simple pole of $Y(s)$ and its residue k_u can be computed as

$$k_u = Y(s)(s - j\omega_0)|_{s - j\omega_0 = 0} = aH(s)|_{s = j\omega_0} = aH(j\omega_0)$$

If $H(s)$ is stable, then all its poles have negative real parts and their time responses all approach 0 as $t \to \infty$. Thus we conclude that if $H(s)$ is stable and if $u(t) = ae^{j\omega_0 t}$, then we have

$$y_{ss}(t) := \lim_{t \to \infty} y(t) = aH(j\omega_0)e^{j\omega_0 t} \tag{9.50}$$

It is the steady-state response of stable $H(s)$ excited by $u(t) = ae^{j\omega_0 t}$, for $t \geq 0$.

Substituting (9.47) into (9.50) yields

$$\begin{aligned} y_{ss}(t) &= aA(\omega_0)e^{j\theta(\omega_0)}e^{j\omega_0 t} = aA(\omega_0)e^{j(\omega_0 t + \theta(\omega_0))} \\ &= aA(\omega_0)\left[\cos(\omega_0 t + \theta(\omega_0)) + j\sin(\omega_0 t + \theta(\omega_0))\right] \end{aligned} \tag{9.51}$$

We list some special cases of (9.50) or (9.51) as a theorem.

Theorem 9.5 Consider a system with proper rational transfer function $H(s)$. If the system is BIBO stable, then

$$u(t) = ae^{j\omega_0 t} \text{ for } t \geq 0 \quad \rightarrow \quad y_{ss}(t) = \lim_{t \to \infty} y(t) = aH(j\omega_0)e^{j\omega_0 t}$$

and, in particular,

$$u(t) = a \quad \text{for } t \geq 0 \quad \rightarrow \quad y_{ss}(t) = aH(0)$$
$$u(t) = a \sin \omega_0 t \quad \text{for } t \geq 0 \quad \rightarrow \quad y_{ss}(t) = a|H(j\omega_0)| \sin(\omega_0 t + \measuredangle H(j\omega_0))$$
$$u(t) = a \cos \omega_0 t \quad \text{for } t \geq 0 \quad \rightarrow \quad y_{ss}(t) = a|H(j\omega_0)| \cos(\omega_0 t + \measuredangle H(j\omega_0))$$

□

The steady-state response of $H(s)$ in (9.50) is excited by the input $u(t) = ae^{j\omega_0 t}$. If $\omega_0 = 0$, then the input is a step function with amplitude a, and the output approaches a step function with amplitude $aH(0)$. If $u(t) = a \sin \omega_0 t = \text{Im } ae^{j\omega_0 t}$, where Im stands for the imaginary part, then the output approaches the imaginary part of (9.51) or $aA(j\omega) \sin(\omega_0 t + \theta(\omega_0))$. Using the real part of $ae^{j\omega_0 t}$, we can obtain the third equation. In conclusion, if we apply a sinusoidal input to a system, then the output will approach a sinusoidal signal with the same frequency, but its amplitude will be modified by $A(\omega_0) = |H(j\omega_0)|$ and its phase by $\theta(\omega_0) = \measuredangle H(j\omega_0)$. We stress that the system must be stable. If it is not, then the output generally grows unbounded and $H(j\omega_0)$ has no physical meaning as we will demonstrate shortly.

Example 9.7.3 Consider a system with transfer function $H(s) = 2/(s+2)$ and consider the signal

$$u(t) = \cos 0.2t + 0.2 \sin 25t \tag{9.52}$$

The signal consists of two sinusoids, one with frequency 0.2 rad/s and the other 25 rad/s. We study the eventual effect of the system on the two sinusoids.

In order to apply Theorem 9.5, we must compute

$$H(j0.2) = \frac{2}{2 + j0.2} = 0.995e^{-j0.0997}$$

and

$$H(j25) = \frac{2}{2 + j25} = 0.079e^{-j1.491}$$

They can also be read from Figure 9.10 but the reading cannot be very accurate. Note that the angles must be expressed in radians. Then Theorem 9.5 implies

$$y_{ss}(t) = 0.995 \cos(0.2t - 0.0997) + 0.0158 \sin(25t - 1.491) \tag{9.53}$$

We see that the system attenuates greatly the high-frequency component (from 0.2 to 0.0159) and passes the low-frequency component with only a small attenuation (from 1 to 0.995). Thus the system is called a low-pass filter. □

Example 9.7.4 Consider a system with transfer function $H(s) = 2/(s - 2)$. We compute its step response. If $u(t) = 1$, for $t \geq 0$, then $U(s) = 1/s$ and

$$Y(s) = H(s)U(s) = \frac{2}{s - 2}\frac{1}{s} = \frac{1}{s - 2} - \frac{1}{s}$$

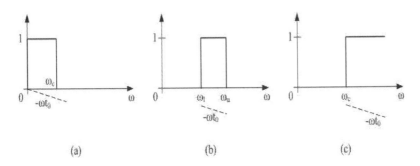

Figure 9.11: (a) Ideal lowpass filter with cutoff frequency ω_c. (b) Ideal bandpass filter with upper and lower cutoff frequencies ω_u and ω_l. (c) Ideal highpass filter with cutoff frequency ω_c.

Thus its step response is

$$y(t) = e^{2t} - 1 = e^{2t} + H(0)$$

for $t \geq 0$, where we have used $H(0) = -1$. Even though the output contains the step function with amplitude $H(0)$, it also contains the exponentially increasing function e^{2t}. As $t \to \infty$, the former is buried by the latter. Thus the output approaches ∞ as $t \to \infty$ and Theorem 9.5 does not hold. In conclusion, the stability condition is essential in using Theorem 9.5. Moreover the frequency response of unstable $H(s)$ has no physical meaning. $\square$

In view of Theorem 9.5, if we can design a stable system with magnitude response as shown in Figure 9.11(a) with solid line and phase response with dotted line, then the system will pass sinusoids with frequency $|\omega| < \omega_c$ and stop sinusoids with frequency $|\omega| > \omega_c$. We require the phase response to be linear to avoid distortion as we will discuss in a later section. We call such a system an *ideal lowpass filter* with cutoff frequency ω_c. The frequency range $[0, \omega_c)$ is called the *passband*, and $[\omega_c, \infty)$ is called the *stopband*. Figures 9.11(b) and (c) show the characteristics of ideal bandpass and highpass filters. The ideal bandpass filter will pass sinusoids with frequencies lying inside the range $[\omega_l, \omega_u]$, where ω_l and ω_u are the lower and upper cutoff frequencies, respectively. The ideal highpass filter will pass sinusoids with frequencies larger than the cutoff frequency ω_c. They are called *frequency selective filters* and are special types of systems.

The impulse response $h(t)$ of the ideal lowpass filter with linear phase $-\omega t_0$ can be computed as

$$h(t) = \frac{\sin[\omega_c(t - t_0)]}{\pi(t - t_0)}$$

for all t in $(-\infty, \infty)$. See Problem 9.20. It is plotted in Figure 2.10(b) with $\omega_c = 1/a = 10$ and $t_0 = 4$. We see that $h(t)$ is nonzero for $t < 0$, thus the ideal lowpass filter is not causal and cannot be built in the real world. In practice, we modify the magnitude responses in Figure 9.11 to the ones shown in Figure 9.12. We insert a *transition band* between the passband and stopband. Furthermore, we specify a *passband tolerance* and

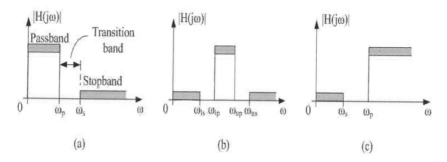

Figure 9.12: Specifications of practical (a) lowpass filter, (b) bandpass filter, and (c) highpass filter.

stopband tolerance as shown with shaded areas. The transition band is generally not specified and is the "don't care" region. We also introduce the *group delay* defined as

$$\text{Group delay} = \tau(\omega) := -\frac{d\theta(\omega)}{d\omega} \tag{9.54}$$

For an ideal filter with linear phase such as $\theta(\omega) = -t_0\omega$, its group delay is t_0, a constant. Thus instead of specifying the phase response to be a linear function of ω, we specify the group delay in the passband to be roughly constant or $t_0 - \epsilon < \tau(\omega) < t_0 + \epsilon$, for some constants t_0 and ϵ. Even with these more relaxed specifications on the magnitude and phase responses, if we specify both of them, it is still difficult to find causal filters to meet both specifications. Thus in practice, we often specify only magnitude responses as shown in Figure 9.12. The design problem is then to find a proper stable rational function of a degree as small as possible to have its magnitude response lying inside the specified region. See, for example, Reference [C7].

9.7.1 Plotting frequency responses

Plotting frequency responses from transfer functions is important in designing filters. We use an example to discuss its plotting. Consider the transfer function[9]

$$H_1(s) = \frac{1000s^2}{s^4 + 28.28s^3 + 5200s^2 + 67882.25s + 5760000} \tag{9.55}$$

Similar to Program 9.2, we type in an edit window the following

```
%Program 9.3 (f913.m)
w=0:0.01:200;s=j*w;
H=1000.*s.^2./(s.^4+28.28.*s.^3+5200.*s.^2+67882.25.*s+5760000);
plot(w,abs(H))
```

If we save the program as f913.m, then typing in the command window

```
>> f913
```

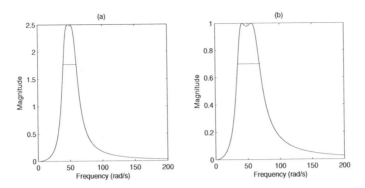

Figure 9.13: (a) Magnitude response and bandwidth of (9.55). (b) Magnitude response and bandwidth of (9.57).

will generate Figure 9.13(a) in a figure window.

The frequency response of (9.55) can be more easily obtained by calling the MATLAB function freqs. Typing in the command window

```
%program 9.3
>> w=0:0.01:200;
>> n=[1000 0 0];d=[1 28.28 5200 67882.25 576000];
>> H=freqs(n,d,w);
>> plot(w,abs(H))
```

will also generate Figure 9.13(a). Note that n and d are the numerator's and denominator's coefficients of (9.55). Thus H=freqs(n,d,w) computes the frequency response of the CT transfer function at the specified frequencies and stores them in H. Note that the last character 's' in freqs denotes the Laplace transform variable s. If we continue to type in the command window

```
>> freqs(n,d)
```

then it will generate the plots in Figure 9.14. Note that freqs(n,d) contains no output H and no frequencies w. In this case, it automatically selects 200 frequencies in $[0, \infty)$ and then generates the plots as shown. Note that the scales used in frequency and magnitude are not linear; they are in logarithmic scales. Such plots are called *Bode plots*. Hand plotting of Bode plots is discussed in some texts on signals and systems because for a simple $H(s)$ with real poles and real zeros, its frequency response can be approximated by sections of straight lines. We will not discuss their plotting because they can now be plotted exactly using a computer.[10]

---

[9]It is a Butterworth filter. See Reference [C7].

[10]Bode plots are discussed in every control text and used in designing control systems. The design method however is complicated and is applicable only to a very limited class of LTI lumped systems. On the other hand, Bode plots are useful in identification: finding transfer functions from measured plots. See Subsection 9.7.5 and Reference [C5].

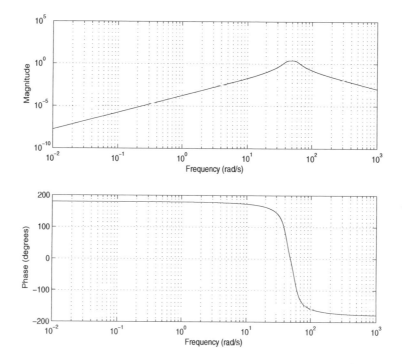

Figure 9.14: Top: Magnitude response of (9.55). Bottom: Phase response of (9.55).

9.7.2 Bandwidth of frequency selective filters

This subsection introduces the concept of bandwidth for frequency-selective filters. This concept is different from the bandwidth of signals. See Subsection 4.4.3. The *bandwidth* of a filter is defined as the width of its passband in the positive frequency range. The bandwidth is ω_c for the ideal lowpass filter in Figure 9.11(a), $\omega_u - \omega_l$ for the ideal bandpass filter in Figure 9.11(b), and infinity for the ideal highpass filter in Figure 9.11(c). Clearly, for highpass filters, the passband cutoff frequency is more meaningful than bandwidth.

For practical filters, different specification of passband tolerance will yield different bandwidth. The passband tolerance is usually defined using the unit of decibel (dB). The dB is the unit of $20 \log_{10} |H(j\omega)|$. Because the bandwidth is defined with respect to peak magnitude, we define

$$a = 20 \log_{10} \left(\frac{|H(j\omega)|}{H_{max}} \right)$$

where H_{max} is the peak magnitude. Note that if $H_{max} = 1$, the system carries out only filtering. If $H_{max} > 1$, the system carries out filtering as well as amplification. If $|H(j\omega)| = H_{max}$, then a is 0 dB. If

$$|H(j\omega)| = 0.707 H_{max} \tag{9.56}$$

then

$$a = 20 \log_{10} 0.707 = -3 \text{ dB}$$

The 3-dB *bandwidth* of a passband is then defined as the width of the frequency range in $[0, \infty)$ in which the magnitude response of $H(s)$ is -3 dB or larger. Note that we can also define a 2-dB or 1-dB bandwidth. We mention that the 3-dB bandwidth is also called the *half-power* bandwidth. The power of $y(t)$ is $y^2(t)$ and, consequently, is proportional to $|H(j\omega)|^2$. If the power at H_{max} is A, then the power at $0.707 H_{max}$ is $(0.707)^2 A = 0.5A$. Thus the 3-dB bandwidth is also called the *half-power bandwidth*.

We give some examples. Consider $H(s) = 2/(s+2)$ whose magnitude response is shown in Figure 9.10(a). Its peak magnitude H_{max} is 1 as shown. From the plot we see that the magnitude is 0.707 or larger for ω in $[0, 2]$. Thus the 3-dB bandwidth of $2/(s+2)$ is 2 rad/s. We call $\omega_p = 2$ the passband edge frequency. Consider the transfer functions in (9.55). Its magnitude response is shown in Figure 9.13(a). Its peak magnitude H_{max} is 2.5. From the plot, we see that for ω in $[40, 60]$, the magnitude is $0.707 \times 2.5 = 1.77$ or larger as shown. We call 40 the lower passband edge frequency and 60 the upper passband edge frequency. Thus the bandwidth of the filter is $60 - 40 = 20$ rad/s.

We give one more example. Consider the transfer function[11]

$$H_2(s) = \frac{921.27s^2}{s^4 + 38.54s^3 + 5742.73s^2 + 92500.15s + 5760000} \tag{9.57}$$

Its magnitude response is plotted, using MATLAB, in Figure 9.13(b). For the transfer function in (9.57), we have $H_{2max} = 1$ and the magnitude response is 0.707 or larger in the frequency range $[35, 70]$. Thus its bandwidth is $70 - 35 = 35$ rad/s.

9.7.3   Non-uniqueness in design

Consider a filter with transfer function

$$H(s) = \frac{a}{s + a} \tag{9.58}$$

where a is a real positive constant. The problem is to use the filter to pass the dc signal 2 and to eliminate $0.2 \sin 80t$ in

$$u(t) = 2 + 0.2 \sin 80t \tag{9.59}$$

Will the design of the filter or the selection of a be unique?

The filter has its magnitude response of the form shown in Figure 9.10(a). Its peak magnitude is $H(0) = a/a = 1$ and its frequency response at $\omega = a$ is

$$H(ja) = \frac{a}{ja + a} = \frac{1}{1 + j} = \frac{1}{1.414 e^{j\pi/4}} = 0.707 e^{-j\pi/4}$$

Thus its 3-dB bandwidth is a rad/s. Because, for $a << 80$,

$$H(j80) = \frac{a}{j80 + a} \approx \frac{a}{j80} = 0.0125a e^{-j\pi/2}$$

---

[11]It is a type I Chebyshev filter. See Reference [C7]

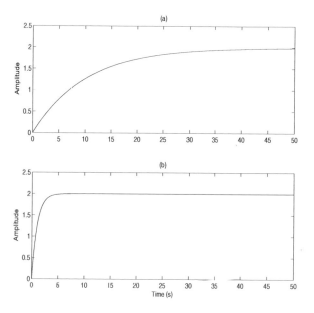

Figure 9.15: (a) The output of (9.58) excited by (9.59) with $a = 0.1$. (b) With $a = 1$.

the steady-state response of the filter excited by $u(t)$ is

$$y_{ss}(t) = 2H(0) + 0.2|H(j80)| \sin(80t + \angle H(j80)) = 2 + 0.0025a \sin(80t - \pi/2)$$

We see that the smaller a is, the more severely the signal $\sin 80t$ is attenuated. However, the time constant of the filter is $1/a$. Thus the smaller a is, the longer for the response to reach steady state. For example, if we select $a = 0.1$, then the amplitude of $\sin 80t$ is reduced to 0.00025, but it takes $5/0.1 = 50$ seconds to reach steady state as shown in Figure 9.15(a). If we select $a = 1$, then the amplitude of $\sin 80t$ is reduced to 0.0025, but it takes only $5/1 = 5$ seconds to reach steady state as shown in Figure 9.15(b). In conclusion, if the speed of response is also important, then the selection of a must be a compromise between filtering and the response time. Other practical issues such as cost and implementation may also play a role. Thus most designs are not unique in practice.

9.7.4 Frequency domain and transform domain

This subsection discusses the roles of frequency and transform domains in designing CT or analog filters. The design problem is to find a stable proper rational transfer function $H(s)$ so that its frequency response will meet some design specifications. As discussed in Section 9.7, if we specify both the magnitude response and phase response, then it will be difficult to find a $H(s)$ to meet both specifications. Thus in practice, we often specify only the magnitude response as shown in Figure 9.12. In other words, we find a stable $H(s)$ whose magnitude response $|H(j\omega)|$ lies inside the shaded areas shown in Figure 9.12. Because the specification is given in the frequency domain, it is

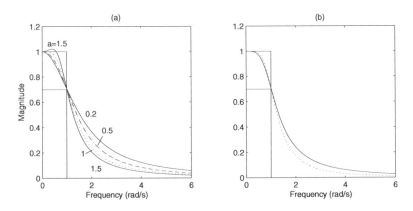

Figure 9.16: (a) Magnitude of (9.51) with $a = 0.2, 0.5, 1, 1.5$. (b) Plots of (9.60) (solid line) and (9.62) (dotted line).

simpler to search a positive real-valued $M(\omega)$ to meet the specification and then find a $H(s)$ to meet $|H(j\omega)| = M(\omega)$.

We use an example to discuss the selection of $M(\omega)$. Suppose we wish to design a lowpass filter with passband edge frequency $\omega_p = 1$ and 3-dB passband tolerance. Let us consider

$$M(\omega) = \frac{1}{\sqrt{a\omega^4 + (1 - a)\omega^2 + 1}} \tag{9.60}$$

where a is a positive real number. Clearly, we have $M(0) = 1$ or $20 \log M(0) = 0$ dB and

$$M(1) = \frac{1}{\sqrt{a + (1 - a) + 1}} = \frac{1}{\sqrt{2}} = 0.707$$

or $20 \log M(1) = -3$ dB, for any a. We plot $M(\omega)$ in Figure 9.16(a) for various a. If $a = 1.5$, then $M(\omega)$ goes outside the permissible passband region. If $a = 0.2, 0.5$, and 1, then all $M(\omega)$ remain inside the permissible passband region. However for $a = 1$, $M(\omega)$ rolls off most steeply to zero. Thus we conclude that

$$M_1(\omega) = \frac{1}{\sqrt{\omega^4 + 1}} \tag{9.61}$$

with $a = 1$, is the best among all (9.60) with various a. We plot in Figure 9.16(b) $M_1(\omega)$ in (9.61) with a solid line and

$$M_2(\omega) = \frac{1}{\sqrt{\omega^5 + 1}} \tag{9.62}$$

with a dotted line. We see that $M_2(\omega)$ rolls off more steeply than $M_1(\omega)$ after $\omega_p = 1$ and is a better magnitude response.

The preceding search of $M_i(\omega)$ is carried out entirely in the frequency domain. This is however not the end of the design. The next step is to find a stable proper rational function $H(s)$ so that $|H(j\omega)| = M_i(\omega)$. It turns out that no proper rational function $H(s)$ exits to meet $|H(j\omega)| = M_2(\omega)$. However we can find

$$H_1(s) = \frac{1}{s^2 + \sqrt{2}s + 1} = \frac{1}{s^2 + 1.414s + 1} \qquad (9.63)$$

to meet $|H_1(j\omega)| = M_1(\omega)$ or $H_1(j\omega)H_1(-j\omega) = M_1^2(\omega)$. Indeed we have

$$\frac{1}{-\omega^2 + j\sqrt{2}\omega + 1} \times \frac{1}{-\omega^2 - j\sqrt{2}\omega + 1} = \frac{1}{(1-\omega^2)^2 + 2\omega^2} = \frac{1}{\omega^4 + 1}$$

Such a $M_1(\omega)$ is said to be *spectral factorizable*. This completes the design. See Reference [C7].

The purpose of this subsection is to show the roles of frequency and transform domains in filter design. The search of $M(\omega)$, limited to those spectral factorizable, is carried out in the frequency domain. Once such an $M(\omega)$ is found, we must find a proper stable rational transfer function $H(s)$ to meet $H(j\omega)H(-j\omega) = M^2(\omega)$. We can then find a minimal realization of $H(s)$ and implement it using an op-amp circuit. We cannot implement a filter directly from the magnitude response in (9.61).

9.7.5 Identification by measuring frequency responses

If the internal structure of an LTI lumped system is known, we can apply physical laws to develop a transfer function or an ss equation to describe it.[12] If the internal structure is not known, can we develop a mathematical equation, from measurements at its input and output terminals, to describe the system? If affirmative, the equation must be an external description (convolution, high-order differential equation, or transfer function), and cannot be an internal description (ss equation).

To develop a convolution, we need the impulse response which in theory can be obtained by measurement. We apply an impulse at the input terminal, then the output measured at the output terminal is the impulse response. Unfortunately, we cannot generate an impulse in practice and there is no impulse response to be measured. It is however simple to generate a step function. Let $y_q(t)$ be the step response of a system. Then the impulse response of the system is given by

$$h(t) = \frac{dy_q(t)}{dt}$$

See Problem 8.33. However differentiation will amplify high-frequency noise which often exists in measurement. Thus in practice, it is difficult to measure impulse responses and to develop convolutions.

The other two external descriptions are high-order differential equations and transfer functions. We discuss only the latter because they are essentially equivalent. See Subsection 8.7.1. One way to develop a transfer function is to measure its frequency response. If a system is stable,[13] and if we apply the input $\sin \omega_0 t$, then Theorem 9.5

---

[12]This subsection may be skipped without loss of continuity.
[13]This can be relaxed to permit a simple pole at $s = 0$.

states that the output will approach $|H(j\omega_0)|\sin(\omega_0 t + \sphericalangle H(j\omega_0))$. Thus we can measure $|H(j\omega_0)|$ and $\sphericalangle H(j\omega_0)$ at frequency ω_0. This measures only two values and the noise will be less an issue. After measuring frequency responses at a number of frequencies, we can sketch the magnitude and phase responses. Clearly, the more frequencies we measure, the more accurate the responses.

The next step is to find a transfer function from the magnitude and phase responses. If the magnitude response is plotted using linear scales, it is difficult to develop a transfer function. However if it is plotted using logarithmic scales for both frequency and magnitude as in Figures 9.14 and 10.2(a), and if the transfer function has only real poles and real zeros, then its magnitude response often can be approximated by straight lines as in Figure 10.2(a). The intersections of those straight lines will yield the locations of poles and zeros. However, if the transfer function has complex poles as in Figure 9.14, then the situation will be more complex. See Reference [C5]. In conclusion, it is possible to develop transfer functions from measurements for simple systems. Commercial devices such as HP 3563A, Control system Analyzer, can carry out this task automatically.

9.7.6   Parametric identification

In identifying a transfer function, if we assume its form to be known but with unknown parameters, the method is called a parametric identification.[14] We discuss in this subsection such a method. Before proceeding, we mention that the method discussed in the preceding subsection is called a non-parametric method in which we assume no knowledge of the form of its transfer function.

Consider an armature-controlled DC motor driving a load. Its transfer function from the applied voltage $u(t)$ to the motor shaft's angular position $y(t)$ can be approximated by

$$H(s) = \frac{Y(s)}{U(s)} = \frac{k_m}{s(\tau_m s + 1)}$$

where k_m is called the motor gain constant and τ_m the motor time constant. It is obtained by ignoring some pole. See Reference [C5] and Subsection 10.2.1. The parameters k_m and τ_m can be obtained from the specification of the motor and the moment of inertia of the load. Because the load such as an antenna may not be of regular shape, computing analytically its moment of inertia may not be simple. Thus we may as well obtain k_m and τ_m directly by measurement. The preceding transfer function is from applied voltage to the motor shaft's angular position. Now if we consider the motor shaft's angular velocity $v(t) = dy(t)/dt$ or $V(s) = sY(s)$ as the output, then the transfer function becomes

$$H_v(s) = \frac{V(s)}{U(s)} = \frac{sY(s)}{U(s)} = \frac{k_m}{\tau_m s + 1}$$

If we apply a step input, that is, $u(t) = 1$, for $t \geq 0$, then the output in the transform domain is

$$V(s) = \frac{k_m}{\tau_m s + 1}\frac{1}{s} = \frac{k_m}{s} - \frac{k_m \tau_m}{\tau_m s + 1} = \frac{k_m}{s} - \frac{k_m}{s + 1/\tau_m}$$

---

[14]This subsection may be skipped without loss of continuity.

Thus the output in the time domain is

$$v(t) = k_m - k_m e^{-t/\tau_m}$$

for $t \geq 0$. The velocity will increase and finally reach the steady state which yields k_m. The time it reaches steady state equals five time constants or $5\tau_m$. Thus from the step response, we can obtain k_m and τ_m.

Our discussion ignores completely the noise problem. Identification under noise is a vast subject area.

9.8 Laplace transform and Fourier transform

This section discusses the relationship between the Laplace transform and the Fourier transform. We consider in this section only positive-time signals. For two-sided signals, see Problem 9.26.

Let $x(t)$ be a positive-time signal, that is, $x(t) = 0$ for all $t < 0$. Then its Laplace transform is

$$X(s) = \mathcal{L}[x(t)] = \int_{t=0}^{\infty} x(t)e^{-st} dt \tag{9.64}$$

and its Fourier transform is

$$\bar{X}(\omega) = \mathcal{F}[x(t)] = \int_{t=0}^{\infty} x(t)e^{-j\omega t} dt \tag{9.65}$$

Note the use of different notations, otherwise confusion may arise. See Section 3.4. From (9.64) and (9.65), we see immediately

$$\bar{X}(\omega) := \mathcal{F}[x(t)] = \mathcal{L}[x(t)]\big|_{s=j\omega} =: X(s)\big|_{s=j\omega} = X(j\omega) \tag{9.66}$$

Does this equation hold for all $x(t)$? The answer is negative as we discuss next.

Consider the positive function $x(t) = e^{2t}$, for $t \geq 0$. Its Laplace transform is $X(s) = 1/(s-2)$. Its Fourier transform however, as discussed in Example 4.3.1, does not exist and thus cannot equal $X(j\omega) = 1/(j\omega - 2)$. Next we consider the step function $q(t) = 1$, for $t \geq 0$. Its Laplace transform is $Q(s) = 1/s$. Its Fourier transform is, as discussed in (4.51),

$$\bar{Q}(\omega) = \pi\delta(\omega) + \frac{1}{j\omega} \tag{9.67}$$

It contains an impulse at $\omega = 0$ and is different from $Q(j\omega)$. Thus (9.66) does not hold for a step function. Note that both e^{2t} and $q(t) = 1$, for $t \geq 0$, are not absolutely integrable. Thus (9.66) does not hold if $x(t)$ is not absolutely integrable.

On the other hand, if $x(t)$ is absolutely integrable, then (9.66) does hold. In this case, its Fourier transform $\bar{X}(\omega)$ is bounded and continuous for all ω, as discussed in Section 4.3, and its Laplace transform can be shown to contain the $j - \omega$ axis in its region of convergence. Thus replacing s by $j\omega$ in $X(s)$ will yield $\bar{X}(\omega)$. Because tables of Laplace transform pairs are more widely available, we may use (9.66) to compute frequency spectra of signals.

Example 9.8.1 Consider

$$x(t) = 3e^{-0.2t}\cos 10t \tag{9.68}$$

for $t \geq 0$ and $x(t) = 0$ for $t < 0$. It is positive time. Because

$$\int_0^\infty |3e^{-0.2t}\cos 10t|dt < 3\int_0^\infty e^{-0.2t}dt = \frac{3}{0.2} < \infty$$

it is absolutely integrable. Its Laplace transform is, using Table 9.1,

$$X(s) = \frac{3(s+0.2)}{(s+0.2)^2 + 100} = \frac{3s+0.6}{s^2 + 0.4s + 100.04} \tag{9.69}$$

Thus the Fourier transform or the frequency spectrum of (9.68) is

$$X(j\omega) = \frac{3s+0.6}{(s+0.2)^2 + 100}\bigg|_{s=j\omega} = \frac{3j\omega + 0.6}{(j\omega + 0.2)^2 + 100} \tag{9.70}$$

□

9.8.1   Why Fourier transform is not used in system analysis

This subsection compares the Laplace and Fourier transforms in system analysis. Consider an LTI lumped system described by

$$y(t) = \int_{\tau=0}^\infty h(t-\tau)u(\tau)d\tau$$

Its Laplace transform is

$$Y(s) = H(s)U(s) \tag{9.71}$$

and its Fourier transform can be computed as

$$\bar{Y}(\omega) = \bar{H}(\omega)\bar{U}(\omega) \tag{9.72}$$

Thus the Fourier transform can also be used to carry out system analysis. However, the Fourier transform is, as we demonstrate in this subsection, less general, less revealing, and more complicated. Thus we do not discuss in this text Fourier analysis of systems.

Equation (9.71) is applicable regardless of whether the system is stable or not and regardless of whether the frequency spectrum of the input is defined or not. For example, consider $h(t) = u(t) = e^t$, for $t \geq 0$. Their Laplace transforms both equal $1/(s-1)$. Thus the output is given by

$$Y(s) = H(s)U(s) = \frac{1}{s-1}\frac{1}{s-1} = \frac{1}{(s-1)^2}$$

and, using Table 9.1, $y(t) = te^t$, for $t \geq 0$. The output grows unbounded. For this example, the Fourier transforms of $h(t)$, $u(t)$, and $y(t)$ are not defined and (9.72) is not applicable. Thus the Laplace transform is more general.

Next we consider a system with transfer function

$$H(s) = \frac{s}{(s+1)(s+2)} \tag{9.73}$$

It is stable and its impulse response (inverse Laplace transform of $H(s)$) is positive time and absolutely integrable. Thus its Fourier transform is, using (9.66),

$$\bar{H}(\omega) = H(j\omega) = \frac{j\omega}{(j\omega+1)(j\omega+2)} = \frac{j\omega}{2 - \omega^2 + 3j\omega} \tag{9.74}$$

From the poles of $H(s)$, we can determine immediately the general form of its time response, its time constant, and, consequently, its speed of response. Although the poles and zero of the system are embedded in its Fourier transform $\bar{H}(\omega)$ in (9.74), it requires some effort to find its poles and zero (see Reference [C5]). Thus $\bar{H}(\omega)$ is less revealing than $H(s)$.

We next compute the step responses of (9.73) and (9.74). The step response of (9.73) is given by

$$Y(s) = \frac{s}{(s+1)(s+2)} \frac{1}{s} = \frac{1}{(s+1)(s+2)} = \frac{1}{s+1} - \frac{1}{s+2}$$

which implies $y(t) = e^{-t} - e^{-2t}$, for $t \geq 0$. If we use (9.72), then the output of (9.74) is given by, using (9.67),

$$\bar{Y}(\omega) = \frac{j\omega}{2 - \omega^2 + 3j\omega} \left[\pi\delta(\omega) + \frac{1}{j\omega} \right]$$

To find its inverse Fourier transform is complicated. Therefore there seems no reason to use the Fourier transform in system analysis.

9.8.2 Phasor analysis

Phasors are often used in texts on circuit analysis to carry out *sinusoidal steady-state analysis*. Consider the cosine function

$$A\cos(\omega t + \theta) = \text{Re}\left[Ae^{j\theta}e^{j\omega t} \right]$$

where $A \geq 0$. For a given ω, the cosine function is uniquely specified by A and θ and the cosine function can be uniquely represented by the *phasor*

$$A \angle \theta$$

Consider a system, in particular, an RLC circuit with transfer function $H(s)$. If $H(s)$ is stable, and if the input is $u(t) = a\cos\omega_0 t$, then the output approaches

$$y_{ss}(t) = a|H(j\omega_0)|\cos(\omega_0 t + \angle H(j\omega_0))$$

In other words, if the input is a cosine function with phasor $a \angle 0$, then the output approaches a cosine function with phasor $a|H(j\omega_0)| \angle H(j\omega_0)$. Thus phasors can be

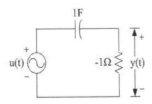

Figure 9.17: Unstable network.

used to study sinusoidal steady-state responses. However, the system must be stable. Otherwise, the result will be incorrect.

Example 9.8.2 Consider the network shown in Figure 9.17. It consists of a capacitor with capacitance 1F and a resistor with resistance -1Ω. Note that such a negative resistance can be generated using an op-amp circuit. See Problem 9.25. The transfer function of the network is $H(s) = s/(s-1)$. If we apply $u(t) = \cos 2t$, then its output can be computed as

$$y(t) = 0.2e^t + |H(j2)| \cos(2t + \angle H(j2))$$

for $t \geq 0$. Even though the output contains the phasor $|H(j2)| \angle H(j2)$, it will be buried by the exponentially increasing function $0.2e^t$. Thus the circuit will burn out or saturate and the phasor analysis cannot be used.□

In conclusion, real-world RLC circuits with positive R, L, and C, are, as discussed earlier, automatically stable. In this case, phasor analysis can be employed to compute their sinusoidal steady-state responses. Furthermore, their impedances can be defined as R, $j\omega L$, and $1/j\omega C$. However, if a system is not stable, phasor analysis cannot be used.

9.8.3   Conventional derivation of frequency responses

This subsection discusses a different way of developing frequency responses. Consider an LTI system described by the convolution

$$y(t) = \int_0^\infty h(t - \tau) u(\tau) d\tau \tag{9.75}$$

In this equation, the system is assumed to be initially relaxed at $t = 0$ and the input is applied from $t = 0$ onward. Now if the system is assumed to be initially relaxed at $t = -\infty$ and the input is applied from $t = -\infty$, then (9.75) must be modified as

$$y(t) = \int_{\tau=-\infty}^\infty h(t - \tau) u(\tau) d\tau \tag{9.76}$$

This is the equation introduced in most texts on signals and systems.

Let us introduce a new variable $\alpha := t - \tau$ for (9.76). Note that t is fixed and τ is the variable. Then we have $d\alpha = -d\tau$ and (9.76) becomes

$$y(t) = \int_{\alpha=\infty}^{-\infty} h(\alpha)u(t-\alpha)d(-\alpha) = \int_{\alpha=-\infty}^{\infty} h(\alpha)u(t-\alpha)d\alpha$$

which becomes, after renaming α as τ,

$$y(t) = \int_{\tau=-\infty}^{\infty} h(t-\tau)u(\tau)d\tau = \int_{\tau=-\infty}^{\infty} h(\tau)u(t-\tau)d\tau \qquad (9.77)$$

Because the role of h and u can be interchanged, (9.77) is said to have a *commutative* property. Note that (9.75) does not have the commutative property as in (9.77) and cannot be used in subsequent development. This is one of the reasons of extending the time to $-\infty$.

Now if we apply $u(t) = e^{j\omega_0 t}$ from $t = -\infty$, then the second form of (9.77) becomes

$$
\begin{aligned}
y(t) &= \int_{\tau=-\infty}^{\infty} h(\tau)e^{j\omega_0(t-\tau)}d\tau = \left(\int_{\tau=-\infty}^{\infty} h(\tau)e^{-j\omega_0 \tau}d\tau \right) e^{j\omega_0 t} \\
&= \left(\int_{\tau=0}^{\infty} h(\tau)e^{-j\omega_0 \tau}d\tau \right) e^{j\omega_0 t} = H(j\omega_0)e^{j\omega_0 t} \qquad (9.78)
\end{aligned}
$$

where we have used $h(t) = 0$ for $t < 0$ and (9.64) with s replaced by $j\omega_0$. This equation is similar to (9.50) with $a = 1$ and is the way of introducing frequency response in most texts on signals and systems.

Even though the derivation of (9.78) appears to be simple, the equation holds, as discussed in the preceding sections, only if the system is stable or $h(t)$ is absolutely integrable. This stability condition is not used explicitly in its derivation and is often ignored. More seriously, the equation in (9.78) holds for all t in $(-\infty, \infty)$. In other words, there is no transient response in (9.78). Thus (9.78) describes only steady-state responses. In reality, it is not possible to apply an input from time $-\infty$. An input can be applied only from some finite time where we may call it time zero. Recall that time zero is a relative one and is defined by us. Thus there is always a transient response. In conclusion, the derivation of (9.50), that uses explicitly the stability condition and is valid only for $t \to \infty$, is more revealing than the derivation of (9.78). It also describes real-world situation.

9.9   Frequency responses and frequency spectra

We showed in Section 9.7 that the output of a stable system with transfer function $H(s)$ excited by the pure sinusoid $u(t) = ae^{j\omega_0 t}$ approaches the pure sinusoid $aH(j\omega_0)e^{j\omega_0 t}$. Real-world signals are rarely pure sinusoids. We now discuss the general case.

The input and output of a system with transfer function $H(s)$ are related by

$$Y(s) = H(s)U(s) \qquad (9.79)$$

This is a general equation, applicable regardless of whether the system is stable or not and regardless of whether the spectrum of the input is defined or not. Let us replace

s in (9.79) by $j\omega$ to yield

$$Y(j\omega) = H(j\omega)U(j\omega) \tag{9.80}$$

If the system is not stable or if the input grows unbounded, then one of $H(j\omega)$ and $U(j\omega)$ does not have any physical meaning and, consequently, neither does their product $Y(j\omega)$. However, if the system is BIBO stable and if the input is absolutely integrable, then $H(j\omega)$ is the frequency response of the system and $U(j\omega)$ is, as shown in (9.66), the frequency spectrum of the input. Now the question is: Does their product have any physical meaning?

To answer the question, we show that if $H(s)$ is stable (its impulse response is absolutely integrable in $[0, \infty)$) and if $u(t)$ is absolutely integrable in $[0, \infty)$, so is $y(t)$. Indeed we have

$$y(t) = \int_{\tau=0}^{\infty} h(t-\tau)u(\tau)d\tau$$

which implies

$$|y(t)| = \left| \int_{\tau=0}^{\infty} h(t-\tau)u(\tau)d\tau \right| \leq \int_{\tau=0}^{\infty} |h(t-\tau)||u(\tau)|d\tau$$

Thus we have

$$\int_{t=0}^{\infty} |y(t)|dt \leq \int_{t=0}^{\infty} \left(\int_{\tau=0}^{\infty} |h(t-\tau)||u(\tau)|d\tau \right) dt = \int_{\tau=0}^{\infty} \left(\int_{t=0}^{\infty} |h(t-\tau)|dt \right) |u(\tau)|d\tau$$

$$= \int_{\tau=0}^{\infty} \left(\int_{\bar{t}=-\tau}^{\infty} |h(\bar{t})|d\bar{t} \right) |u(\tau)|d\tau = \left(\int_{\bar{t}=0}^{\infty} |h(\bar{t})|d\bar{t} \right) \left(\int_{\tau=0}^{\infty} |u(\tau)|d\tau \right)$$

where we have interchanged the order of integrations, introduced a new variable $\bar{t} = t - \tau$, and used the causality condition $h(\bar{t}) = 0$ for $\bar{t} < 0$. Thus if $h(t)$ and $u(t)$ are absolutely integrable, so is $y(t)$. Consequently the spectrum of the output is well defined and equals the product of the frequency response of the system and the frequency spectrum of the input, that is

(Frequency spectrum of output signal)

$= $ (Frequency response of LTI stable system) $\times$ (Frequency spectrum of input signal)

This is an important equation.

Consider the magnitude and phase responses of an ideal lowpass filter with cutoff frequency ω_c shown in Figure 9.11(a) or

$$H(j\omega) = \begin{cases} 1 \cdot e^{-j\omega t_0} & \text{for } |\omega| \leq \omega_c \\ 0 & \text{for } |\omega| > \omega_c \end{cases} \tag{9.81}$$

where we have also introduced a linear phase $-\omega t_0$, for some constant t_0, in the passband. Now if $u(t) = u_1(t) + u_2(t)$ and if the magnitude spectra of $u_1(t)$ and $u_2(t)$ are as shown in Figure 9.18, then the output's frequency spectrum is given by

$$Y(j\omega) = H(j\omega)U(j\omega) = H(j\omega)U_1(j\omega) + H(j\omega)U_2(j\omega) = U_1(j\omega)e^{-j\omega t_0} \tag{9.82}$$

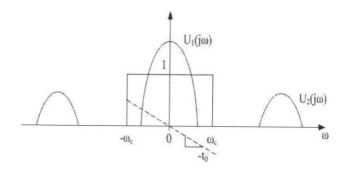

Figure 9.18: Spectra of $u_1(t)$ and $u_2(t)$.

where $H(j\omega)U_2(j\omega)$ is identically zero because their nonzero parts do not overlap. As derived in Problem 4.12, $U_1(j\omega)e^{-j\omega t_0}$ is the frequency spectrum of $u_1(t - t_0)$. Thus the output of the ideal lowpass filter is

$$y(t) = u_1(t - t_0) \tag{9.83}$$

That is, the filter stops the signal $u_2(t)$ and passes $u_1(t)$ with only a delay of t_0 seconds. This is called a *distortionless transmission* of $u_1(t)$. Note that t_0 is the group delay of the ideal lowpass filter defined in (9.54). We see that filtering is based entirely on (9.80).

9.9.1 Why modulation is not an LTI process

Modulation is fundamental in communication and most texts on communication start with the discussion of LTI systems and then discuss various modulation schemes. See, for example, References [C10, H2, L2]. The simplest modulation, as discussed in Subsection 4.4.1, is to shift the spectrum $\bar{X}(\omega)$ of a signal $x(t)$ to $\bar{X}(\omega \pm \omega_c)$, where ω_c is the carrier frequency.

Consider an LTI stable system with transfer function $H(s)$. Its input and output are related by $Y(s) = H(s)U(s)$ or

$$Y(j\omega) = H(j\omega)U(j\omega)$$

If $U(j\omega) = 0$, for ω in some frequency range, then the equation implies $Y(j\omega) = 0$ for the same frequency range. It means that *the output of an LTI stable system cannot contain nonzero frequency components other than those contained in the input* or, equivalently, *an LTI system can only modify the nonzero frequency spectrum of an input but cannot generate new frequency components.* Modulation however always create new frequencies and thus cannot be a linear time-invariant process. Recall that the modulation discussed in Section 4.4.1 is a linear but time-varying process.

Then why bother to discuss LTI systems in communication texts? In modulation, the frequency spectrum of a signal will be shifted to new locations. This is achieved using mostly nonlinear devices. The output of a nonlinear device will generate the desired modulated signal as well as some undesired signals. The undesired signals must be eliminated using LTI lumped filters. Demodulators such as envelop detectors are nonlinear devices and must include LTI lowpass filters. Thus filters are integral parts of all communication systems. Communication texts however are concerned only with the effects of filtering on frequency spectra based on (9.80). They are not concerned with actual design of filters. Consequently many communication texts introduce only the Fourier transform (without introducing the Laplace transform) and call $H(j\omega)$ or $\bar{H}(\omega)$ a transfer function.

9.9.2   Resonance – Time domain and frequency domain

In this subsection[15] we show that the phenomenon of resonance can be easily explained in the frequency domain but not in the time domain.

Consider a system with transfer function

$$H(s) = \frac{20}{s^2 + 0.4s + 400.04} = \frac{20}{(s + 0.2)^2 + 20^2} \tag{9.84}$$

It is stable and has its impulse response (inverse Laplace transform of H(s)) shown in Figure 9.19(a) and its magnitude response (magnitude of $H(j\omega)$) shown in Figure 9.19(aa). The magnitude response shows a narrow spike at frequency roughly $\omega_r = 20$ rad/s. We call ω_r the *resonance frequency*.

Let us study the outputs of the system excited by the inputs

$$u_i(t) = e^{-0.5t} \cos \omega_i t$$

with $\omega_1 = 5$, $\omega_2 = 20$, and $\omega_3 = 35$. The inputs $u_i(t)$ against t are plotted in Figures 9.19(b), (c), and (d). Their magnitude spectra $|U_i(j\omega)|$ against ω are plotted in Figures 9.19(bb), (cc), and (dd). Figures 9.19(bbb), (ccc), and (ddd) show their excited outputs $y_i(t)$. Even though the three inputs have the same peak magnitude 1 and roughly the same total amount of energy[16], their excited outputs are very different as shown in Figure 9.19(bbb), (ccc), and (ddd). The second output has the largest peak magnitude and the most energy among the three outputs. This will be difficult to explain in the time domain. However it becomes transparent in the frequency domain. The nonzero portion of the magnitude response of the system and the nonzero portions of the magnitude spectra of the first and third inputs do not overlap. Thus their products are practically zero for all ω and the corresponding outputs are small. On the other hand, the nonzero portion of the magnitude response of the system and the nonzero portion of the magnitude spectrum of the second input coincide. Thus their product is nonzero in the neighborhood of $\omega = 20$ rad/s and the corresponding output is large as shown.

No physical system is designed to be completely rigid. Every structure or mechanical system will vibrate when it is subjected to a shock or an oscillating force. For example, the wings of a Boeing 747 vibrate fiercely when it flies into a storm. Thus in designing a system, if its magnitude response has a narrow spike, it is important that

---

[15]This subsection may be skipped without loss of continuity.
[16]Because their magnitude spectra have roughly the same form and magnitude, it follows from (4.36) that the three inputs have roughly the same total amount of energy.

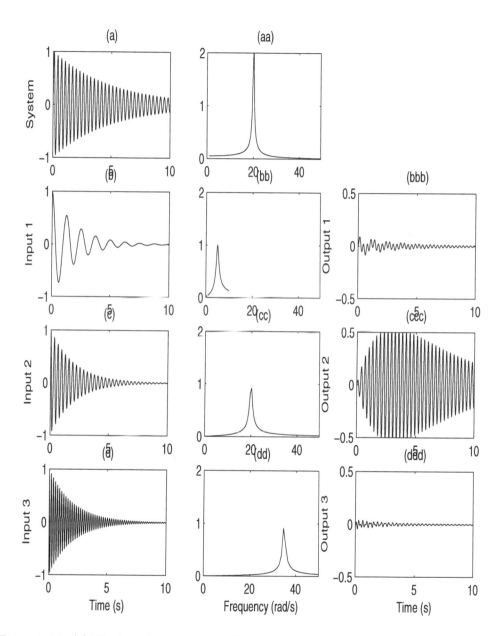

Figure 9.19: (a) The impulse response $h(t)$ of the system in (9.84). (aa) The magnitude response $|H(j\omega)|$ of the system in (9.84). (b) $u_1 = e^{-0.5t}\cos 5t$. (bb) Its magnitude spectrum. (bbb) Its excited output $y_1(t)$. (c) $u_2 = e^{-0.5t}\cos 20t$. (cc) Its magnitude spectrum. (ccc) Its excited output $y_2(t)$. (d) $u_3 = e^{-0.5t}\cos 35t$. (dd) Its magnitude spectrum. (ddd) Its excited output $y_3(t)$.

the spike should not coincide with the most significant part of the frequency spectrum of possible external excitation. Otherwise, excessive vibration may occur and cause eventual failure of the system. The most infamous such failure was the collapse of the first Tacoma Narrows Suspension Bridge in Seattle in 1940 due to wind-induced resonance.[17]

To conclude this section, we discuss how the plots in Figures 9.19 are generated. The inverse Laplace transform of (9.84) is $h(t) = e^{-0.2t} \sin 20t$. Typing

```
t=0:0.05:10;h=exp(-0.2*t).*sin(20*t);plot(t,h)
```

yields the impulse response of (9.84) in Figure 9.19(a). Typing

```
n=20;d=[1 0.4 400.04]; [H,w]=freqs(n,d);plot(w,abs(H))
```

yields the magnitude response of (9.84) in Figure 9.19(aa). Note that the function [H,w]=freqs(n,d) selects automatically 200 frequencies, denoted by w, and then computes the frequency responses at those frequencies and stores them in H. Typing

```
t=0:0.05:10;u1=exp(-0.5*t).*cos(5*t);plot(t,u1)
```

yields the input u_1 in Figure 9.19(b). Because the Laplace transform of $u_1(t)$ with $\omega_1 = 5$ is

$$U_1(s) = (s+0.5)/((s+0.5)^2 + 25) = (s+0.5)/(s^2 + s + 25.25)$$

the frequency spectrum in Figure 9.19(bb) can be obtained by typing

```
n1=[1 0.5];d1=[1 1 25.25];[H1,w1]=freqs(n1,d1);plot(w1,abs(H1))
```

The output of (9.84) excited by $u_1(t)$, shown in Figure 9.19(bbb), is obtained by typing

```
y1=lsim(n,d,u1,t);plot(t,y1)
```

Recall that the computation is carried out using the control-form ss equation realization of (9.84). The rest of Figure 9.19 are similarly generated.

9.10   Reasons for not using ss equations in design

We introduced for transfer functions the concepts of coprimeness, poles, zeros, and general form of step responses. The corresponding concepts in ss equations are controllability, observability, eigenvalues, transmission zeros. To develop the general form of responses, we must transform an ss equation into Jordan form using similarity transformations. Developing these results in ss equations requires some matrix theory and is much more complicated and less transparent than in transfer functions. Thus its discussion is omitted. The interested reader is referred to Reference [C6].

Now we give reasons for not using ss equations in design.

---

[17]Some texts, for example Reference [D1], use the collapse of the bridge to illustrate the concept of stability. It is however more a resonance issue than a stability issue.

1. A proper rational transfer function of degree n has at most $2(n + 1)$ nonzero parameters. For a general ss equation of dimension n, the matrix $\mathbf{A}$ is of order $n \times n$ and has n^2 entries. The vectors $\mathbf{b}$ and $\mathbf{c}$ each have n entries. Thus the set $\{\mathbf{A}, \mathbf{b}, \mathbf{c}, d\}$ has a total of $n^2 + 2n + 1$ parameters. For example, if $n = 5$, then a transfer function may have 12 parameters, whereas an ss equation may have 36 parameters. Thus the use of transfer functions in design is considerably simpler than the use of ss equations.

2. Filters to be designed are specified in the frequency domain as shown in Figure 9.12. The design consists of searching a proper stable transfer function of a smallest possible degree so that its magnitude response meets the specifications. In this searching, ss equations cannot be used. Thus we use exclusively transfer functions in filter design.

3. In control systems, design often involves feedback. Transfer functions of feedback systems can be easily computed as in Figures 7.8(c) and 7.11. Note that those formulas are directly applicable to the CT case if z is replaced by s. State-space equations describing feedback systems are complicated. Thus it is simpler to use transfer functions to design control systems. See the next chapter and References [C5, C6].

In conclusion, we encounter in this and the preceding chapters three domains:

- *Transform domain*: where we carry out design. In filters design, we search a transfer function whose frequency response meets the specification given in the frequency domain. In control system design, we search a compensator, a transfer function, so that the feedback system meets the specifications in the time domain such as overshoot and response time. Once a transfer function is found, we use its minimal realization to carry out implementation.

- *Time domain*: where actual processing is carried out. The state-space equation is most convenient in real-time processing. Systems' frequency responses and signals' frequency spectra do not play any role in actual real-time processing.

- *Frequency domain*: where the specification is developed from frequency spectra of signals to be processed.

It is important to understand the role of these three domains in engineering.

9.10.1 A brief history

This section discusses briefly the advent of transfer functions and state-space equations in ECE programs. We discuss first some historical events which are relevant to our discussion.

The simplest control system is to control the water level of the storage tank of a flush toilet. After each flush, the water level in the storage tank will return to a desired or preset level. It is achieved using a float to control the opening of the valve.[18] It

---

[18] If you never see its operation, go home and open the cover of the storage tank of a toilet. Flush it and see how water comes back automatically to a fixed level.

is an *automatic control* because it does not requires any human's intervention. It is a *feedback* system because the float is controlled by the *actual* water level. This type of water level control can be traced all the way back to water clocks developed in ancient times, circa BC. A water clock consisted of a big tank with a opening through which a small stream would flow into a long tube with a linear scale. If the water level of the tank was kept constant, then the speed of the stream would be constant and the amount of water accumulated in the tube could be used to tell time. The water level in the tank was controlled using a float just as in flush toilets. Thus automatic feedback control systems have been in existence for over two thousand years.

In 1786, James Watt developed an efficient steam engine by introducing a condenser; he also introduced a centrifugal flyball governor to better control the engine speed. In addition, improved temperature regulators and pressure regulators were also developed. These devices all used feedback to achieve automatic control to counter variations in the pressure and temperature of steam and in load. Note that steam engines were invented in the late seventeenth century (1601-1700) in order to pump water out of coal mines. Prior to their uses in steam engines, temperature regulators were used to control the temperature of incubators for hatching chickens and the centrifugal governors were widely used in windmills to control the spacing and pressure of millstones. Thus an invention, in reality, is often incremental or improvement of exiting ones. We mention that feedback control is indispensable in refrigerators which first appeared in 1858. We also mention that Gustaf Dalén, a Swedish inventor and industrialist, received the Nobel Prize in Physics 1912 "for his invention of automatic regulators for use in conjunction with gas accumulators for ... ". The regulators were probably gas pressure regulators.

The invention of three-element vacuum tubes by Lee De Forest in 1907 was a cornerstone for speech transmission and radio broadcasting. Using vacuum tubes to amplify fading speech, telephone transmission from New York to San Francisco became operational in 1915. Filters were also developed to eliminate transmission noise. Using sinusoidal signals generated by vacuum tubes to modulate audio signals, radio broadcasting by more than one station became possible. Amplifiers for audio signals and for radio signals, power amplifiers, tuners, filters, and oscillators were developed to build radio transmitters and receivers. By 1922, there were more than 500 ratio stations in the US and a large number of transmitters and receivers. By then, it is fair to say that the design of electronic circuits had become routine. The design was probably all carried out using empirical methods.

Oliver Heaviside (1850-1925) used p to denote differentiation and introduced resistive operators R, Lp, and $1/Cp$ to study electric circuits in 1887. This ushered the concepts of impedances and transfer functions. Heaviside computed the step response of a simple pole using an infinite power series and then derived the step response of a transfer function with no zero and a number of simple poles excluding the pole at the origin using partial fraction expansion. Heaviside's operational calculus however lacked an inversion formula and was incomplete. It was recognized around 1940 that the Laplace transform encompassed the Heaviside method. The theory of Laplace transform, first introduced by Pierre-Simon Laplace in 1782, was complete and rigorous. See References [C2, C10, D0]. Because of its simplicity, the Laplace transform and the transfer or system function have become an indispensable tool in studying

LTI systems. See Reference [G2]. We mention that filters were widely used by 1922. However the Butterworth filter was first suggested in 1930 and Chebyshev and Elliptic filters appeared later. They were all based on transfer functions.

In order to explain the dancing of flyballs and jerking of machine shafts in steam engines, James Maxwell developed in 1868 a third-order linearized differential equation to raise the issue of stability. He concluded that in order to be stable, the characteristic polynomial of the differential equation must be CT stable. This earmarked the beginning of mathematical study of control systems. In 1895, Edward Routh developed the Routh test (Section 9.5.3). In 1932, Henry Nyquist developed a graphical method of checking the stability of a feedback system from its open-loop system. The method however is fairly complex and is not suitable for design. In 1940, Hendrick Bode simplified the method to use the phase margin and gain margin of the open-loop system to check stability. However the method is applicable only to a small class of open-loop systems. Moreover, the relationship between phase and gain margins and system performances is vague. In 1948, W. R. Evans developed the root-locus method to carry out design of feedback systems. The method is general but the compensator used is essentially limited to degree 0. The aforementioned methods are all based on transfer functions and constitute the entire bulk of most texts on control systems published before 1970.

State-space equations first appeared in the engineering literature in the early 1960s. The formulation is precise: it first gives definitions, and then develops conditions and finally establishes theorems. Moreover its formulation for SISO and MIMO systems are the same and all results for SISO systems can be extended to MIMO systems. The most celebrated results are: If an ss equation is controllable, then state feedback can achieve arbitrary pole-placement. If an ss equation is observable, then a state estimator with any desired poles can be constructed. By 1980, ss equations and designs were introduced into many undergraduate texts on control.

With the impetus of ss results, researchers took a fresh look in the 1970s into transfer functions. By considering a rational function of a ratio of two polynomials, the polynomial-fraction approach was born. By so doing, results in SISO systems can also be extended to MIMO systems. The important concept in this approach is coprimeness. Under the coprimeness assumption, it is possible to achieve pole-placement and model-matching designs. The method is simpler and the results are more general than those based on ss equations. See Reference [C5, C6].

From the preceding discussion, we see that mathematical methods of designing feedback control systems started to appear in the 1940s. By then feedback control systems had been in existence for over 150 years counting from the feedback control of steam engines. Because the aforementioned design methods are applicable only to LTI lumped systems, whereas all practical systems have gone through many iterations of improvement, are generally very good, and cannot be described by simple mathematical equations. Thus it is not clear to this author how much of control text's design methods are used in practice.

The role of transfer functions has been firmly established in ECE curricula. Now transfer functions appear in texts on circuit analysis, analog filter design, microelectronic circuits, digital signal processing, and, to a lesser extent, communication. The role of ss equations however is still not clear. State-space equations are now introduced in every undergraduate control texts. See, for example, Reference [D1]. They stress analytical solutions and discuss some design.

The class of systems studied in most texts on signals and systems is limited to linear time-invariant and lumped systems. Such a system can be described by a convolution, higher-order differential equation, ss equation, and rational transfer function. The first three are in the time domain and the last one is in the frequency domain. Many existing texts stress convolutions and Fourier analysis of systems and ignore ss equations. This text gives reasons for not doing so. We stress the use of ss equations in computer computation, real-time processing, and op-amp circuit implementations and the use of transfer functions in qualitative analysis of systems. This selection of topics may be more useful in practice.

Problems

9.1 What are the poles and zeros of the following transfer functions:

$$H_1(s) = \frac{s^2 - 1}{3s^2 + 3s - 6}$$

$$H_2(s) = \frac{2s + 5}{3s^2 + 9s + 6}$$

$$H_3(s) = \frac{s^2 - 2s + 5}{(s + 0.5)(s^2 + 4s + 13)}$$

Plot the poles and zeros of $H_3(s)$ on a complex s-plane. Note that you have the freedom in selecting the scales of the plot.

9.2 Find the impulse and step responses of a system with transfer function

$$H(s) = \frac{2s^2 - 10s + 1}{s^2 + 3s + 2}$$

9.3 Verify that if $H(s)$ is proper, then its impulse response equals the step response of $sH(s)$. Note that in computer computation, a transfer functions is first realized as an ss equation and then carried out computation. If $H(s)$ is biproper, then $sH(s)$ is improper and cannot be realized in a standard-form ss equation. Thus if we use a computer to compute the impulse response of $H(s)$, we require $H(s)$ to be strictly proper which implies $d = 0$ in its ss-equation realization. If $H(s)$ is biproper, the MATLAB function `impulse` will simply ignore d. For example, for $H_1(s) = 1/(s + 1)$ and $H_2(s) = H_1(s) + 2 = (2s + 3)/(s + 1)$, typing in MATLAB `n1=1;d=[1 1];impulse(n1,d)` and `n2=[2 3];d=[1 1];impulse(n2,d)` will yield the same result. See also Problem 8.6.

9.4* Consider the heating system shown in Figure 9.20. Let $y(t)$ be the temperature of the chamber and $u(t)$ be the amount of heat pumping into the chamber. Suppose they are related by

$$\dot{y}(t) + 0.0001y(t) = u(t)$$

If no heat is applied and if the temperature is 80°, how long will it take for the temperature to drop to 70°? [Hint: Use (8.43).]

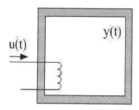

Figure 9.20: Heating system.

9.5 What is the general form of the step response of a system with transfer function

$$H(s) = \frac{10(s-1)}{(s+1)^3(s+0.1)}$$

9.6 Consider the transfer function

$$H(s) = \frac{N(s)}{(s+2)^4(s+0.2)(s^2+2s+10)}$$

where $N(s)$ is a polynomial of degree 7 or less and has $N(0) = 320$. What is the form of its step response?

9.7 Consider a system with transfer function

$$H(s) = \frac{s+3}{(s+1)(s-1)}$$

If we apply the input $u(t)$ with Laplace transfer

$$U(s) = \frac{s-1}{s(s+3)}$$

what is its output? Is every pole of $H(s)$ excited by $u(t)$? In general, the output of an unstable system excited by any input will be unbounded. This problem shows an exception.

9.8 Can we define a system to be stable if a bounded input excites a bounded output?

9.9 Is it true that if a system is not stable only if all its poles lie inside the right-half s-plane or on the $j\omega$-axis?

9.10 Use the Routh test to check the CT stability for each of the following polynomials.

1. $s^5 + 3s^3 + 2s^2 + s + 1$
2. $s^5 + 4s^4 - 3s^3 + 2s^2 + s + 1$
3. $s^5 + 4s^4 + 3s^3 + 2s^2 + s + 1$
4. $s^5 + 6s^4 + 23s^3 + 52s^2 + 54s + 20$

9.11 Show that a polynomial of degree 1 or 2 is a CT stable polynomial if and only if all coefficients are of the same sign.

9.12 Show that the polynomial

$$s^3 + a_1 s^2 + a_2 s + a_3$$

is a CT stable polynomial if and only if $a_1 > 0$, $a_2 > 0$, and $a_1 a_2 > a_3 > 0$.

9.13 Verify from their transfer functions that the two circuits in Figures 9.7(b) and (c) are stable.

9.14 What are the steady-state and transient responses of the system in Problem 9.5 excited by a step input? How many seconds will it take for the transient response to die out or, equivalently, for the response to reach steady state?

9.15 What are the steady-state and transient responses of the system in Problem 9.6 excited by a step input? How many seconds will it take for the transient response to die out or, equivalently, for the response to reach steady state?

9.16* Consider the system with transfer function $1/(s+1)^3$. Compute its step response. What is its transient response? Verify that the transient response decreases to less than 1% of its peak value in nine time constants.

9.17 Consider the transfer function

$$H(s) = \frac{-10}{s + 10}$$

Compute $H(j\omega)$ at $\omega = 0, 10$, and 100 and then sketch roughly its magnitude and phase responses for ω in $[-100, 100]$.

9.18 Verify that $H_1(s) = 2/(s+2)$ and $H_2(s) = 2/(s-2)$ have the identical magnitude response.

9.19 What is the steady-state response of a system with transfer function

$$H(s) = \frac{s - 0.2}{s^2 + s + 100}$$

excited by the input

$$u(t) = 2 + \cos 10t - \sin 100t$$

How long will it take to reach steady state? Is it a lowpass, bandpass, or highpass filter?

9.20 The ideal lowpass filter in Figure 9.11(a) can be expressed as

$$H(\omega) = \begin{cases} e^{-j\omega t_0} & \text{for } |\omega| \le \omega_c \\ 0 & \text{for } \omega| > \omega_c \end{cases}$$

Verify that its inverse Fourier transform defined in (4.22) is given by

$$h(t) = \frac{\sin[\omega_c(t - t_0)]}{\pi(t - t_0)}$$

for all t in $(-\infty, \infty)$. Because $h(t)$ is not identically zero for all $t < 0$, the ideal lowpass filter is not causal and cannot be built in the real world.

9.21 Consider the transfer function

$$H(s) = \frac{k}{s+a}$$

where a and k are real constants with $a > 0$. Show that it is a lowpass filter with 3-dB passband $[0, \ a]$ and bandwidth a. Is the condition $a > 0$ necessary? Do we need $k > 0$?

9.22 Consider the network shown in Figure 9.21(a). Compute its transfer function.

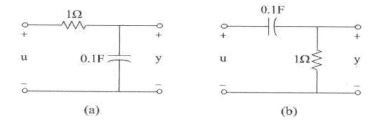

Figure 9.21: Two *RC* circuits.

Is it of the form in Problem 9.21? Is it lowpass? What is its 3-dB bandwidth? What is its steady state response excited by the input

$$u(t) = \sin 0.1t + \sin 100t$$

9.23 Consider the network shown in Figure 9.21(b). Compute its transfer function and plot their magnitude and phase responses. What is its steady state response excited by the input

$$u(t) = \sin 0.1t + \sin 100t$$

Is it lowpass or highpass? What is its 3-dB bandwidth? What is its 3-dB passband edge frequency?

9.24 Verify that for a properly designed transfer function of degree 2, a lowpass, bandpass, and highpass filter must assume respectively the following form

$$
\begin{aligned}
H_l(s) &= \frac{b}{s^2 + a_2 s + a_3} \\
H_b(s) &= \frac{bs + d}{s^2 + a_2 s + a_3} \\
H_h(s) &= \frac{bs^2 + cs + d}{s^2 + a_2 s + a_3}
\end{aligned}
$$

with $b \neq 0$ and $|d| << a_3$.

9.25 Consider the op-amp circuit shown in Figure 9.22. Using the ideal model to verify that the voltage $v(t)$ and current $i(t)$ shown are related by $v(t) = -Ri(t)$. Thus the circuit can be used to implement a resistor with a negative resistance.

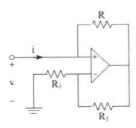

Figure 9.22: Negative impedance converter.

9.26* Let $x(t)$ be two sided and absolutely integrable in $(-\infty, \infty)$. Let $x_+(t) = x(t)$, for $t \geq 0$ and $x_+(t) = 0$, for $t < 0$. In other words, $x_+(t)$ denotes the positive-time part of $x(t)$. Let $x_-(t) = x(t)$, for $t < 0$ and $x_-(t) = 0$, for $t \geq 0$. In other words, $x_-(t)$ denotes the negative-time part of $x(t)$. Verify

$$\mathcal{F}[x(t)] = \mathcal{L}[x_+(t)]\big|_{s=j\omega} + \mathcal{L}[x_-(-t)]\big|_{s=-j\omega}$$

Using this formula, the Fourier transform of a two-sided signal can be computed using the (one-sided) Laplace transform. Note that $x_-(t)$ is negative time, but $x_-(-t)$ is positive time.

9.27 Let $X(s)$ be the Laplace transform of $x(t)$ and be a proper rational function. Show that $x(t)$ approaches a constant (zero or nonzero) if and only if all poles, except possibly a simple pole at $s = 0$, of $X(s)$ have negative real parts. In this case, we have

$$\lim_{t \to \infty} x(t) = \lim_{s \to 0} sX(s)$$

This is called the *final-value theorem*. To use the theorem, we first must check essentially the stability of $X(s)$. It is not as useful as Theorem 9.5.

Chapter 10

Model reduction and some feedback Designs

10.1 Introduction

This chapter introduces three independent topics.[1] The first topic, consisting of Sections 10.2 and 10.3, discusses model reduction which includes the concept of dominant poles as a special case. Model reduction is widely used in engineering and yet rarely discussed in most texts. Because it is based on systems' frequency responses and signals' frequency spectra, its introduction in this text is most appropriate. We introduce the concept of operational frequency ranges of devices and apply it to op-amp circuits, seismometers, and accelerometers.

The second topic, consisting of Sections 10.4 and 10.5, discusses composite systems, in particular, feedback systems. Feedback is used in refrigerators, ovens, and homes to maintain a set temperature and in auto-cruise of automobiles to maintain a set speed. We use examples to demonstrate the necessity and advantage of using feedback. We then discuss pole-placement design and feedback design of inverse systems.

The last topic discusses the Wien-bridge oscillator. The circuit will generate a sustained oscillation once it is excited and the input is removed. We first design it directly. We then develop a feedback model for an op-amp circuit and then use it to design the oscillator. We also relate the conventional oscillation condition with pole location.

10.2 Op-amp circuits based on a single-pole model

We introduce model reduction by way of answering the question posed in Section 6.7.2. Consider the op-amp circuits shown in Figure 6.9 and repeated in Figure 10.1. We showed in Chapter 6 that the two circuits, using LTI memoryless models with finite and infinite open-loop gains, can be described by the same equation $v_o(t) = v_i(t)$.

---

[1]The order of their studies can be changed.

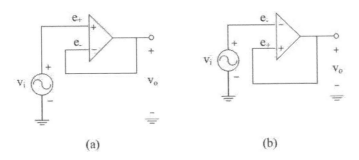

Figure 10.1: (a) Voltage follower. (b) Useless system.

However, the circuit in Figure 10.1(a) can be used as a voltage follower or buffer, but not the one in Figure 10.1(b). We now give the reason.

If an op amp is modeled as LTI memoryless with open-loop gain A, then its transfer function is A and its frequency response is independent of frequency. In reality, typical magnitude and phase responses of an op amp are as shown in Figure 10.2. It is plotted using logarithmic scales for frequencies and for magnitudes. But the phase is plotted with linear scale. The magnitude plot can be approximated by three sections of straight lines (one horizontal, one with slope $-45°$, and one with an even larger slope). They intersect roughly at $f = 8$ and 3×10^6 in Hz or $\omega = 16\pi$ and $6\pi \times 10^6$ in rad/s. The horizontal line passes through the gain at 2×10^5. Thus the transfer function of the op amp is

$$A(s) = \frac{2 \times 10^5}{\left(1 + \frac{s}{16\pi}\right)\left(1 + \frac{s}{6\pi \cdot 10^6}\right)} \tag{10.1}$$

See Subsection 9.7.5 and References [C5, C8]. It has two poles, one at -16π and the other at $-6\pi \times 10^6$. The response due to the latter pole will vanish much faster than that of the former, thus the response of (10.1) is dominated by the pole -16π and the transfer function in (10.1) can be reduced to or approximated by

$$A(s) = \frac{2 \times 10^5}{\left(1 + \frac{s}{16\pi}\right)} = \frac{32\pi \cdot 10^5}{s + 16\pi} \approx \frac{10^7}{s + 50.3} \tag{10.2}$$

This is called a *single-pole* or *dominate-pole* model of the op amp. This simplification or model reduction will be discussed in the next subsection.

If an op amp is modeled as memoryless and operates in its linear region, then its inputs and output are related by

$$v_o(t) = A[e_+(t) - e_-(t)]$$

in the time-domain, or

$$V_o(s) = A[E_+(s) - E_-(s)] \tag{10.3}$$

in the transform-domain, where A is a constant and capital letters are the Laplace transforms of the corresponding lower-case letters. Now if the op amp is modeled to have memory and has the transfer function $A(s)$, then (10.3) must be modified as

$$V_o(s) = A(s)[E_+(s) - E_-(s)] \tag{10.4}$$

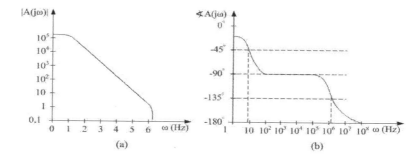

Figure 10.2: (a) Magnitude response of an op amp. (b) Its frequency response.

Now we use this equation with $A(s)$ in (10.2) to study the stability of the op-amp circuits. The use of (10.1) will yield the same conclusion but the analysis will be more complicated.

Consider the circuit in Figure 10.1(a). Substituting $V_i(s) = E_+(s)$ and $V_o(s) = E_-(s)$ into (10.4) yields

$$V_o(s) = A(s)(V_i(s) - V_o(s))$$

which implies

$$(1 + A(s))V_o(s) = A(s)V_i(s)$$

Thus the transfer function of the circuit in Figure 10.1(a) is

$$H(s) = \frac{V_o(s)}{V_i(s)} = \frac{A(s)}{A(s) + 1} \tag{10.5}$$

or, substituting (10.2),

$$H(s) = \frac{\frac{10^7}{s+50.3}}{1 + \frac{10^7}{s+50.3}} = \frac{10^7}{s + 50.3 + 10^7} \approx \frac{10^7}{s + 10^7} \tag{10.6}$$

It is stable. If we apply a step input, the output $v_o(t)$, as shown in Theorem 9.5, will approach $H(0) = 1$. If we apply $v_i(t) = \sin 10t$, then, because

$$H(j10) = \frac{10^7}{j10 + 10^7} \approx 1e^{j \cdot 0}$$

the output will approach $\sin 10t$. Furthermore, because the time constant is $1/10^7$, it takes roughly 5×10^{-7} second to reach steady state. In other words, the output will follow the input almost instantaneously. Thus the circuit is a very good voltage follower.

Now we consider the op-amp circuit in Figure 10.1(b). Substituting $V_i(s) = E_-(s)$ and $V_o(s) = E_+(s)$ into (10.4) yields

$$V_o(s) = A(s)(V_o(s) - V_i(s))$$

which implies

$$(1 - A(s))V_o(s) = -A(s)V_i(s)$$

Thus the transfer function of the circuit is

$$H(s) = \frac{V_o(s)}{V_i(s)} = \frac{-A(s)}{1 - A(s)} = \frac{A(s)}{A(s) - 1} \tag{10.7}$$

or, substituting (10.2),

$$H(s) = \frac{\frac{10^7}{s+50.3}}{\frac{10^7}{s+50.3} - 1} = \frac{10^7}{10^7 - s - 50.3} \approx \frac{10^7}{10^7 - s} \tag{10.8}$$

It has a pole in the RHP. Thus the circuit is not stable and its output will grow unbounded when an input is applied. Thus the circuit will either burn out or run into a saturation region and cannot be used as a voltage follower.

To conclude this section, we mention that the instability of the circuit in Figure 10.1(b) can also be established using a memoryless but nonlinear model of the op amp. See Reference [C9, pp. 190–193]. But our analysis is simpler.

10.2.1   Model reduction – Operational frequency range

Before discussing model reduction, we give a theorem.

Theorem 10.1 Consider a transfer function $H(s)$ and a simpler transfer function $H_s(s)$. If $H(s)$ is stable and if its frequency response equals the frequency response of $H_s(s)$ in some frequency range B, then $H(s)$ can be reduced to $H_s(s)$ for input signals whose nonzero frequency spectra lie inside B. We call B the *operational frequency range* of $H_s(s)$.□

Proof Let $y(t)$ and $y_s(t)$ be the outputs of $H(s)$ and $H_s(s)$ excited by the same $u(t)$. Then we have

$$Y(j\omega) = H(j\omega)U(j\omega) \tag{10.9}$$

and

$$Y_s(j\omega) = H_s(j\omega)U(j\omega)$$

As discussed in Section 9.9, (10.9) has physical meaning only if the system is stable. If $H(j\omega) = H_s(j\omega)$ for ω in B, then $Y(j\omega) = Y_s(j\omega)$ for ω in B. If the spectrum of $u(t)$ is zero outside B or $U(j\omega) = 0$ for ω lying outside B, then $Y(j\omega) = Y_s(j\omega) = 0$ for ω outside B. Thus we have $Y(j\omega) = Y_s(j\omega)$ for all ω. Consequently we conclude $y(t) = y_s(t)$ for all t. This establishes the theorem. □

In practical application we require only $|H(j\omega)| \approx |H_s(j\omega)|$ for ω in B and $U(j\omega) \approx 0$ for ω outside B. Here we consider only the magnitude responses without considering the phase responses, just as in designing filters discussed at the end of Section 9.7. Because every condition involves approximation, there is no need to specify B precisely. We now discuss its application.

Consider the voltage follower in Figure 10.1(a) with transfer function in (10.6). The magnitude response of (10.6) is plotted in Figure 10.3(a) against $\log_{10}\omega$ for ω from $10^0 = 1$ to 10^{11}. We see that the magnitude response is 1 in the frequency range $[0, 10^6]$. In this frequency range, the magnitude responses of (10.6) and $H_s(s) = 1$ are the same. Thus if the frequency spectra of input signals lie inside the range, the

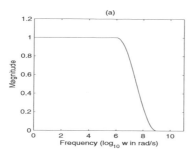
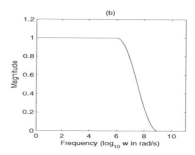

Figure 10.3: (a) Magnitude response of (10.6). (b) Magnitude response of (10.8).

transfer function in (10.6) can be simplified as $H_s(s) = 1$. Indeed, the frequency spectra of the step input and $v_i(t) = \sin 10t$ lie inside the range and their outputs of (10.6) and $H_s(s) = 1$ are indistinguishable as discussed earlier. However if the spectrum of a signal lies outside the range, then the simplified model cannot be used. For example, consider $u(t) = \cos 10^{20}t$ whose nonzero spectrum lies outside $[0, 10^6]$. For this input, the output of $H_s(s) = 1$ is $y_s(t) = \cos 10^{20}t$, for $t \geq 0$. To find the output of $H(s)$ in (10.6), we compute

$$H(j10^{20}) = \frac{10^7}{j10^{20} + 10^7} \approx \frac{10^7}{j10^{20}} \approx 0 \cdot e^{-j\pi/2}$$

Thus the output of $H(s)$ approaches $0 \times \cos(10^{20}t - \pi/2) = 0$ as $t \to \infty$. This output is clearly different from the output of $H_s(s) = 1$.

The transfer function in (10.6) is essentially a lowpass filter with 3-dB passband $[0, 10^7]$ (see Problem 9.21). As mentioned earlier, there is no need to specify B exactly, thus it is convenient to select the operational frequency range B as the 3-dB passband. In conclusion, if the input of a voltage follower is of low frequency or, more precisely, its frequency spectrum lies inside the operational frequency range $[0, 10^7]$, then there is no need to use the more realistic model in (10.6). We can use the simplified model $H_s(s) = 1$. This is widely used in practice. Note that the stability of $H(s)$ is essential in this simplification. For example, the magnitude response of (10.8) is shown in Figure 10.3(b) which is identical to the one in Figure 10.3(a). But the unstable transfer function in (10.8) cannot be simplified as $H_s(s) = 1$.

We give a different example. Consider the transfer function

$$C(s) = k_p + \frac{k_i}{s} + k_d s$$

where k_p is a *proportional* gain, k_i is the parameter associated with $1/s$, an *integrator*, and k_d is associated with s, a *differentiator*. Thus the transfer function is called a *PID controller* and is widely used in control systems. See Reference [C5].

The differentiator $k_d s$ is actually a simplified model. In reality, it is designed as

$$H(s) = \frac{k_d s}{1 + s/N} \tag{10.10}$$

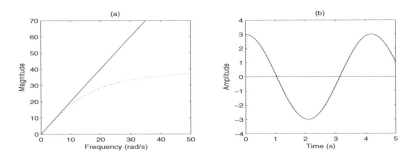

Figure 10.4: (a) Magnitude responses of $2s$ (solid line) and (10.10) with $N = 20$ (dotted line). (b) Outputs of $2s$ (solid line) and (10.10) (dotted line).

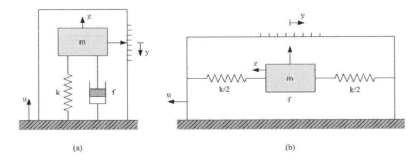

Figure 10.5: (a) Seismometer. (b) Accelerometer.

where N is a constant, called a *taming factor*. The transfer function is biproper and can be implemented without using any differentiator. We plot in Figure 10.4(a) with a solid line the magnitude response of $k_d s$ with $k_d = 2$ and with a dotted line the magnitude response of (10.10) with $N = 20$. We see that they are close for low frequencies or for frequency in $[0, 10]$. Now if we apply the signal $\cos 1.5t$ whose spectrum lies inside the range, then the outputs of (10.10) and $k_d s$ are indistinguishable as shown in Figure 10.4(b) except in the neighborhood of $t = 0$. Note that the transfer function in (10.10) with $N = 20$ has time constant $1/20 = 0.05$ and it takes roughly 0.25 second for the output of (10.10) to reach steady-state as shown in Figure 10.4(b). Thus the transfer function in (10.10) acts as a differentiator for low frequency signals.

10.3   Seismometers

A seismometer is a device to measure and record vibratory movements of the ground caused by earthquakes or man-made explosion. There are many types of seismometers. We consider the one based on the model shown in Figure 10.5(a). A block with mass m, called a *seismic mass*, is supported inside a case through a spring with spring constant k and a dashpot as shown. The dashpot generates a viscous friction with viscous friction coefficient f. See the discussion pertaining Figure 8.2. The case is rigidly attached to the ground. Let u and z be, respectively, the displacements of the case and seismic mass relative to the inertia space. They are measured from the equilibrium position. By this, we mean that if $u = 0$ and $z = 0$, then the gravity of the seismic mass and the spring force cancel out and the mass remains stationary. The input u in Figure 10.5(a) is the movement of the ground.

Now if the ground vibrates and $u(t)$ becomes nonzero, the spring will exert a force to the seismic mass and cause it to move. If the mass is rigidly attached to the case, then $z(t) = u(t)$. Otherwise, we generally have $z(t) < u(t)$. Let us define $y(t) := u(t) - z(t)$. It is the displacement of the mass with respect to the case and can be read from the scale on the case as shown. It can also be transformed into a voltage signal using, for example, a potentiometer. Note that the direction of $y(t)$ in Figure 10.5(a) is opposite to that of $u(t)$. Now the acceleration force of m must be balanced out by the spring force $ky(t)$ and the viscous friction $f\dot{y}(t)$. Thus we have

$$ky(t) + f\dot{y}(t) = m\ddot{z}(t) = m(\ddot{u}(t) - \ddot{y}(t)) \tag{10.11}$$

or

$$m\ddot{y}(t) + f\dot{y}(t) + ky(t) = m\ddot{u}(t) \tag{10.12}$$

Applying the Laplace transform and assuming zero initial conditions, we obtain

$$ms^2Y(s) + fsY(s) + kY(s) = ms^2U(s)$$

Thus the transfer function from u to y of the seismometer is

$$H(s) = \frac{Y(s)}{U(s)} = \frac{ms^2}{ms^2 + fs + k} =: \frac{s^2}{s^2 + \bar{f}s + \bar{k}} \tag{10.13}$$

where $\bar{f} := f/m$ and $\bar{k} := k/m$. The system is always stable for any positive $\bar{f}$ and $\bar{k}$ (see Problem 9.11).

The reading $y(t)$ of a seismometer should be proportional to the actual ground movement $u(t)$. This can be achieved if (10.13) can be reduced or simplified as

$$H_s(s) = \frac{s^2}{s^2} = 1$$

Thus the design problem is to search $\bar{f}$ and $\bar{k}$ so that the frequency range B in which the magnitude response of $H(s)$ equals that of $H_s(s)$ is large.

In order for (10.13) to reduce to $H_s(s) = s^2/s^2$, we require $|j\omega\bar{f} + \bar{k}| \ll |(j\omega)|^2$. Thus a smaller $\bar{k}$ may yield a larger frequency range B. We plot in Figure 10.6(a) the magnitude responses of (10.13) with $\{\bar{f} = 0.63, \bar{k} = 0.2\}$ (dotted line), $\{\bar{f} = 2, \bar{k} = 5\}$ (dot-and-dashed line), and $\{\bar{f} = 4, \bar{k} = 10\}$ (dashed line), and the magnitude response of $H_s(s) = 1$ (horizontal solid line). They are obtained in MATLAB by typing

```
w=0:0.01:5;n=[1 0 0];
d1=[1 0.63 0.2];H1=freqs(n,d1,w);d2=[1 2 5];H2=freqs(n,d2,w);
d3=[1 4 10];H3=freqs(n,d3,w);
plot(w,abs(H1),':',w,abs(H2),'-.',w,abs(H3),'--',[0 15],[1 1])
```

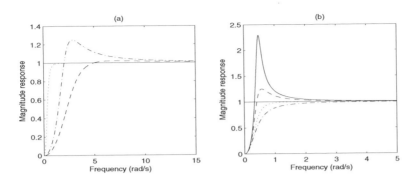

Figure 10.6: (a) Magnitude responses of (10.13) for $\{\bar{k} = 0.2,\ \bar{f} = 0.63\}$ (dotted line), $\{\bar{k} = 5,\ \bar{f} = 2\}$ (dot-and-dashed line), and $\{\bar{k} = 10,\ \bar{f} = 4\}$ (dashed line). (b) Magnitude responses of (10.13) for $\bar{k} = 0.2$ and $\bar{f} = 0.2$ (solid line), 0.4 (dashed line), 0.63 (dotted line), 0.8 (dash-and-dotted line).

Indeed the smallest $\bar{k} = 0.2$ yields the largest $B = [1.25,\ \infty)$. The selection of $\bar{f}$ is carried out by simulation. For the same $\bar{k} = 0.2$, we plot in Figure 10.6(b) the magnitude responses for $\bar{f} = 0.2, 0.4, 0.63$, and 0.8. We see that the magnitude response for $\bar{f} = 0.63$ (dotted line) approaches 1 at the smallest ω or has the largest frequency range $\omega > 1.25$ in which it equals 1. Thus we conclude that (10.13) with $\bar{f} = 0.63$ and $\bar{k} = 0.2$ is a good seismometer. However, it will yield accurate results *only* for signals whose nonzero frequency spectra lie inside the range $B = [1.25,\ \infty)$. In other words, in the frequency range B, the transfer function in (10.13) can be reduced as

$$H_s(s) = \frac{Y_s(s)}{U(s)} = \frac{s^2}{s^2} = 1$$

which implies $y_s(t) = u(t)$, for all t. This is illustrated by an example.

Consider the signal $u_1(t) = e^{-0.3t} \sin t \cos 20t$, for $t \geq 0$ shown in Figure 10.7(a). Its magnitude spectrum can be computed using FFT, as discussed in Chapter 5, and is plotted in Figure 10.8(a). We see that its magnitude spectrum lies entirely inside $B = [1.25, \infty)$. The output of the seismometer excited by $u_1(t)$ is shown in Figure 10.7(aa). It is obtained in MATLAB as follows:

```
n=[1 0 0];d=[1 0.63 0.2];t=0:0.1:10;
u=exp(-0.3*t).*sin(t).*cos(20*t);
y=lsim(n,d,u,t);plot(t,y)
```

We see that $y(t) = u_1(t)$, for all $t \geq 0$. Indeed the transfer function in (10.13) can be reduced to $H_s(s) = 1$.

Next we consider the signal $u_2(t) = e^{-0.3t} \sin t$ shown in Figure 10.7(b). Its magnitude spectrum is shown in Figure 10.8(b). It is significantly different from zero outside B. Thus we cannot expect the output to equal $u_2(t)$. This is indeed the case as shown in Figure 10.7(bb). This shows the importance of the concepts of operational frequency ranges of systems and frequency spectra of signals.

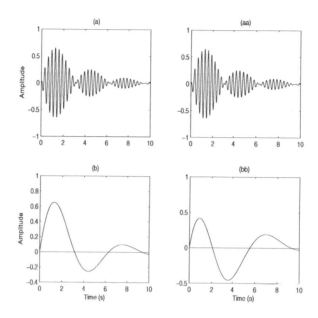

Figure 10.7: (a) $u_1(t) = e^{-0.3t} \sin t \cos 20t$. (aa) Output of the seismometer excited by u_1. (b) $u_2(t) = e^{-0.3t} \sin t$. (bb) Output of the same seismometer excited by u_2.

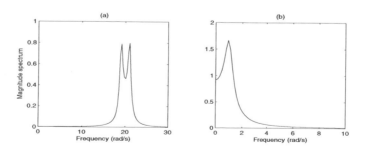

Figure 10.8: (a) Magnitude spectrum of $u_1(t) = e^{-0.3t} \sin t \cos 20t$. (b) Magnitude spectrum of $u_2(t) = e^{-0.3t} \sin t$.

To conclude this section, we mention that even for the simple model in Figure 10.5(a), we may raise many design questions. Can we obtain a larger B by selecting a smaller $\bar{k}$? What are the real-world constraints on $\bar{f}$ and $\bar{k}$? Once a seismometer is designed and installed, it will always give some reading even if the input's frequency spectrum is not inside B. Thus we may also question the importance of B. In conclusion, a real-world design problem is much more complicated than the discussion in this section.

10.3.1   Accelerometers

Consider next the system shown in Figure 10.5(b). It is a model of a type of accelerometers. An accelerometer is a device that measures the acceleration of an object to which the device is attached. By integrating the acceleration twice, we obtain the velocity and distance (position) of the object. It is used in inertial navigation systems on airplanes and ships. It is now widely used to trigger airbags in automobiles.

The model consists of a block with seismic mass m attached to a case through two springs as shown in Figure 10.5(b). The case is filled with oil to create viscous friction and is attached rigidly to an object such as an airplane. Let u and z be, respectively, the displacements of the case and the mass with respect to the inertia space. Because the mass is floating inside the case, u may not equal z. We define $y := u - z$. It is the displacement of the mass with respect to the case and can be transformed into a voltage signal. Note that the direction of $y(t)$ in Figure 10.5(b) is opposite to that of $u(t)$. The input of the accelerometer is the displacement u and the output is acceleration y. Let the spring constant of each spring be $k/2$ and let the viscous friction coefficient be f. Then we have, as in (10.12) and (10.13),

$$ky(t) + f\dot{y}(t)] = m\ddot{z}(t) = m\ddot{u}(t) - m\ddot{y}(t)$$

and

$$H(s) = \frac{Y(s)}{U(s)} = \frac{ms^2}{ms^2 + fs + k} =: \frac{s^2}{s^2 + \bar{f}s + \bar{k}} \tag{10.14}$$

where $\bar{f} := f/m$ and $\bar{k} := k/m$. This system is always stable for any positive $\bar{f}$ and $\bar{k}$ (see Problem 9.11). Note that the transfer function in (10.14) is identical to (10.13) which is designed to act as a seismometer. Now for the same transfer function, we will design it so that it acts as an accelerometer.

The reading $y(t)$ of an accelerometer should be proportional to the acceleration of the case or $d^2u(t)/dt^2$. This can be achieved if (10.14) can be simplified as $H_s(s) = s^2/\bar{k}$. Indeed, the output $y_s(t)$ of $H_s(s)$ excited by $u(t)$ is

$$H_s(s) = \frac{Y_s(s)}{U(s)} = \frac{s^2}{\bar{k}} \tag{10.15}$$

or, in the time domain,

$$y_s(t) = \frac{1}{\bar{k}} \frac{d^2u(t)}{dt^2} \tag{10.16}$$

Thus the output $y_s(t)$ is proportional to the acceleration of the case.

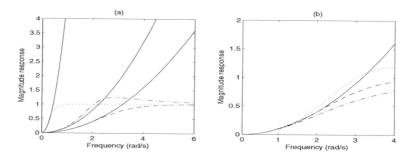

Figure 10.9:   (a) Magnitude responses of (10.14) and (10.15) for $\{\bar{k} = 0.2,\ \bar{f} = 0.63\}$(dotted line), $\{\bar{k} = 10,\ \bar{f} = 4\}$ (dashed line), and $\{\bar{k} = 5,\ \bar{f} = 2\}$.   (b) Magnitude responses of (10.14) for $\bar{k} = 10$ and $\bar{f} = 3$ (dotted line), 4 (dashed line), 5 (dash-and-dotted line).

In order for (10.14) to reduce to (10.15), we require $|(j\omega)^2 + j\omega\bar{f}| \ll \bar{k}$. Thus a larger $\bar{k}$ may yield a larger B. We plot in Figure 10.9(a) the magnitude responses of (10.14) for $\{\bar{f} = 0.63,\ \bar{k} = 0.2\}$ (dotted line), $\{\bar{f} = 2,\ \bar{k} = 5\}$ (dot-and-dashed line), and $\{\bar{f} = 4,\ \bar{k} = 10\}$ (dashed line), and the corresponding magnitude responses of (10.15) (all solid lines). Indeed the largest $\bar{k} = 10$ yields the largest B. We then plot in Figure 10.9(b) the magnitude responses of (10.14) with $\bar{k} = 10$ for $\bar{f} = 3$, 4, and 5. We see that the magnitude response of (10.14) for $\bar{f} = 4$ (dashed line) is very close to the magnitude response of (10.15) with $\bar{k} = 10$ in the frequency range roughly $B = [0,\ 2.3]$. Thus in this frequency range, the system in Figure 10.5(b) described by (10.14) with $\bar{f} = 4$ and $\bar{k} = 10$ can function as an accelerometer.

If the movement of an object, such as a commercial airplane or big ship, is relatively smooth, then the frequency spectrum of the input signal is of low frequency and the accelerometer can be used. Note that an accelerometer will always give out a reading, even if it is not accurate. Consequently, its velocity and distance (or position) may not be accurate. Fortunately, the reading of the position of an airplane can be reset or corrected using GPS (global positioning system) signals or signals from ground control stations.

We see that the same transfer function can be designed to function as a seismometer or as an accelerometer. The design involves the search of parameters of (10.13) or (10.14) so that its magnitude response will match a desired magnitude response in a large frequency range. The device will yield a correct result only if the frequency spectrum of an input lies inside the range. Thus the design involves three concepts: frequency responses of systems, frequency spectra of signals, and operational frequency ranges of devices. However, once a devise is designed, it will generate an output no matter what input is applied. It pays no attention to the spectrum of the input. If the spectrum lies inside the operational frequency range, the result will be accurate. If not, the result will be less accurate as shown in Figures 10.7(b) and (bb). Thus although the concept of operational frequency range is important, its precise definitions is not essential.

The subject of sensors, including seismometers and accelerometers, is a vast one.

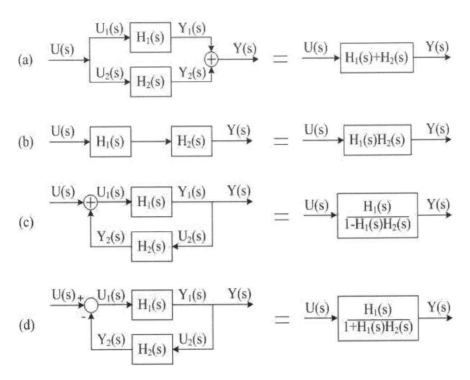

Figure 10.10: (a) Parallel connection. (b) Tandem connection. (c) Negative-feedback connection. (d) Positive-feedback connection.

There are piezoelectric, micro-mechanical, microthermal, magnetic, capacitive, optical, chemical and biological sensors and transducers. See, for example, Reference [F1]. Mathematical modeling of those real-world devices will be complicated. The models in Figure 10.5 are used as examples in many books. However, their discussions often stop after developing the transfer function. No reasons are given why it can act as a seismometer or an accelerometer.

10.4   Composite systems – Loading problem

A system is often built by connecting two or more subsystems. There are basically four types of connections as shown in Figure 10.10. They are parallel, tandem, and positive or negative feedback connections. As discussed in Section 7.8, computing their time-domain descriptions is difficult. Thus we compute only their transfer functions. The procedure is purely algebraic and is identical to the one in Section 7.8.

Let the subsystems in Figure 10.10 be described by

$$Y_1(s) = H_1(s)U_1(s) \quad \text{and} \quad Y_2(s) = H_2(s)U_2(s) \qquad (10.17)$$

where $H_i(s)$, for $i = 1, 2$, are proper rational functions. In the *parallel* connection shown in Figure 10.10(a), we have $U_1(s) = U_2(s) = U(s)$ and $Y(s) = Y_1(s) + Y_2(s)$.

By direct substitution, we have

$$Y(s) = H_1(s)U_1(s) + H_2(s)U_2(s) = (H_1(s) + H_2(s))U(s) =: H_p(s)U(s)$$

Thus the overall transfer function of the parallel connection is simply the sum of the two individual transfer functions or $H_p(s) = H_1(s) + H_2(s)$.

In the *tandem* or *cascade* connection shown in Figure 10.10(b), we have $U(s) = U_1(s)$, $Y_1(s) = U_2(s)$, and $Y_2(s) = Y(s)$. Its overall transfer function is the multiplication of the two individual transfer functions or $H_t(s) = H_1(s)H_2(s) = H_2(s)H_1(s)$.

In the *positive-feedback* connection shown in Figure 10.10(c), we have $U_1(s) = U(s) + Y_2(s)$ and $Y_1(s) = Y(s) = U_2(s)$. By direct substitution, we have

$$Y_2(s) = H_2(s)U_2(s) = H_2(s)Y_1(s) = H_2(s)H_1(s)U_1(s)$$

and

$$U_1(s) = U(s) + Y_2(s) = U(s) + H_2(s)H_1(s)U_1(s)$$

which implies

$$[1 - H_2(s)H_1(s)]U_1(s) = U(s) \quad \text{and} \quad U_1(s) = \frac{U(s)}{1 - H_2(s)H_1(s)}$$

Because $Y(s) = Y_1(s) = H_1(s)U_1(s)$, we have

$$H_{pf}(s) := \frac{Y(s)}{U(s)} = \frac{H_1(s)}{1 - H_1(s)H_2(s)} \tag{10.18}$$

This is the transfer function of the positive feedback system in Figure 10.10(c). For the *negative-feedback* connection shown in Figure 10.10(d), using the same procedure, we can obtain its overall transfer function as

$$H_{nf}(s) := \frac{Y(s)}{U(s)} = \frac{H_1(s)}{1 + H_1(s)H_2(s)} \tag{10.19}$$

This can also be obtained from (10.18) by changing the negative feedback to positive feedback using Figure 7.10.

The preceding manipulation is correct mathematically, but may not be so in engineering. The next example illustrates this fact.

Example 10.4.1 Consider the two RC circuits shown in Figure 10.11(a). Using impedances, we can find their transfer functions as

$$H_1(s) = \frac{1/s}{1 + 1/s} = \frac{1}{s+1} \quad \text{and} \quad H_2(s) = \frac{1}{1 + 1/2s} = \frac{2s}{2s+1} \tag{10.20}$$

Let us connect them in tandem as shown with dashed lines. We compute its transfer function from u to y. The impedance of the parallel connection of $1/s$ and $1 + 1/2s$ is

$$Z_1(s) = \frac{(1/s)(1 + 1/2s)}{(1/s) + (1 + 1/2s)} = \frac{2s+1}{s(2s+3)}$$

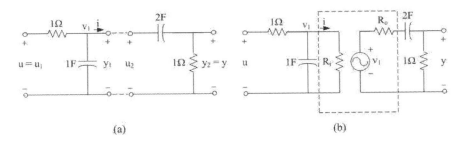

(a)   (b)

Figure 10.11: (a) Two circuits. (b) Inserting an isolating amplifier.

Thus the voltage $V_1(s)$ shown is given by

$$V_1(s) = \frac{Z_1(s)}{1 + Z_1(s)} U(s) = \frac{2s + 1}{2s^2 + 5s + 1} U(s)$$

and the output $Y(s)$ is given by

$$Y(s) = \frac{1}{1 + 1/2s} V_1(s) = \frac{2s}{2s + 1} \frac{2s + 1}{2s^2 + 5s + 1} U(s)$$

Thus the transfer function from u to y of the tandem connection is

$$H_t(s) = \frac{2s}{2s^2 + 5s + 1}$$

Does it equal the product of $H_1(s)$ and $H_2(s)$? We compute

$$H_1(s)H_2(s) = \frac{1}{s + 1} \frac{2s}{2s + 1} = \frac{2s}{2s^2 + 3s + 1}$$

which is different from $H_t(s)$. Thus the transfer function of the tandem connection of the two networks does not equal the product of the two individual transfer functions.
□

   The preceding example shows that $H_t(s) = H_1(s)H_2(s)$ may not hold in practice. If this happens, we say that the connection has a loading problem. The loading in Figure 10.11(a) is due to the fact that the current $i(t)$ shown is zero before connection and becomes nonzero after connection. This provides a method of eliminating the loading.
   Let us insert an amplifier with gain 1 or a voltage follower between the two circuits as shown in Figure 10.11(b). If the amplifier has a very large input resistance R_i as is usually the case, the current $i(t)$ will remain practically zero. If the output resistance R_o is zero or very small, then there will be no internal voltage drop in the amplifier. Thus the amplifier is called an *isolating amplifier* or a *buffer* and can eliminate or reduce the loading problem. In conclusion, in electrical systems, loading can often be eliminated by inserting voltage followers as shown in Figure 6.10. Because op amps have large input resistances, the loading problem in op-amp circuits is generally negligible.

10.4.1 Complete characterization

The overall transfer function of any composite system in Figure 10.10 can be readily computed as shown in the preceding section. It is then natural to ask: Does the overall transfer function so obtained characterize completely the composite system? This will be discussed in this subsection.

Recall from Section 8.11.1 that a system is completely characterized by its transfer function if its number of energy storage elements equals the degree of its transfer function. The degree of a proper transfer function $H(s) = N(s)/D(s)$ is defined as the degree of $D(s)$ if $N(s)$ and $D(s)$ are coprime or have no common root. Thus in determining the degree of $N(s)/D(s)$, we must first cancel out all common roots between $N(s)$ and $D(s)$.

Consider $H_i(s) = N_i(s)/D_i(s)$. It is assumed that $N_i(s)$ and $D_i(s)$ are coprime. Then the degree of $H_i(s)$ equals the degree of $D_i(s)$, denoted by N_i. If every subsystem in Figure 10.10 is completely characterized by its transfer function, then every composite system has a total number of $N_1 + N_2$ energy storage elements. Let $H_o(s)$ be the transfer function of an overall system. Then the overall system is completely characterized by $H_o(s)$ if and only if $H_o(s)$ has a degree equal to $N_1 + N_2$. This condition can be stated in terms of the poles and zeros of $H_i(s)$ as in the next theorem.

Theorem 10.2 Consider two systems S_1 and S_2 which are completely characterized by its proper rational transfer functions $H_1(s)$ and $H_2(s)$.

1. The parallel connection of S_1 and S_2 is completely characterized by its transfer function $H_p = H_1(s) + H_2(s)$ if and only if $H_1(s)$ and $H_2(s)$ has no pole in common.

2. The tandem connection of S_1 and S_2 is completely characterized by its transfer function $H_t = H_1(s)H_2(s)$ if and only if there is no pole-zero cancellation between $H_1(s)$ and $H_2(s)$.

3. The feedback connection of S_1 and S_2 is completely characterized by its transfer function $H_f = H_1(s)/(1 \pm H_2(s))$ if and only if no pole of $H_2(s)$ is canceled by any zero of $H_1(s)$.

We first use examples to demonstrate the case for feedback connection.

Example 10.4.2 Consider two systems S_i with transfer functions

$$H_1(s) = \frac{2}{s-1} \quad \text{and} \quad H_2(s) = \frac{s-1}{(s+3)(s+1)}$$

where no pole of $H_2(s)$ is canceled by any zero of $H_1(s)$. Thus the feedback system is completely characterized by its overall transfer function. Indeed, the transfer function of their positive-feedback connection is

$$
\begin{aligned}
H_{pf}(s) &= \frac{\frac{2}{s-1}}{1 - \frac{2}{s-1} \cdot \frac{s-1}{(s+3)(s+1)}} = \frac{\frac{2}{s-1}}{1 - \frac{2}{(s+3)(s+1)}} \\
&= \frac{\frac{2}{s-1}}{\frac{(s+3)(s+1)-2}{(s+3)(s+1)}} = \frac{2(s+1)(s+3)}{(s-1)(s^2 + 4s + 1)}
\end{aligned}
$$

It has degree 3, sum of the degrees of $H_1(s)$ and $H_2(s)$. Thus the feedback system is completely characterized by $H_{pf}(s)$. Note that the cancellation of the zero of $H_2(s)$ by the pole of $H_1(s)$ does not affect the complete characterization. □

Example 10.4.3 Consider two systems S_i with transfer functions

$$H_1(s) = \frac{s-1}{(s+1)(s+2)} \quad \text{and} \quad H_2(s) = \frac{2}{s-1}$$

where the pole of $H_2(s)$ is canceled by the zero of $H_1(s)$. Thus the feedback system is not completely characterized by its overall transfer function. Indeed, the transfer function of their positive-feedback connection is

$$
\begin{aligned}
H_{pf}(s) &= \frac{\frac{s-1}{(s+1)(s+3)}}{1 - \frac{2}{s-1} \cdot \frac{s-1}{(s+1)(s+3)}} = \frac{\frac{s-1}{(s+1)(s+3)}}{1 - \frac{2}{(s+1)(s+3)}} \\
&= \frac{\frac{s-1}{(s+1)(s+3))}}{\frac{s^2+4s+3-2}{(s+1)(s+3)}} = \frac{s-1}{s^2+4s+1}
\end{aligned}
$$

It has degree 2, less than the sum of the degrees of $H_1(s)$ and $H_2(s)$. Thus the feedback system is not completely characterized by $H_{pf}(s)$. □

The preceding two examples show the validity of the feedback part of Theorem 10.2. We now argue formally the first part of Theorem 10.2. The overall transfer function of the parallel connection is

$$H_p(s) = \frac{N_1(s)}{D_1(s)} + \frac{N_2(s)}{D_2(s)} = \frac{N_1(s)D_2(s) + N_2(s)D_1(s)}{D_1(s)D_2(s)} \qquad (10.21)$$

If $D_1(s)$ and $D_2(s)$ have the same factor $s+a$, then the same factor will also appear in $(N_1(s)D_2(s) + N_2(s)D_1(s))$. Thus the degree of $H_p(s)$ is less than the sum of the degrees of $D_1(s)$ and $D_2(s)$ and the parallel connection is not completely characterized by $H_p(s)$. Suppose $D_1(s)$ has the factor $s+a$ but $D_2(s)$ does not. Then the only way for $N_1(s)D_2(s) + N_2(s)D_1(s)$ to have the factor $s+a$ is that $N_1(s)$ contains $s+a$. This is not possible because $D_1(s)$ and $N_1(s)$ are coprime. Thus we conclude that the parallel connection is completely characterized by $H_p(s)$ if and only if $H_1(s)$ and $H_2(s)$ have no common poles. The cases for tandem and feedback connections can be similarly argued.

10.4.2   Necessity of feedback

Feedback is widely used in practice to achieve automation (such as in water level control) and to counter load variations and external disturbances (as in auto cruise). We show in this subsection that, for any unstable system, the only way to stabilize it is to use feedback.

Consider a system with transfer function

$$H_1(s) = \frac{1}{s-1} = \frac{N_1(s)}{D_1(s)} \qquad (10.22)$$

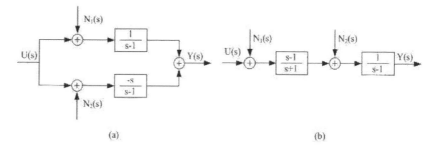

Figure 10.12: (a) Parallel connection. (b) Tandem connection.

It has one pole in the RHP (right-half plane) and is not stable. Is it possible to find a transfer function $H_2(s) = N_2(s)/D_2(s)$ so that their parallel connection is stable? This is discussed in the next example.

Example 10.4.4 Consider the parallel connection shown in Figure 10.12(a). The overall transfer function is

$$H_p(s) = \frac{1}{s-1} + \frac{-s}{s-1} = \frac{1-s}{s-1} = -1 \qquad (10.23)$$

For this example, $H_p(s)$ does not contain the poles of the original subsystems and appears to be a good system. However such a design is not acceptable in practice and in theory as we discuss next.

Suppose we need one resistor with 10 kΩ to implement each subsystem. Because all resistors have some tolerances, it is unlikely to have two 10 kΩ resistors with identical resistance. See the last paragraph of Subsection 6.8.1. Thus in practice, it is difficult to achieve exact cancelation. Even if exact cancelation is possible, noise often exists in physical systems. We assume that noise $n_i(t)$ or $N_i(s)$ enters system $H_i(s)$ as shown in Figure 10.12(a). Then the output of the parallel connection will grow unbounded unless $n_1(t) = n_2(t)$, for all t which is unlikely. In conclusion, any design which involves exact cancelation of unstable poles is not acceptable in practice.

The design in Figure 10.12(a) is neither acceptable in theory. We have assumed that each subsystem is completely characterized by its transfer function of degree 1. Then the parallel connection has two energy storage elements. However its transfer function $H_p(s) = -1$ has degree 0. Thus $H_p(s)$ does not characterize completely the parallel connection, as discussed in the preceding subsection. In conclusion, we cannot introduce a parallel connection to make $H_1(s)$ stable.□

Next we consider the tandem connection of $H_2(s) = N_2(s)/D_2(s)$ and $H_1(s)$. Its overall transfer function is

$$H_t(s) = H_1(s)H_2(s) = \frac{N_1(s)N_2(s)}{D_1(s)D_2(s)} \qquad (10.24)$$

We see that the pole of $H_1(s)$ remains in $H_t(s)$ unless it is canceled by a zero of $H_2(s)$ as we discuss in the next example.

Example 10.4.5 Consider the tandem connection shown in Figure 10.12(b). The overall transfer function is

$$H_t(s) = \frac{s-1}{s+1}\frac{1}{s-1} = \frac{1}{s+1} \tag{10.25}$$

and is a stable transfer function. However all discussion in the preceding example applies directly to this design. That is, the design is not acceptable in practice and in theory. □

Next we consider the negative-feedback system shown in Figure 10.10(d). Its overall transfer function is

$$H_{nf}(s) = \frac{H_1(s)}{1 + H_1(s)H_2(s)}$$

which becomes, after substituting $H_i(s) = N_i(s)/D_i(s)$,

$$
\begin{aligned}
H_{nf}(s) &= \frac{N_1(s)/D_1(s)}{1 + N_1(s)N_2(s)/(D_1(s)D_2(s))}\\
&= \frac{N_1(s)D_2(s)}{D_1(s)D_2(s) + N_1(s)N_2(s)}
\end{aligned}
\tag{10.26}
$$

We see that the poles of (10.26) are different from the poles of $H_i(s)$, for $i = 1, 2$. Thus feedback will introduce new poles. This is in contrast to the parallel and tandem connections where the poles of $H_i(s)$ remain unchanged as we can see from (10.21) and (10.24).

Now for the transfer function $H_1(s)$ in (10.22), if we introduce a negative feedback with $H_2(s) = 2$, then the feedback transfer function is

$$H_{nf}(s) = \frac{H_1(s)}{1 + H_1(s)H_2(s)} = \frac{1/(s-1)}{1 + 2/(s-1)} = \frac{1}{(s-1)+2} = \frac{1}{s+1} \tag{10.27}$$

The unstable pole of $H_1(s)$ at $s = 1$ is being *shifted* to $s = -1$ in $H_{nf}(s)$. Thus the feedback system is stable. The design does not involve any cancelation. Furthermore the feedback system is completely characterized by (10.27) because its degree equals the total number of energy-storage elements in the system. In conclusion, the only way to stabilize an unstable system is to introduce feedback.

More generally, suppose we are required to design a composite system to improve the performance of a system which has some undesirable poles. If we use the parallel or tandem connections, the only way to remove those poles is by direct cancelation. Such design however is not acceptable as discussed earlier. If we introduce feedback, then those poles, as we will shown in a later section, can be shifted to any desired positions. Thus the only way to improve the performance of the system is to introduce feedback.

10.4.3   Advantage of feedback

We showed in the preceding subsection the necessity of introducing feedback. Even if feedback is not necessary, we often use it to improve the robustness of a system as we discuss in this subsection.

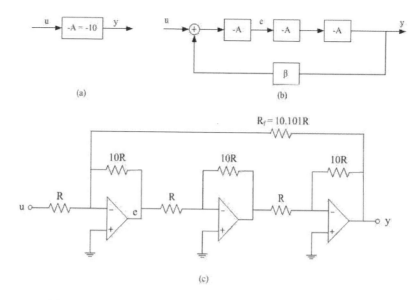

Figure 10.13: (a) Inverting amplifier. (b) Inverting feedback amplifier. (c) Op-amp implementation of (b).

Suppose we are asked to design an inverting amplifier with gain $A = 10$. Such an amplifier can be easily built using the circuit in Figure 6.13(a) with $R_2 = 10R_1$ and is denoted by the box in Figure 10.13(a). Now consider the arrangement shown in Figure 10.13(b). It consists of three boxes, each with gain $-A = -10$ and a positive feedback with gain β.

All systems in Figure 10.13 are memoryless. The transfer function of a memoryless system with gain A is simply A. Let $-A_f$ be the gain from u to y of the positive-feedback system in Figure 10.13(b). Then we have, using (10.18),

$$-A_f = \frac{(-A)^3}{1 - \beta(-A)^3} = -\frac{A^3}{1 + \beta A^3}$$

or

$$A_f = \frac{A^3}{1 + \beta A^3} \tag{10.28}$$

Next we will find a β so that $A_f = 10$. We solve

$$10 = \frac{10^3}{1 + 10^3 \beta}$$

which implies $10 + 10^4 \beta = 10^3$ and

$$\beta = \frac{10^3 - 10}{10^4} = 0.099$$

In other words, if $\beta = 0.099$, then the feedback system in Figure 10.13(b) is also an inverting amplifier with gain 10. The feedback gain β can be implemented as shown in Figure 10.13(c) with $R_f = R/\beta = 10.101R$ (Problem 10.9).

n	1	2	3	4	5	6	7	8	9	10
A	10	9.0	8.1	7.29	6.56	5.9	5.3	4.78	4.3	3.87
A_f	10	9.96	9.91	9.84	9.75	9.63	9.46	9.25	8.96	8.6

Table 10.1: Gains for Inverting Open and Feedback Amplifiers.

Even though the inverting feedback amplifier in Figure 10.13(b) uses three times more components than the one in Figure 10.13(a), it is the preferred one. We give the reason. To *dramatize* the effect of feedback, we assume that A decreases 10% each year due to aging or whatever reason. In other words, A is 10 in the first year, 9 in the second year, and 8.1 in the third year as listed in the second row of Table 10.1. Next we compute A_f from (10.28) with $\beta = 0.099$ and $A = 9$:

$$A_f = \frac{9^3}{1 + 0.099 \times 9^3} = 9.963$$

We see that even though A decreases 10%, A_f decreases only $(10 - 9.963)/10 = 0.0037$ or less than 0.4%. If $A = 8.1$, then

$$A_f = \frac{8.1^3}{1 + 0.099 \times 8.1^3} = 9.913$$

and so forth. They are listed in the third row of Table 10.1. We see that the inverting feedback amplifier is much less sensitive to the variation of A.

If an amplifier is to be replaced when its gain decreases to 9 or less, then the open-loop amplifier in Figure 10.13(a) lasts only one year; whereas the feedback amplifier in Figure 10.13(b) lasts almost 9 years. Thus even though the feedback amplifier uses three times more components, it lasts nine times longer. Thus it is more cost effective. Not to mention the inconvenience and cost of replacing the open-loop amplifier every year. In conclusion, a properly designed feedback system is much less sensitive to parameter variations and external disturbances. Thus feedback is widely used in practice.

10.5   Design of control systems – Pole placement

In addition to the feedback configurations shown in Figures 10.10(c) and (d), there are many other possible feedback configurations. The one shown in Figure 10.14, called the *unity feedback* configuration, is widely used to design control systems. A typical design problem is: Give a plant with proper rational transfer function $P(s)$, design a compensator with proper rational transfer function $C(s)$ and a gain A so that the plant output $y(t)$ will track the reference or desired input $r(t)$. The design depends on the type of signal $r(t)$ to be tracked. The more complicated $r(t)$, the more complicated the design. The simplest problem assumes $r(t)$ to be a step function with amplitude $a \neq 0$. Setting a desired temperature in a heating system or a desired speed in auto cruise of an automobile belongs to this problem.

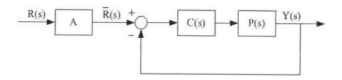

Figure 10.14: Unity-feedback system.

We discuss two approaches in designing such systems. The first approach is to select a $C(s)$ with open parameters and then search the parameters and hope that the resulting system will achieve the design. This is basically a trial-and-error method and as such, we start with the simplest compensator or a compensator of degree 0 ($C(s) = k$) to carry out the search. If not successful, we then try a compensator of degree 1. The conventional root-locus and Bode-plot methods take this approach. The second approach is to search an overall transfer function $H_f(s)$ which achieve the design and then compute the required compensator $C(s)$. If we call $H_f(s)$ a desired model, then this is a model matching problem. The quadratic optimal design method takes this approach. See Reference [C5].

In this section we use the first approach to carry out the design. The method is different from the aforementioned methods and is purely algebraic. We use an example to illustrate the procedure. Consider a plant with transfer function

$$P(s) = \frac{s - 2}{(s + 0.5)(s - 1)} \tag{10.29}$$

It has a pole in the right half s-plane and is unstable. The problem is to find a compensator $C(s)$ and a gain A in Figure 10.14 so that its output will track any step reference input.

The first step in design is to find a $C(s)$ to make the feedback system stable. If not, the output $y(t)$ will grow unbounded for any applied $r(t)$ and cannot track any step reference input. We first try $C(s) = k$, where k is a real constant, a compensator of degree 0. Then the transfer function from r to y is

$$
\begin{aligned}
H_f(s) &= A\frac{C(s)P(s)}{1 + C(s)P(s)} = A\frac{k\frac{s-2}{(s+0.5)(s-1)}}{1 + k\frac{s-2}{(s+0.5)(s-1)}} \\
&= A\frac{k(s - 2)}{(s + 0.5)(s - 1) + k(s - 2)} \\
&= A\frac{k(s - 2)}{s^2 + (k - 0.5)s - 0.5 - 2k} \tag{10.30}
\end{aligned}
$$

The condition for $H_f(s)$ to be stable is that the polynomial

$$D(s) = s^2 + (k - 0.5)s - 0.5 - 2k \tag{10.31}$$

be CT stable. The conditions for (10.31) to be stable are $k - 0.5 > 0$ and $-0.5 - 2k > 0$ (See Problem 9.11). Any $k > 0.5$ meeting the first inequality will not meet the second inequality. Thus the feedback system cannot be stabilized using a compensator of degree 0.

Next we try a compensator of degree 1 or

$$C(s) = \frac{N_1 s + N_0}{D_1 s + D_0} \tag{10.32}$$

where N_i and D_i are real numbers. Using this compensator, $H_f(s)$ becomes, after some simple manipulation,

$$
\begin{aligned}
H_f(s) &= A \frac{(N_1 s + N_0)(s - 2)}{(D_1 s + D_0)(s + 0.5)(s - 1) + (N_1 s + N_0)(s - 2)} \\
&=: \frac{A(N_1 s + N_0)(s - 2)}{D_f(s)}
\end{aligned} \tag{10.33}
$$

where

$$
\begin{aligned}
D_f(s) &= D_1 s^3 + (D_0 - 0.5 D_1 + N_1) s^2 + (-0.5 D_1 - 0.5 D_0 - 2 N_1 + N_0) s \\
&\quad + (-0.5 D_0 - 2 N_0)
\end{aligned} \tag{10.34}
$$

Now $H_f(s)$ is stable if the polynomial of degree 3 in (10.34) is CT stable. Because the four parameters of the compensator in (10.32) appear in the four coefficients of $D_f(s)$, we can easily find a compensator to make $H_f(s)$ stable. In fact we can achieve more. We can find a $C(s)$ to place the poles of $H_f(s)$ or the roots of $D_f(s)$ in any positions so long as complex-conjugate poles appear in pairs. For example, we can select the poles arbitrarily at $-1, -1, -1$ or, equivalently, select $D_f(s)$ as

$$D_f(s) = (s + 1)^3 = s^3 + 3s^2 + 3s + 1 \tag{10.35}$$

Equating its coefficients with those of (10.34), we obtain

$$
\begin{aligned}
D_1 &= 1 \\
D_0 - 0.5 D_1 + N_1 &= 3 \\
-0.5 D_1 - 0.5 D_0 - 2 N_1 + N_0 &= 3 \\
-0.5 D_0 - 2 N_0 &= 1
\end{aligned} \tag{10.36}
$$

Substituting $D_1 = 1$ into the other equations and then carrying out elimination, we can finally obtain

$$D_1 = 1 \quad D_0 = 8.8 \quad N_1 = -5.3 \quad N_0 = -2.7$$

Thus the compensator is

$$C(s) = \frac{-5.3s - 2.7}{s + 8.8} \tag{10.37}$$

For this compensator, the overall transfer function $H_f(s)$ in (10.33) becomes

$$H_f(s) = \frac{A(s - 2)(-5.3s - 2.7)}{s^3 + 3s^2 + 3s + 1} \tag{10.38}$$

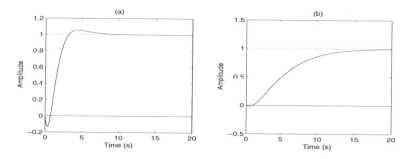

Figure 10.15: (a) Step response of Figure 10.14 with all three poles selected at -1. (b) With all three poles selected at -0.4.

Because $H_f(s)$ is stable, if we apply $r(t) = a$, for $t \geq 0$, then its output approaches $aH_f(0)$ as $t \to \infty$ (Theorem 9.5). Thus we select A in (10.38) to make $H_f(0) = 1$ or $A(-2)(-2.7)/1 = 1$ which yield

$$A = \frac{1}{5.4} = 0.1852 \tag{10.39}$$

This completes the design. In other words, for the plant transfer function in (10.29), if we use the compensator in (10.37) and the gain in (10.39), then the output of the feedback system in Figure 10.14 will track any step reference input. To verify this, we plot in Figure 10.15(a) the output of the feedback system excited by $r(t) = 1$, for $t \geq 0$. The output indeed tracks the reference input. The response has a small overshoot. Its response also goes negative when the input is applied as shown. It is called *undershoot* and is due to the RHP zero of $P(s)$. Overall, it is a satisfactory design.

The preceding pole-placement design method is a general one. For a plant with a strictly proper transfer function of degree N, if we introduce a compensator $C(s)$ of degree $N - 1$, then the $(2N - 1)$ number of poles of the unity-feedback system can be arbitrarily assigned. The design procedure involves only matching up coefficients. Even though the method is simple, straightforward and powerful, it is applicable only to LTI lumped systems with a known rational transfer function. If a plant is not LTI and lumped or if we use a nonlinear compensator, then the method is not applicable.

Before proceeding, we mention that for the plant in (10.29), it will be difficult to use the root-locus method to carry out the design. Neither can the Bode-plot method be used because (10.29) has an unstable pole at $s = 1$. If we find an ss realization of $P(s)$, we can use state feedback and state estimator to carry the design. But the procedure and the concepts involved will be much more complicated. Note that pole-placement will introduce some zeros as shown in (10.33) and (10.38) which we have no control. In model matching, we can control poles as well as zeros. However we must use a more complicated feedback configuration. The interested reader is referred to References [C5, C6].

10.5.1   Is the design unique?

There are two types of specifications in designing a control system: steady-state and transient. The steady-state specification is concerned with accuracy; it is the final reading discussed in Subsection 9.1.1. For the design problem in Figure 10.14, if $r(t) = a$ and if the steady-state output is $y_{ss}(t) = 0.9a$, then the tracking has 10% error $((a - 0.9a)/a = 0.1)$. If $y_{ss}(t) = a$, then the system can track any step reference input $r(t) = a$ without any error. This design can be easily achieved by selecting the gain A in Figure 10.14 so that $H_f(0) = 1$ as in (10.39). Thus accuracy is easy to achieve in designing control systems.

The transient specification is concerned with response time and overshoot. For the pole-placement design in the preceding section, by selecting all three poles at -1, we obtain the response shown in Figure 10.15(a). If we select all three poles at -0.4, then the step response of the resulting system will be as shown in Figure 10.15(b). The response has no overshoot; it has a smaller undershoot but a larger response time than the one in Figure 10.15(a). Which design shall we select depends on which is more important: response time or overshoot.

The selection of all three poles at -1 or at -0.4 is carried out by computer simulation. If we select a real pole and a pair of complex-conjugate poles in the neighborhood of -1 or -0.4, we will obtain probably a comparable result. Thus the design is not unique. How to select a set of desired poles is not a simple problem. See Reference [C5].

10.6   Inverse systems

Consider a system with transfer function $H(s)$. If we know the input $u(t)$, we can compute its output $y(t)$ (the inverse Laplace transform of $H(s)U(s)$). Now if we know the output $y(t)$, can we compute its input $u(t)$?

Consider a system with transfer function $H(s)$. We call the system with transfer function $H_{in}(s) := H^{-1}(s)$ the *inverse system* of $H(s)$. If the inverse $H_{in}(s)$ of $H(s)$ exists, we apply $y(t)$ to $H_{in}(s)$, then its output $\bar{y}(t)$ equals $u(t)$ for

$$\bar{y}(t) = \mathcal{L}^{-1}\left[H_{in}(s)Y(s)\right] = \mathcal{L}^{-1}\left[H^{-1}(s)H(s)U(s)\right] = \mathcal{L}^{-1}[U(s)] = u(t)$$

This problem may find application in oil exploration: To find the source from measured data.

Every system, including every inverse system, involves two issues: realizability and stability. If the transfer function of a system is not a proper rational function, then its implementation requires the use of differentiators. If the transfer function is not stable, then its output will grow unbounded for any input. Thus we require every inverse system to have a proper rational transfer function and to be stable.

Consider a system with transfer function $H(s)$. If $H(s)$ is biproper, then its inverse $H_{in}(s) = H^{-1}(s)$ is also biproper. If all zeros of $H(s)$ lie inside the left half s-plane,[2] then $H^{-1}(s)$ is stable. Thus if a stable biproper transfer function has all its zeros lying inside the LHP, then its inverse system exists and can be easily built. If it has one or

---

[2]Such a transfer function is called a *minimum-phase transfer function*. See Reference [C5].

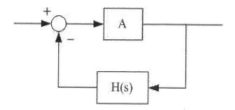

Figure 10.16: Feedback implementation of inverse system.

more zeros inside the RHP or on the $j\omega$-axis, then its inverse system is not stable and cannot be used.

If $H(s)$ is strictly proper, then $H^{-1}(s)$ is improper and cannot be realized without using differentiators. In this case, we may try to implement its inverse system as shown in Figure 10.16 where A is a very large positive gain. The overall transfer function of the feedback system in Figure 10.16 is

$$H_o(s) = \frac{A}{1 + AH(s)}$$

Now if A is very large such that $|AH(j\omega)| \gg 1$, then the preceding equation can be approximated by

$$H_o(s) = \frac{A}{1 + AH(s)} \approx \frac{A}{AH(s)} = H^{-1}(s) \tag{10.40}$$

Thus the feedback system can be used to implement approximately an inverse system. Furthermore, for A very large, $H_o(s)$ is practically independent on A, thus the feedback system is insensitive to the variations of A. Thus it is often suggested in the literature to implement an inverse system as shown in Figure 10.16. See, for example, Reference [O1, pp. 820-821].

The problem is not so simple as suggested. It is actually the model reduction problem discussed in Section 10.2.1. In order for the approximation in (10.40) to hold, the overall system $H_o(s)$ must be stable. Moreover, the approximation holds only for signals whose spectra are limited to some frequency range. This is illustrated with examples.

Example 10.6.1 Consider $H(s) = 0.2/(s+1)$. Because $H^{-1}(s) = 5(s+1)$ is improper, it cannot be implemented without using differentiators. If we implement its inverse system as shown in Figure 10.16, then the overall transfer function is

$$H_o(s) = \frac{A}{1 + AH(s)} = \frac{A(s+1)}{s + 1 + 0.2A}$$

It is stable for any $A \geq -5$. Thus if A is very large, $H_o(s)$ can be reduced to

$$H_{os}(s) \approx \frac{A(s+1)}{0.2A} = 5(s+1) = H^{-1}(s)$$

in the frequency range $|j\omega + 1| \ll 0.2A$. Thus the feedback system can be used to implement approximately the inverse system of $H(s)$. The approximation, however, is valid only for low-frequency signals as demonstrated in the following.

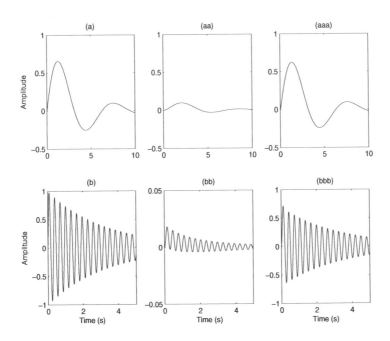

Figure 10.17: (a) Input $u_1(t)$ of $H(s)$. (aa) Output $y_1(t)$ of $H(s)$. (aaa) Output of $H_o(s)$ excited by $y_1(t)$. (b) Input $u_2(t)$ of $H(s)$. (bb) Output $y_2(t)$ of $H(s)$. (bbb) Output of $H_o(s)$ excited by $y_2(t)$.

Consider the signal $u_1(t) = e^{-0.3t} \sin t$ shown in Figure 10.17(a). Its nonzero magnitude spectrum, as shown in Figure 10.8(b), lies inside $[0, 5]$. If we select $A = 100$, then we have $|j\omega + 1| \ll 0.2A$ for ω in $[0, 5]$. The output $y_1(t)$ of $H(s)$ excited by $u_1(t)$ is shown in Figure 10.17(aa). The output of $H_o(s)$ with $A = 100$ excited by $y_1(t)$ is shown in Figure 10.17(aaa). It is close to $u_1(t)$. Thus the feedback system in Figure 10.16 implements the inverse of $H(s)$ for $u_1(t)$.

Next we consider $u_2(t) = e^{-0.3t} \sin 20t$ and select $A = 100$. The corresponding results are shown in Figures 10.17(b), (bb), and (bbb). The output of the feedback system has the same wave form as the input $u_2(t)$ but its amplitude is smaller. This is so because we do not have $|j\omega + 1| \ll 0.2A = 20$. Thus the feedback system is not exactly an inverse system of $H(s)$ for high-frequency signals.□

Example 10.6.2 Consider

$$H(s) = 1/(s^3 + 2s^2 + s + 1)$$

Because $H^{-1}(s) = s^3 + 2s^2 + s + 1$ is improper, it cannot be implemented without using differentiators. If we implement its inverse system as in Figure 10.16, then the overall transfer function is

$$H_o(s) = \frac{A}{1 + AH(s)} = \frac{A(s^3 + 2s^2 + s + 1)}{s^3 + 2s^2 + s + 1 + A} \qquad (10.41)$$

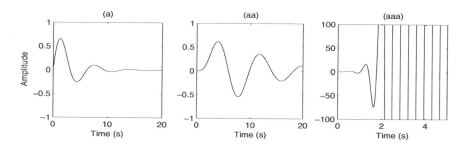

Figure 10.18: (a) Input $u(t)$ of $H(s)$ in (10.41). (b) Its output $y(t)$. (c) Output of $H_o(s)$ excited by $y(t)$.

Using the Routh test, we can show that $s^3 + 2s^2 + s + 1 + A$ is not a CT stable polynomial for any $A \geq 1$ (Problem 10.14). Thus $H_o(s)$ is not stable for A large and cannot be reduced to $H^{-1}(s)$. To verify this assertion, we apply $u(t) = e^{-0.3t}\sin t$ shown in Figure 10.18(a) to $H(s)$. Its output is shown in Figure 10.18(b). Now we apply this output to the inverse system $H_o(s)$ in (10.41) with $A = 1000$. The output of $H_o(s)$ is shown in Figure 10.18(c). It grows unbounded and does not resemble $u(t)$. In conclusion, the feedback system in Figure 10.16 cannot be used to implement the inverse system of (10.41).□

Even though feedback implementation of inverse systems is often suggested in the literature, such implementation is not always possible as shown in the preceding examples. Thus care must be exercised in using the implementation.

10.7 Wien-bridge oscillator

A sinusoidal generator is a device that will maintain a sinusoidal oscillation once it is excited and the excitation is removed. We discuss one simple such circuit in this section.

Consider the op-amp circuit, called the *Wien-bridge oscillator*, shown in Figure 10.19(a) where the op amp is modeled as ideal. We use transform impedances to compute its transfer function from the input v_i to the output v_o. Let $Z_3(s)$ and $Z_4(z)$ be, respectively, the impedances of the parallel and series connections of R and C shown in Figure 10.19(a). Then we have

$$Z_3 = \frac{R(1/Cs)}{R + (1/Cs)} = \frac{R}{RCs + 1} \qquad (10.42)$$

and

$$Z_4 = R + (1/Cs) \qquad (10.43)$$

The voltage $E_+(s)$ at the noninverting terminal is, using $I_+(s) = 0$,

$$E_+(s) = \frac{Z_3}{Z_3 + Z_4}V_o(s)$$

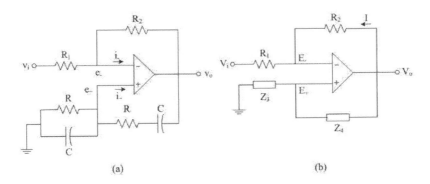

Figure 10.19: (a) Wien-bridge oscillator. (b) Using impedances.

The current flowing from v_o to v_i is, using $I_-(s) = 0$,

$$I(s) = \frac{V_o(s) - V_i(s)}{R_1 + R_2}$$

Thus the voltage $E_-(s)$ at the inverting terminal with respect to the ground is

$$
\begin{aligned}
E_-(s) &= V_i(s) + I(s)R_1 = V_i(s) + \frac{R_1}{R_1 + R_2}(V_o(s) - V_i(s)) \\
&= \frac{R_1 V_o(s) + R_2 V_i(s)}{R_1 + R_2}
\end{aligned}
$$

Equating $E_+(s) = E_-(s)$ for the ideal model yields

$$\frac{Z_3}{Z_3 + Z_4}V_o(s) = \frac{R_1 V_o(s) + R_2 V_i(s)}{R_1 + R_2}$$

which implies

$$(R_1 + R_2)Z_3 V_o(s) = (Z_3 + Z_4)R_1 V_o(s) + (Z_3 + Z_4)R_2 V_i(s)$$

and

$$(R_2 Z_3 - R_1 Z_4)V_o(s) = (Z_3 + Z_4)R_2 V_i(s)$$

Thus the transfer function from v_i to v_o is

$$H(s) = \frac{V_o(s)}{V_i(s)} = \frac{(Z_3 + Z_4)R_2}{R_2 Z_3 - R_1 Z_4} \tag{10.44}$$

Substituting (10.42) and (10.43) into (10.44) and after simple manipulation, we finally obtain the transfer function as

$$H(s) = \frac{-((RCs)^2 + 3RCs + 1)R_2}{R_1(RCs)^2 + (2R_1 - R_2)RCs + R_1} \tag{10.45}$$

It has two poles and two zeros.

We now discuss the condition for the circuit to maintain a sustained oscillation once it is excited and then the input is removed. If $H(s)$ has one or two poles inside the RHP, its output will grow unbounded once the circuit is excited. If the two poles are inside the LHP, then the output will eventually vanish once the input is removed. Thus the condition for the circuit to maintain a sustained oscillation is that the two poles are on the $j\omega$-axis. This is the case if

$$2R_1 = R_2$$

Under this condition, the denominator of (10.45) becomes $R_1[(RC)^2 s^2 + 1]$ and its two roots are located at $\pm j\omega_0$ with

$$\omega_0 := \frac{1}{RC}$$

They are pure imaginary poles of $H(s)$. After the circuit is excited and the input is removed, the output $v_o(t)$ is of the form

$$v_o(t) = k_1 \sin(\omega_0 t + k_2)$$

for some constants k_1 and k_2. It is a sustained oscillation with frequency $1/RC$ rad/s. Because of its simplicity in structure and design, the Wien-bridge oscillator is widely used.

Although the frequency of oscillation is fixed by the circuit, the amplitude k_1 of the oscillation depends on how it is excited. Different excitation will yield different amplitude. In order to have a fixed amplitude, the circuit must be modified. First we select a R_2 slightly larger than $2R_1$. Then the transfer function in (10.45) becomes unstable. In this case, even though no input is applied, once the power is turned on, the circuit will start to oscillate due to thermal noise or power-supply transient. The amplitude of oscillation will increase with time. When it reaches A, a value predetermined by a nonlinear circuit, called a limiter, the circuit will maintain $A\sin\omega_r t$, where ω_r is the imaginary part of the unstable poles and is very close to ω_0. See Reference [S1]. Thus an actual Wien-bridge oscillator is more complex than the one shown in Figure 10.19. However, the preceding linear analysis does illustrate the basic design of the oscillator.[3]

10.8   Feedback model of general op-amp circuit

The Wien-bridge oscillator was designed directly in the preceding section. Most electronics texts carry out its design using a feedback model. In order to do so, we develop a feedback model for a general op-amp circuit.

Consider the op-amp circuit shown in Figure 10.20(a) in which the op amp is modeled as $I_-(s) = -I_+(s) = 0$ and $V_o(s) = A(s)[E_+(s) - E_-(s)]$. Here we do not assume $E_+(s) = E_-(s)$. Let $Z_i(s)$, for $i = 1:4$, be any transform impedances. We develop a block diagram for the circuit.

---

[3]We mention that the Wien-bridge oscillator is not BIBO stable. An oscillator is not designed to process signals; it is designed to maintain an oscillation once it is excited and the input is removed. Thus its stability should be defined for its natural responses. See footnote No. 5 in page 223. The oscillator is said to be marginally stable.

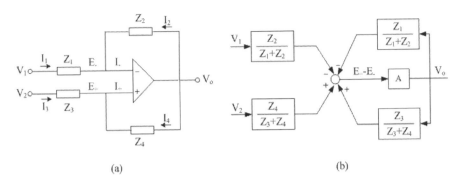

Figure 10.20: (a) Op-amp circuit. (b) Its block diagram.

The current I_1 passing through the impedance Z_1 is $(V_1 - E_-)/Z_1$, and the current I_2 passing through Z_2 is $(V_o - E_-)/Z_2$. Because $I_- = 0$, we have $I_1 = -I_2$ or

$$\frac{V_1 - E_-}{Z_1} = -\frac{V_o - E_-}{Z_2}$$

which implies

$$E_- = \frac{Z_2 V_1 + Z_1 V_o}{Z_1 + Z_2} = \frac{Z_2}{Z_1 + Z_2} V_1 + \frac{Z_1}{Z_1 + Z_2} V_o \qquad (10.46)$$

This equation can also be obtained using a linearity property. If $V_o = 0$ or the output terminal is grounded, then the voltage E_- at the inverting terminal excited by V_1 is $[Z_2/(Z_1 + Z_2)]V_1$. If $V_1 = 0$ or the inverting terminal is grounded, then the voltage E_- excited by V_o is $[Z_1/(Z_1 + Z_2)]V_o$. Using the additivity property of a linear circuit, we obtain (10.46).

Likewise, because $I_+ = 0$, we have, at the noninverting terminal,

$$\frac{V_2 - E_+}{Z_3} = -\frac{V_o - E_+}{Z_4}$$

which implies

$$E_+ = \frac{Z_4 V_2 + Z_3 V_o}{Z_3 + Z_4} = \frac{Z_4}{Z_3 + Z_4} V_2 + \frac{Z_3}{Z_3 + Z_4} V_o \qquad (10.47)$$

This equation can also be obtained using the additivity property as discussed above. Using (10.46), (10.47), and $V_o = A(E_+ - E_-)$, we can obtain the block diagram in Figure 10.20(b). It is a feedback model of the op-amp circuit in Figure 10.20(a). If $V_1 = 0$ and $V_2 = 0$, the feedback model reduces to the one in Reference [H4, p. 847].

10.8.1   Feedback model of Wien-bridge oscillator

Applying Figure 10.20(b) to Figure 10.19(b), we can obtain for the Wien-bridge oscillator the feedback model shown in Figure 10.21(a) with Z_3 and Z_4 given in (10.42) and (10.43). Note that the adder in Figure 10.20(b) has been plotted as two adders in Figure 10.21(a). This is permitted because the signals at the input of the gain A

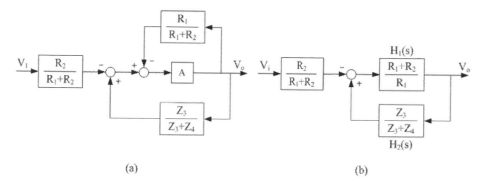

(a) (b)

Figure 10.21: (a) Feedback model of Wien-bridge oscillator with finite A. (b) With $A = \infty$.

are the same. If A is very large or infinite, the negative-feedback loop implements the inverse of $R_1/(R_1 + R_2)$. Note that the negative-feedback system is memoryless and is always stable. Thus Figure 10.21(a) can be reduced as shown in Figure 10.21(b). Let us define

$$H_1(s) := \frac{R_1 + R_2}{R_1} \tag{10.48}$$

and

$$H_2(s) := \frac{Z_3}{Z_3 + Z_4} = \frac{RCs}{(RCs)^2 + 3RCs + 1} \tag{10.49}$$

where we have substituted (10.42) and (10.43). Because the input V_i enters into the inverting terminal and feedback enters into the noninverting terminal, the adder has the negative and positive signs shown. This is essentially the feedback model used in References [H4, S1]. Note that the system in Figure 10.21(b) is a positive feedback system.

The transfer function of the feedback system in Figure 10.21(b) can be computed as

$$H(s) = \frac{V_o(s)}{V_i(s)} = \frac{-R_2}{R_1 + R_2} \times \frac{H_1(s)}{1 - H_1(s)H_2(s)} \tag{10.50}$$

Substituting (10.48) and (10.49) into (10.50), we will obtain the same transfer function in (10.45) (Problem 10.17).

We discuss an oscillation condition, called the *Barkhausen criterion*. The criterion states that if there exists an ω_0 such that

$$H_1(j\omega_0)H_2(j\omega_0) = 1 \tag{10.51}$$

then the feedback system in Figure 10.21(b) will maintain a sinusoidal oscillation with frequency ω_0. Note that once the circuit is excited, the input is removed. Thus the oscillation condition depends only on $H_1(s)$ and $H_2(s)$. Their product $H_1(s)H_2(s)$ is called the *loop gain*. It is the product of the transfer functions along the loop.

The criterion in (10.51) can be established as follows. Suppose the feedback system in Figure 10.21(b) maintains the steady-state oscillation $\mathrm{Re}(ae^{j\omega_0 t})$ at the output after the input is removed ($v_i = 0$). Then the steady-state output of $H_2(s)$ is

$\text{Re}(aH_2(j\omega_0)e^{j\omega_0 t})$ (See (9.50)). This will be the input of $H_1(s)$ because $v_i = 0$. Thus the steady-state output of $H_1(s)$ is

$$\text{Re}(H_1(j\omega_0)aH_2(j\omega_0)e^{j\omega_0 t})$$

If the condition in (10.51) is met, then this output reduces to $\text{Re}(ae^{j\omega_0 t})$, the sustained oscillation. This establishes (10.51). The criterion is widely used in designing Wien-bridge oscillators. See References [H4, S1].

We mention that the criterion is the same as checking whether or not $H(s)$ in (10.50) has a pole at $j\omega_0$. We write (10.51) as

$$1 - H_1(j\omega_0)H_2(j\omega_0) = 0$$

which implies $H_1(j\omega_0) \neq 0$. Thus we have $H(j\omega_0) = \infty$ or $-\infty$, that is, $j\omega_0$ is a pole of $H(s)$ in (10.50). In other words, the criterion in (10.51) is the condition for $H(s)$ to have a pole at $j\omega_0$. Because $H(s)$ has only real coefficients, if $j\omega_0$ is a pole, so is $-j\omega_0$. Thus the condition in (10.51) checks the existence of a pair of complex conjugate poles at $\pm j\omega_0$ in $H(s)$. This is the condition used in Section 10.7. In conclusion, the Wien-bridge oscillator can be designed directly as in Section 10.7 or using a feedback model as in this subsection.

Problems

10.1 Consider the voltage follower shown in Figure 10.1(a). What is its overall transfer function if the op amp is modeled to have memory with transfer function

$$A(s) = \frac{10^5}{s + 100}$$

Is the voltage follower a lowpass filter? Find its 3-dB passband. We may consider the passband as its operational frequency range.

10.2 Consider the positive-feedback op-amp circuit shown in Figure 6.17 with $R_2 = 10R_1$. Show that it has transfer function $H_s(s) = -10$ if its op amp is modeled as ideal. Suppose the op amp is modeled to have memory with transfer function in (10.2). What is its transfer function $H(s)$? Verify that for ω in $[0, 1000]$, the frequency responses of $H(s)$ and $H_s(s)$ are almost identical. For low-frequency signals with spectra lying inside $[0, 1000]$, can $H(s)$ be reduced to $H_s(s)$? What are the step responses of $H(s)$ and $H_s(s)$? Note that although the circuit in Figure 6.17 cannot be used as a *linear* amplifier; it can be used to build a very useful nonlinear device, called Schmitt trigger.

10.3 Consider the noninverting amplifier shown in Figure 6.18 with $R_2 = 10R_1$. What is its transfer function if the op amp is modeled as ideal? What is its transfer function if the op amp is modeled to have memory with transfer function in (10.2)? Is it stable? What is its time constant? What is its operational frequency range in which the op amp can be modeled as ideal?

10.4 Find a realization of (10.10) with $k_d = 2$ and $N = 20$ and then implement it using an op-amp circuit.

10.5 Consider the transfer function in (10.13). If $m = 1$ and $k = 2$, find the f so that the operational frequency range for (10.13) to act as a seismometer is the largest. Is this operational frequency range larger or smaller than $[1.25, \infty)$ computed for $m = 1$, $k = 0.2$, and $f = 0.63$?

10.6 Repeat Problem 10.5 for $m = 1$ and $k = 0.02$.

10.7 Consider the negative feedback system shown in Figure 10.10(d) with

$$H_1(s) = \frac{10}{s+1} \quad \text{and} \quad H_2(s) = -2$$

Check the stability of $H_1(s)$, $H_2(s)$, and the feedback system. Is it true that if all subsystems are stable, then its negative feedback system is stable? Is the statement that negative feedback can stabilize a system necessarily correct?

10.8 Consider the positive feedback system shown in Figures 10.10(c) with

$$H_1(s) = \frac{-2}{s-1} \quad \text{and} \quad H_2(s) = \frac{3s+4}{s-2}$$

Check the stability of $H_1(s)$, $H_2(s)$, and the feedback system. Is it true that if all subsystems are unstable, then its positive feedback is unstable? Is the statement that positive feedback can destabilize a system necessarily correct?

10.9 Verify that if $R_f = R/\beta$, then Figure 10.13(c) implements the β in Figure 10.13(b).

10.10 Consider the unity feedback system shown in Figure 10.14 with

$$P(s) = \frac{2}{s(s+1)}$$

Can you find a gain $C(s) = k$ so that the poles of the feedback system are located at $-0.5 \pm j2$? Can you find a gain $C(s) = k$ so that the poles of the feedback system are located at $-1 \pm j2$?

10.11 For the problem in Problem 10.10, find a proper rational function $C(s)$ of degree 1 to place the three poles of the feedback system at $-2, -1 + j2$, and $-1 - j2$. What is the gain A in order for the output to track any step reference input?

10.12 Consider the op-amp circuit shown in Figure 10.22 where the op amp is modeled as $V_o(s) = A(s)[E_+(s) - E_-(s)]$ and $I_- = -I_+ = 0$. Verify that the transfer function from $V_i(s)$ to $V_o(s)$ is

$$H(s) = \frac{V_o(s)}{V_i(s)} = \frac{-A(s)Z_2(s)}{Z_1(s) + Z_2(s) + A(s)Z_1(s)}$$

Also derive it using the block diagram in Figure 10.20(b).

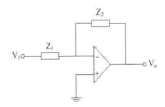

Figure 10.22: Op-amp circuit

10.13 Consider the op-amp circuit in Figure 10.22 with $Z_1 = R$ and $Z_2 = 10R$. Find its transfer functions if $A = 10^5$ and $A = 2 \times 10^5$. The two open-loop gains differ by 100%. What is the difference between the two transfer functions? Is the transfer function sensitive to the variation of A?

10.14 Use the Routh test to verify that the polynomial $s^3 + 2s^2 + s + 1 + A$ is a CT stable polynomial if and only if $-1 < A < 1$.

10.15 Can you use Figure 10.16 to implement the inverse system of $H(s) = (s + 1)/(s^2 + 2s + 5)$? How about $H(s) = (s - 1)/(s^2 + 2s + 5)$?

10.16 Consider the stable biproper transfer function $H(s) = (s - 2)/(s + 1)$. Is its inverse system stable? Can its inverse system be implemented as shown in Figure 10.16?

10.17 Verify that the transfer function in (10.50) equals the one in (10.45).

10.18 Use the ideal model to compute the transfer function from v_i to v_o of the op-amp circuit in Figure 10.23. What is the condition for the circuit to maintain a sinusoidal oscillation once it is excited, and what is its frequency of oscillation? Are the results the same as those obtained in Section 10.7?

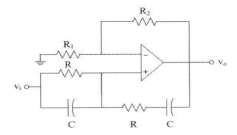

Figure 10.23:

10.19 Use Figure 10.20(b) to develop a feedback block diagram for the circuit in Figure 10.23, and then compute its transfer function. Is the result the same as the one computed in Problem 10.18?

Chapter 11

DT LTI and lumped systems

11.1 Introduction

We introduced in Chapter 7 discrete convolutions for DT LTI systems with finite or infinite memory and then (non-recursive) difference equations, state-space (ss) equations, and rational transfer functions for DT systems with finite memory. Now we will extend the results to DT systems with infinite memory. The extension is possible only for a small subset of such systems. For this class of systems, we will discuss their general properties. The discussion is similar to the CT case in Chapters 8 and 9 and will be brief.

As discussed in Section 6.4.1, in the study of DT systems we may assume the sampling period to be 1. This will be the standing assumption for all sections except the last where DT systems will be used to process CT signals.

Before proceeding, we need the formula in (4.67) or

$$\sum_{n=0}^{\infty} r^n = \frac{1}{1-r} \tag{11.1}$$

where $|r| < 1$ and r can be real or complex. Note that the infinite summation in (11.1) diverges if $|r| > 1$ and $r = 1$ and is not defined if $r = -1$. See Section 3.5.

11.2 Some z-Transform pairs

Consider a DT signal $x[n]$ defined for all n in $(-\infty, \infty)$. Its z-transform is defined in Section 7.7 as

$$X(z) := \mathcal{Z}[x[n]] := \sum_{n=0}^{\infty} x[n]z^{-n} \tag{11.2}$$

where z is a complex variable. Note that it is defined only for the positive-time part of $x[n]$ or $x[n]$, for $n \geq 0$. Because its negative-time part ($x[n]$, for $n < 0$) is not used, we often assume $x[n] = 0$, for $n < 0$, or $x[n]$ to be positive time. Using the definition alone,

347

we have developed the concept of transfer functions and discussed its importance in Section 7.8. In this section, we discuss further the z-transform.

Example 11.2.1 Consider the DT signal $x[n] = 1.3^n$, for $n \geq 0$. This signal grows unbounded as $n \to \infty$. Its z-transform is

$$X(z) = \mathcal{Z}[1.3^n] = \sum_{n=0}^{\infty} 1.3^n z^{-n}$$

This is an infinite power series and is not very useful. Fortunately, for the given $x[n]$, we can cast it into the form of (11.1) with $r = 1.3z^{-1}$ as

$$X(z) = \sum_{n=0}^{\infty} \left(1.3z^{-1}\right)^n = \frac{1}{1 - 1.3z^{-1}} = \frac{z}{z - 1.3} \qquad (11.3)$$

which is a simple rational function of z. This is the z-transform we will use throughout this chapter. However (11.3) holds only if $|1.3z^{-1}| < 1$ or $1.3 < |z|$. For example, if $z = 1$, the infinite sum in (11.3) diverges (approaches ∞) and does not equal $1/(1 - 1.3) = -10/3$. Thus we have

$$\mathcal{Z}[1.3^n] = \frac{1}{1 - 1.3z^{-1}} = \frac{z}{z - 1.3}$$

only if $|z| > 1.3$. We call $z/(z - 1.3)$ the z-transform of 1.3^n and 1.3^n the inverse z-transform of $z/(z - 1.3)$, denoted as

$$\mathcal{Z}^{-1}\left[\frac{z}{z - 1.3}\right] = 1.3^n$$

for $n \geq 0$.$\square$

The z-transform in (11.3) is, strictly speaking, defined only for $|z| > 1.3$, or the region outside the dotted circle with radius 1.3 shown in Figure 11.1. The region is called the *region of convergence*. The region of convergence is important if we use the integration formula

$$x[n] := \mathcal{Z}^{-1}[X(z)] := \frac{1}{2\pi j} \oint X(z) z^{n-1} dz \qquad (11.4)$$

or, in particular, by selecting $z = ce^{j\omega}$,

$$\begin{aligned} x[n] & := \mathcal{Z}^{-1}[X(z)] = \frac{1}{2\pi j} \int_{\omega=0}^{2\pi} X(ce^{j\omega}) \left(ce^{j\omega}\right)^{n-1} cje^{j\omega} d\omega \\ & = \frac{c^n}{2\pi} \int_{\omega=0}^{2\pi} X(ce^{j\omega}) e^{jn\omega} d\omega \end{aligned} \qquad (11.5)$$

to compute the inverse z-transform. For our example, if we select $c = 2$, then the integration contour lies inside the region of convergence and (11.5) will yield

$$x[n] = \begin{cases} 1.3^n & \text{for } n \geq 0 \\ 0 & \text{for } n < 0 \end{cases}$$

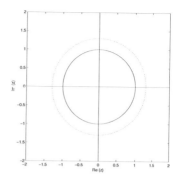

Figure 11.1: Region of convergence.

which is the original positive-time signal. However, if we select $c = 1$, then the integration contour lies outside the region of convergence and (11.5) will yield the following negative-time signal

$$x[n] = \begin{cases} 0 & \text{for } n \geq 0 \\ -1.3^n & \text{for } n < 0 \end{cases}$$

In conclusion, without specifying the region of convergence, the inversion formula in (11.5) may not yield the original $x[n]$. In other words, the relationship between $x[n]$ and $X(z)$ is *not* one-to-one without specifying the region of convergence. Fortunately in practical application, we study only positive-time signals and consider the relationship between $x[n]$ and $X(z)$ to be *one-to-one*. Once we obtain a z-transform $X(z)$ from a positive-time signal, we automatically assume its inverse z-transform to be the original positive-time signal. Thus there is no need to consider the region of convergence and the inversion formula. Indeed when we develop transfer functions in Section 7.7, the region of convergence was not mentioned. Nor will it appear again in the remainder of this chapter.

Before proceeding, we mention that if $x[n]$ is not positive time, we may define its two-sided z-transform as

$$X_{II}(s) := \mathcal{Z}_{II}[x[n]] := \sum_{n=-\infty}^{\infty} x[n]z^{-n}$$

For such a transform, the region of convergence is essential because the same $X_{II}(z)$ may have many different two-sided $x[n]$. Thus its study is much more complicated than the (one-sided) z-transform introduced in (11.2). Fortunately, the two-sided z-transform is rarely, if not never, used in practice. Thus its discussion is omitted.

We now extend 1.3^n to $x[n] = b^n$, for $n \geq 0$, where b can be a real or complex number. Note that b^n is an exponential sequence as discussed in Subsection 2.8.1. Its z-transform is

$$X(z) = \mathcal{Z}[b^n] = \sum_{n=0}^{\infty} b^n z^{-n} = \sum_{0}^{\infty} \left(bz^{-1}\right)^n = \frac{1}{1 - bz^{-1}} = \frac{z}{z - b} \qquad (11.6)$$

If $b = 0$, $x[n] = \delta_d[n]$ is the impulse sequence defined in (2.24) with $n_0 = 0$. If $b = 1$, $x[n] = 1$ is a step sequence. Thus we have

$$\mathcal{Z}[\delta_d[n]] = \frac{1}{1 - 0 \cdot z^{-1}} = \frac{1}{1} = 1$$

and

$$\mathcal{Z}[\text{step sequence}] = \mathcal{Z}[1] = \frac{1}{1 - z^{-1}} = \frac{z}{z - 1}$$

If $b = ae^{j\omega_0}$, where a and ω_0 are real numbers, then

$$\mathcal{Z}[a^n e^{jn\omega_0}] = \mathcal{Z}[(ae^{j\omega_0})^n] = \frac{1}{1 - ae^{j\omega_0} z^{-1}} = \frac{z}{z - ae^{j\omega_0}}$$

Using this, we can compute

$$
\begin{aligned}
\mathcal{Z}[a^n \sin\omega_0 n] &= \mathcal{Z}\left[\frac{a^n e^{j\omega_0 n} - a^n e^{-j\omega_0 n}}{2j}\right] = \frac{1}{2j}\left[\frac{z}{z - ae^{j\omega_0}} - \frac{z}{z - ae^{-j\omega_0}}\right] \\
&= \frac{z}{2j}\left[\frac{z - ae^{-j\omega_0} - z + ae^{j\omega_0}}{z^2 - a(e^{j\omega_0} + e^{-j\omega_0})z + a^2}\right] = \frac{a(\sin\omega_0)z}{z^2 - 2a(\cos\omega_0)z + a^2} \quad (11.7)
\end{aligned}
$$

and

$$\mathcal{Z}[a^n \cos\omega_0 n] = \frac{(z - a\cos\omega_0)z}{z^2 - 2a(\cos\omega_0)z + a^2} \qquad (11.8)$$

If $a = 1$, they reduce to

$$\mathcal{Z}[\sin\omega_0 n] = \frac{(\sin\omega_0)z}{z^2 - 2(\cos\omega_0)z + 1}$$

and

$$\mathcal{Z}[\cos\omega_0 n] = \frac{(z - \cos\omega_0)z}{z^2 - 2(\cos\omega_0)z + 1}$$

Before proceeding , we develop the formula

$$\mathcal{Z}[nx[n]] = -z\frac{dX(z)}{dz}$$

The differentiation of (11.2) with respect to z yields

$$\frac{d}{dz}X(z) = \sum_{n=0}^{\infty}(-n)x[n]z^{-n-1}$$

which becomes, after multiplying its both sides by $-z$,

$$-z\frac{d}{dz}X(z) = \sum_{n=0}^{\infty} nx[n]z^{-n}$$

Its right-hand-side, by definition, is the z-transform of $nx[n]$. This establishes the formula. Using the formula and $\mathcal{Z}[b^n] = z/(z - b)$, we can establish

$$\mathcal{Z}[nb^n] = -z\frac{d}{dz}\left[\frac{z}{z - b}\right] = -z\frac{(z - b) - z}{(z - b)^2} = \frac{bz}{(z - b)^2} = \frac{bz^{-1}}{(1 - bz^{-1})^2}$$

$x[n],\ n \geq 0$	$X(z)$	$X(z)$
$\delta[n]$	1	1
$\delta[n - n_0]$	z^{-n_0}	z^{-n_0}
1 or $q[n]$	$\frac{z}{z-1}$	$\frac{1}{1-z^{-1}}$
b^n	$\frac{z}{z-b}$	$\frac{1}{1-bz^{-1}}$
nb^n	$\frac{bz}{(z-b)^2}$	$\frac{bz^{-1}}{(1-bz^{-1})^2}$
$n^2 b^n$	$\frac{(z+b)bz}{(z-b)^3}$	$\frac{(1+bz^{-1})bz^{-1}}{(1-bz^{-1})^3}$
$b^n \sin \omega_0 n$	$\frac{(\sin \omega_0)bz}{z^2 - 2(\cos \omega_0)bz + b^2}$	$\frac{(\sin \omega_0)bz^{-1}}{1 - 2(\cos \omega_0)bz^{-1} + b^2 z^{-2}}$
$b^n \cos \omega_0 n$	$\frac{(z - b\cos \omega_0)z}{z^2 - 2(\cos \omega_0)bz + b^2}$	$\frac{(1 - (\cos \omega_0)bz^{-1}}{1 - 2(\cos \omega_0)bz^{-1} + b^2 z^{-2}}$

Table 11.1: z-Transform Pairs

Applying the formula once again, we can obtain

$$\mathcal{Z}[n^2 b^n] - \frac{b(z + b)z}{(z - b)^3}$$

We list in Table 11.1 some z-transform pairs. Unlike the Laplace transform where we use exclusively positive-power form, we may encounter both negative-power and positive-power forms in the z-transform. Thus we list both forms in the table.

The z-transforms listed in Table 11.1 are all proper rational functions of z. Note that the definitions in Subsection 8.7.2 are applicable to any rational function, be it a rational function of s or z. If $x[0] \neq 0$, such as in $\delta_d[n], 1, b^n$, and $b^n \cos \omega_0 n$, then its z-transform is biproper. If $x[0] = 0$, such as in nb^n, $n^2 b^n$, and $b^n \sin \omega_0 n$, then its z-transform is strictly proper. The Laplace transforms in Table 9.1 are all strictly proper except the one of $\delta(t)$.

The z-transforms of the DT signals in the preceding examples are all rational functions of z. This is also the case if a DT signal is of finite length. For example, the z-transform of $x[n]$, for $n = 0 : 5$, and $x[n] = 0$, for all $n > 5$, is

$$
\begin{aligned}
X(z) &= x[0] + x[1]z^{-1} + x[2]z^{-2} + x[3]z^{-3} + x[4]z^{-4} + x[5]z^{-5} \\
&= \frac{x[0]z^5 + x[1]z^4 + x[2]z^3 + x[3]z^2 + x[4]z + x[5]}{z^5}
\end{aligned}
$$

It is a proper rational function of z. However if a positive-time sequence is of infinite length, then its z-transform is an infinite power series of z^{-1} and the following situations may occur:

1. The infinite power series has no region of convergence. The sequences e^{n^2} and e^{e^n}, for $n \geq 0$, are such examples. These two sequences are mathematically contrived and do not arise in practice.

2. The infinite power series exists but cannot be expressed in closed form. Most, if not all, randomly generated sequences of infinite length belong to this type.

3. The infinite series can be expressed in closed form but the form is not a rational function of z. For example, if $h[0] = 0$ and $h[n] = 1/n$, for $n \geq 1$, then it can be shown that its z-transform is $H(z) = -\ln(1 - z^{-1})$. It is an irrational function of z. Note that $\sin z$, $\sqrt{z}$, and $\ln z$ are all irrational.

4. Its z-transform exists and is a rational function of z.

We study in this chapter only z-transforms that belong to the last class.

11.3   DT LTI lumped systems – proper rational functions

We introduced in Chapter 7 that every DT LTI system can be described by a discrete convolution

$$y[n] = \sum_{k=0}^{\infty} h[n-k]u[k] \tag{11.9}$$

where $y[n]$ and $u[n]$ are respectively the output and input of the system and n is the time index. The sequence $h[n]$ is the impulse response and is the output of the system excited by $u[n] = \delta_d[n]$, an impulse sequence. The impulse response has the property $h[n] = 0$, for all $n < 0$, if the system is causal. Recall that the convolution describes only forced responses or, equivalently, is applicable only if the system is initially relaxed at $n_0 = 0$.

The application of the z-transform to (11.9) yields, as derived in (7.25),

$$Y(z) = H(z)U(z) \tag{11.10}$$

where

$$H(z) = \sum_{n=0}^{\infty} h[n]z^{-n} \tag{11.11}$$

is the (DT) transfer function of the system. Recall from (7.29) that the transfer function can also be defined as

$$H(z) := \frac{\mathcal{Z}[\text{output}]}{\mathcal{Z}[\text{input}]} = \frac{Y(z)}{U(z)}\bigg|_{\text{Initially relaxed}}$$

We now define a DT system to be a *lumped* system if its transfer function is a rational function of z. It is *distributed* if its transfer function cannot be so expressed. According to this definition, *every* DT LTI FIR system is lumped because its transfer function, as discussed earlier, is a rational function of z. For a DT LTI IIR system, its transfer function (an infinite power series), as listed in the preceding section, may not be defined or be expressed in closed form. Even if it can be expressed in closed form, the form may not be a rational function. Thus the set of DT LTI IIR lumped systems is, as shown in Figure 8.12, a small part of DT LTI systems.

In the remainder of this chapter we study only DT LTI lumped systems or systems that are describable by

$$H(z) = \frac{N(z)}{D(z)} \tag{11.12}$$

where $N(z)$ and $D(z)$ are polynomials of z with real coefficients. Note that all discussion in Section 8.7.2 are directly applicable here. For example, $H(z)$ is improper if $\deg N(z) > \deg D(z)$, proper if $\deg D(z) \geq \deg N(z)$. In the CT case, we study only proper rational functions of s because, as discussed in Section 8.5.2, improper rational functions will amplify high-frequency noise. In the DT case, we also study only proper rational functions of z. But the reason is different. Consider the simplest improper rational function $H(z) = z/1 = z$ or $Y(z) = H(z)U(z) = zU(z)$. As discussed in Subsection 7.7.1, z is an unit-sample advance operator. Thus the inverse z-transform of $Y(z) = zU(z)$ is

$$y[n] = u[n+1]$$

The output at n depends on the future input at $n+1$. Thus the system is not causal. In general, a DT system with an improper rational transfer function is not causal and cannot be implemented in real time (subsection 2.6.1). In conclusion, we study proper rational transfer functions because of the causality reason in the DT case and of noise problem in the CT case.

To conclude this section, we mention that a rational function can be expressed in two forms such as

$$H(z) \quad = \quad \frac{8z^3 - 24z - 16}{2z^5 + 20z^4 + 98z^3 + 268z^2 + 376z} \tag{11.13}$$

$$= \quad \frac{8z^{-2} - 24z^{-4} - 16z^{-5}}{2 + 20z^{-1} + 98z^{-2} + 268z^{-3} + 376z^{-4}} \tag{11.14}$$

The latter is obtained from the former by multiplying z^{-5} to its numerator and denominator. The rational function in (11.13) is said to be in *positive-power* form and the one in (11.14) in *negative-power* form. Either form can be easily obtained from the other. In the CT case, we use exclusively positive-power form as in Chapters 8 and 9. If we use the positive-power form in the DT case, then many results in the CT case can be directly applied. Thus with the exception of the next subsection, we use only positive-power form in this chapter.[1]

11.3.1 Rational transfer functions and Difference equations

This section discusses the relationship between rational transfer functions and linear difference equations with constant coefficients.[2] As shown in Section 7.7.1, if $x[n]$ is a positive-time sequence ($x[n] = 0$ for all $n < 0$), and if $X(z)$ is its z-transform, then we have

$$\mathcal{Z}[x[n-k]] = z^{-k}X(z) \tag{11.15}$$

for any positive integer k. Using (11.15), we can readily obtain a difference equation from a rational transfer function and vise versa. For example, consider

$$H(z) = \frac{Y(z)}{U(z)} = \frac{z^2 + 2z}{2z^3 + z^2 - 0.4z + 0.8} = \frac{z^{-1} + 2z^{-2}}{2 + z^{-1} - 0.4z^{-2} + 0.8z^{-3}} \tag{11.16}$$

---

[1]The negative-power form is widely used in texts on digital signal processing. We may define properness, poles, and zeros directly in negative-power form. See Reference [C7]. It is however simpler to define properness using positive-power form as we did in Subsection 8.7.2.

[2]This subsection may be skipped without loss of continuity.

The negative-power form is obtained from the positive-power form by multiplying z^{-3} to its numerator and denominator. We write (11.16) as

$$(2 + z^{-1} - 0.4z^{-2} + 0.8z^{-3})Y(z) = (z^{-1} + 2z^{-2})U(z)$$

or

$$2Y(z) + z^{-1}Y(z) - 0.4z^{-2}Y(z) + 0.8z^{-3}Y(z) = z^{-1}U(z) + 2z^{-2}U(z)$$

Its inverse z-transform is, using (11.15),

$$2y[n] + y[n-1] - 0.4y[n-2] + 0.8y[n-3] = u[n-1] + 2u[n-2] \qquad (11.17)$$

This is a third-order linear difference equation with constant coefficients or a third-order LTI difference equation. Conversely, given an LTI difference equation, we can readily obtain its transfer function using (11.15).

To conclude this subsection, we give the reason of using the negative-power form in (11.16). As derive in (7.33) and (7.34), we have

$$\mathcal{Z}[x[n+1]] = z[X(z) - x[0]] \qquad (11.18)$$

and

$$\mathcal{Z}[x[n+2]] = z^2 \left[X(z) - x[0] - x[1]z^{-1} \right]$$

where $x[0]$ and $x[1]$ are not necessarily zero. Thus the use of the positive-power form will be more complicated. However the final result will be the same. See Reference [C7, C8].

11.3.2   Poles and zeros

Consider a proper rational transfer function $H(z)$. A finite real or complex number λ is called a *zero* of $H(z)$ if $H(\lambda) = 0$. It is called a *pole* if $|H(\lambda)| = \infty$. This definition is identical to the one in Section 9.2 and all discussion there is directly applicable. Thus the discussion here will be brief.

Consider a proper rational transfer function

$$H(z) = \frac{N(z)}{D(z)}$$

where $N(z)$ and $D(z)$ are two polynomials with real coefficients. If they are coprime or have no roots in common, then all roots of $N(z)$ are the zeros of $H(z)$ and all roots of $D(z)$ are the poles of $H(z)$. For example, consider

$$H(z) = \frac{N(z)}{D(z)} = \frac{8z^3 - 24z - 16}{2z^5 + 20z^4 + 98z^3 + 268z^2 + 376z + 208} \qquad (11.19)$$

It is a ratio of two polynomials. Its numerator's and denominator's coefficients can be represented in MATLAB as n=[8 0 -24 -16] and d=[2 20 98 268 376 208]. Typing in MATLAB >> roots(n) will yield $2, -1, -1$. Thus we can factor $N(z)$ as

$$N(z) = 8(z - 2)(z + 1)(z + 1)$$

Note that $N(z)$, $N(z)/8$, and $kN(z)$, for any nonzero constant k, have the same set of roots, and the MATLAB function `roots` computes the roots of a polynomial with leading coefficient 1. Typing in MATLAB >> `roots(d)` will yield $-2, -2, -2, -2 + j3, -2 - j3$. Thus we can factor $D(z)$ as

$$D(z) = 2(z+2)^3(z+2-3j)(z+2+3j)$$

Clearly, $N(z)$ and $D(z)$ have no common roots. Thus $H(z)$ in (11.19) has zeros at $2, -1, -1$ and poles at $-2, -2, -2, -2 + 3j, -2 - 3j$. Note that the poles and zeros of $H(z)$ can also be obtained using the MATLAB function `tf2zp` as in Section 9.2.

A pole or zero is called *simple* if it appears only once, and *repeated* if it appears twice or more. For example, the transfer function in (11.19) has a simple zero at 2 and two simple poles at $-2 \pm 3j$. It has a repeated zero at -1 with multiplicity 2 and a repeated pole at -2 with multiplicity 3.

We mention that the zeros and poles of a transfer function can also be obtained using the MATLAB function `tf2zp`, an acronym for transfer function to zero/pole. For the transfer function in (11.19), typing in the command window of MATLAB

```
>> n=[8 0 -24 -16];d=[2 20 98 268 376 208];
>> [z,p,k]=tf2zp(n,d)
```

will yield z=[-1 -1 2]; p=[-2 -2 -2 -2-3j -2+3j]; k=4. It means that $H(z)$ in (11.19) can also be expressed as

$$H(z) = \frac{4(z-2)(z+1)^2}{(z+2)^3(z+2+3j)(z+2-3j)} \tag{11.20}$$

It is called the *zero/pole/gain* form. Note that the gain $k = 4$ is the ratio of the leading coefficients of $N(z)$ and $D(z)$.

11.4 Inverse z-transform

In this section, we discuss how to compute the time sequence of a z-transform. We use examples to discuss the procedure.

Example 11.4.1 Consider a system with transfer function

$$H(z) = \frac{2z^2 + 10}{z^2 - 1.2z - 1.6} = \frac{2z^2 + 10}{(z-2)(z+0.8)} \tag{11.21}$$

We compute its step response, that is, the output excited by a step input. If $u[n] = 1$, for $n \geq 0$, then its z-transform is $U(z) = z/(z-1)$. Thus the step response of (11.21) in the transform domain is

$$Y(z) = H(z)U(z) = \frac{2z^2 + 10}{(z-2)(z+0.8)} \cdot \frac{z}{z-1} \tag{11.22}$$

To find its response in the time domain, we must compute its inverse z-transform. The procedure consists of two steps: (1) expanding $Y(z)$ as a sum of terms whose inverse z-transforms are available in a table such as Table 11.1, and then (2) using the table to find the inverse z-transform. In order to use the procedure in Section 9.3.1, we expand $Y(z)/z$, instead of $Y(z)$, as[3]

$$\frac{Y(z)}{z} \ =: \ \bar{Y}(z) = \frac{2z^2 + 10}{(z-2)(z+0.8)(z-1)}$$

$$= \ k_0 + k_1\frac{1}{z-2} + k_2\frac{1}{z+0.8} + k_u\frac{1}{z-1} \qquad (11.23)$$

Then we have $k_0 = \bar{Y}(\infty) = 0$,

$$k_1 \ = \ \bar{Y}(z)(z-2)\big|_{z=2} = \frac{2z^2 + 10}{(z+0.8)(z-1)}\bigg|_{z=2} = \frac{18}{2.8} = 6.43$$

$$k_2 \ = \ \bar{Y}(z)(z+0.8)\big|_{z=-0.8} = \frac{2z^2 + 10}{(z-0.2)(z-1)}\bigg|_{z=-0.8} = \frac{2(-0.8)^2 + 10}{(-2.8)(-1.8)} = 2.24$$

and

$$k_u \ = \ \bar{Y}(z)(z-1)\big|_{z=1} = \frac{Y(z)(z-1)}{z}\bigg|_{z=1} = H(z)|_{z=1} \qquad (11.24)$$

$$= \ H(1) = \frac{2+10}{1-1.2-1.6} = \frac{12}{-1.8} = -6.67$$

Thus $Y(z)/z$ can be expanded as

$$\frac{Y(z)}{z} = 0 + 6.43\frac{1}{z-2} + 2.24\frac{1}{z+0.8} - 6.67\frac{1}{z-1}$$

Its multiplication by z yields

$$Y(z) = 6.43\frac{z}{z-2} + 2.24\frac{z}{z+0.8} - 6.67\frac{z}{z-1}$$

Thus its inverse z-transform is, using Table 11.1,

$$y[n] = 6.43 \cdot 2^n + 2.24 \cdot (-0.8)^n - 6.67 \cdot 1^n = 6.43 \cdot 2^n + 2.24 \cdot (-0.8)^n - 6.67 \ \ (11.25)$$

for $n \geq 0$. This is the step response of the system described by (11.21). Note that the sequences 2^n and $(-0.8)^n$ are due to the poles 2 and -0.8 of the system and the sequence $-6.67(1)^n$ is due to the pole 1 of the input's z-transform.□

Example 11.4.2 The step response of the preceding example can also be obtained by expanding $Y(z)$ as

$$Y(z) \ = \ H(z)U(z) = \frac{2z^2 + 10}{(z-2)(z+0.8)} \cdot \frac{z}{z-1}$$

$$= \ k_0 + k_1\frac{z}{z-2} + k_2\frac{z}{z+0.8} + k_3\frac{z}{z-1}$$

---

[3]The reason of this expansion will be given in the next example.

Note that every term in the expansion is listed in Table 11.1. Thus once k_i are computed, then its inverse z-transform can be directly obtained. However if $z = \infty$, the equation implies $Y(\infty) = k_0 + k_1 + k_2 + k_3$. Thus the procedure in Section 9.3.1 cannot be directly applied. This is the reason of expanding $Y(z)/z$ in the preceding example. Note that we can also expand $Y(z)$ as

$$
\begin{aligned}
Y(z) &= \frac{(2z^2 + 10)z}{(z-2)(z+0.8)(z-1)} = \frac{2 + 10z^{-2}}{(1 - 2z^{-1})(1 + 0.8z^{-1})(1 - z^{-1})} \\
&= \bar{k}_0 + \bar{k}_1 \frac{1}{1 - 2z^{-1}} + \bar{k}_2 \frac{1}{1 + 0.8z^{-1}} + \bar{k}_3 \frac{1}{1 - z^{-1}}
\end{aligned}
$$

This is in negative-power form and every term in the expansion is in Table 11.1. But the procedure in Section 9.3.1 must be modified in order to compute $\bar{k}_i$. In conclusion there are many ways of carrying out expansions. □

A great deal more can be said regarding the procedure. We will not discuss it further for the same reasons given in Subsection 9.3.2. Note that the reasons discussed for CT transfer functions are directly applicable to DT transfer functions.

Before proceeding, we mention that the inverse z-transform of $X(z)$ can also be obtained by expressing $X(z)$ as a negative power series using direct division. For example, consider

$$
X(z) = \frac{3z + 4}{z^3 + 2z^2 - z + 3} \tag{11.26}
$$

Let us carry out the following direct division

$$
\begin{array}{r}
3z^{-2} \quad -2z^{-3} \quad +7z^{-4} \quad +\cdots \\
\hline
z^3 \ +2z^2 \ -z \ +3 \ )\overline{\ 3z \quad\quad +4 \phantom{-3z^{-1}}} \\
3z \quad\quad +6 \quad -3z^{-1} \quad +9z^{-2} \\
\hline
-2 \quad +3z^{-1} \quad -9z^{-2} \\
-2 \quad -4z^{-1} \quad +2z^{-2} \quad -6z^{-3} \\
\hline
+7z^{-1} \quad -11z^{-2} \quad +6z^{-3}
\end{array}
$$

Then we can express $X(z)$ as

$$
X(z) = 0 + 0 \cdot z^{-1} + 3z^{-2} - 2z^{-3} + 7z^{-4} + \cdots
$$

Thus the inverse z-transform of $X(z)$ is $x[0] = 0, x[1] = 0, x[2] = 3, x[4] = -2$, and so forth. This method is difficult to express the inverse in closed form. However it can be used to establish a general property. Consider a proper rational function $X(z) = N(z)/D(z)$ with $M = \text{Deg}(D(z)) - \text{Deg}(N(z)) \geq 0$. The integer M is the degree difference between the denominator and numerator of $X(z)$. Then the inverse z-transform $x[n]$ of $X(z)$ has the property

$$
x[n] = 0, \ x[1] = 0, \ \cdots, \ x[M-1] = 0, \ x[M] = N_0/D_0 \neq 0 \tag{11.27}
$$

where N_0 and D_0 are the leading coefficients of $N(z)$ and $D(z)$. For example, for the rational function in (11.26), we have $M = 2$ and its time sequence has first non-zero value at time index $n = M = 2$ with $x[2] = 3/1 = 3$. In particular, if $X(z)$ is biproper, then $M = 0$ and its time sequence is nonzero at $n = 0$.

11.4.1 Step responses – Roles of poles and zeros

In this subsection, we will develop a general form of step responses. Using the form, we can discuss the roles of poles and zeros of transfer functions. Before proceeding, we discuss some z-transform pairs.

Consider $X(z) = (b_1 z + b_0)/(z - p)$. It is biproper and has one pole at p. Note that most of the Laplace transform pairs in Table 9.1 are strictly proper; whereas most of the z-transform pairs in Table Table 11.1 are biproper. Thus we consider biproper $X(z)$. Using the procedure in Example 11.4.1, we can obtain

$$\mathcal{Z}^{-1}\left[\frac{b_1 z + b_0}{z - p}\right] = k_0 \delta_d[n] + k_1 p^n \tag{11.28}$$

for $n \geq 0$, where $k_0 = -b_0/p$ and $k_1 = (b_1 p + b_0)/p$ (Problem 11.9). Because $k_0 \delta_d[n] = 0$, for all $n \geq 1$, the time response of $X(z)$ is dictated mainly by the pole p. Note that if $b_0 = 0$, then (11.28) reduces to $b_1 p^n$.

Equation (11.28) holds for p real or complex. If p is complex such as $p = \alpha + j\beta = re^{j\theta}$, then its complex conjugate $p^* = \alpha - j\beta = re^{-j\theta}$ is also a pole and the responses of p and p^* can be combined as

$$\mathcal{Z}^{-1}\left[\frac{b_2 z^2 + b_1 z + b_0}{(z - re^{j\theta})(z - re^{-j\theta})}\right] = k_0 \delta_d[n] + k_1 r^n \sin(\theta n + k_2) \tag{11.29}$$

for $n \geq 0$, where k_i can be expressed in terms of b_i, r and θ. The expressions however are complex and will not be derived, because we are interested only in its general form. Note that it is the DT counterpart of (9.15). Note also that in the CT case, the real part of a complex pole governs the envelope of the response and the imaginary part governs the frequency of oscillation. In the DT case, the real and imaginary parts of a complex pole have no physical meaning. Instead, its magnitude governs the envelope of the response and its phase governs the frequency of oscillation.

We next discuss responses of repeated poles. Using Table 11.1, we have

$$\mathcal{Z}^{-1}\left[\frac{b_3 z^3 + b_2 z^2 + b_1 z + b_0}{(z - p)^3}\right] = k_0 \delta_d[n] + k_1 p^n + k_2 n p^n + k_3 n^2 p^n \tag{11.30}$$

for $n \geq 0$, where all parameters are real. This is the DT counterpart of (9.21).

With the preceding discussion, we can develop the general form of step responses. Consider the following transfer functions

$$H_1(z) = \frac{0.5882 z^4}{(z + 0.6)^2 (z - 0.9e^{j0.5})(z - 0.9e^{-j0.5})}$$
$$= \frac{0.5882 z^4}{z^4 - 0.3796 z^3 - 0.7256 z^2 + 0.4033 z + 0.2916} \tag{11.31}$$
$$H_2(z) = \frac{1.1764(z + 0.5) z^3}{z^4 - 0.3796 z^3 - 0.7256 z^2 + 0.4033 z + 0.2916} \tag{11.32}$$
$$H_3(z) = \frac{0.0654(z^2 + 4z + 4) z^2}{z^4 - 0.3796 z^3 - 0.7256 z^2 + 0.4033 z + 0.2916} \tag{11.33}$$

They all have the same set of poles and the property $H_i(1) = 1$. Even though they have different zeros other than those at $z = 0$,[4] their step responses can all be expressed in the z-transform domain as

$$Y_i(z) = H_i(z)U(z) = H_i(z)\frac{z}{z-1}$$

or

$$Y_i(z) = k_0 + \frac{b_2 z^2 + b_1 z + b_0}{(z+0.6)^2} + \frac{\bar{b}_2 z^2 + \bar{b}_1 z + \bar{b}_0}{(z - 0.9e^{j0.5})(z - 0.9e^{-j0.5})} + \frac{k_u z}{z-1} \qquad (11.34)$$

In this equation, we are interested in only the parameter k_u associated with the pole of the step input. If we multiply $(z-1)/z$ to (11.34) and then set $z = 1$, we obtain

$$k_u = \left. Y_i(z)\frac{z-1}{z}\right|_{z=1} = \left. H_i(z)\right|_{z=1} = H_i(1) \qquad (11.35)$$

This equation also follows from the procedure discussed in Example 11.4.1, as derived in (11.24). Note that (11.35) is applicable for any $H_i(z)$ which has no pole at $z = 1$.

The inverse z-transform of (11.34) or the step responses of (11.31) through (11.33) in the time domain are all of the form, using (11.28) through (11.30),

$$y[n] = \bar{k}_0 \delta_d[n] + k_1(-0.6)^n + k_2 n(-0.6)^n + k_3 0.9^n \sin(0.5n + k_4) + H_i(1) \qquad (11.36)$$

for $n \geq 0$. This form is determined solely by the poles of $H_i(z)$ and $U(z)$. The transfer function $H_i(z)$ has a repeated real pole at -0.6, which yields the response $k_1(-0.6)^n + k_2 n(-0.6)^n$, and a pair of complex conjugate poles $0.9e^{\pm j0.5}$, which yields $k_3 0.9^n \sin(0.5n + k_4)$. The z-transform of the step input is $z/(z-1)$. Its pole at $z = 1$ yields the response $H_i(1) \times 1^n = H_i(1)$. We plot in Figure 11.2 their step responses followed by zero-order holds. See the discussion pertaining to Figure 7.5. They can be obtained in MATLAB by typing in an edit window the following

```
%Program 11.1 (f112.m)
n1=0.5882*[1 0 0 0 0];d=[1 -0.3796 -0.7256 0.4033 0.2916];
n2=1.1764*[1 0.5 0 0 0];n3=0.0654*[1 4 4 0 0];
pig1=tf(n1,d,1);pig2=tf(n2,d,1);pig3=tf(n3,d,1);
step(pig1,60)
hold on
step(pig2,':',60)
step(pig3,'-.',60)
```

The first two lines express the numerators and denominators of $H_i(z)$ as row vectors. The function pig1=tf(n1,d,1) uses the transfer function H1=n1/d to define the DT system and names it pig1. It is important to have the third argument $T = 1$ in tf(n1,d,T). Without it, tf(n1,d) defines a CT system. The function step(pig1,60)

---

[4] We introduce zeros at $z = 0$ to make all $H_i(z)$ biproper so that their nonzero step responses all start to appear at $n = 0$ as discussed in the preceding section. Without introducing the zeros at z, the responses in Figure 11.2 will be difficult to visualize and compare.

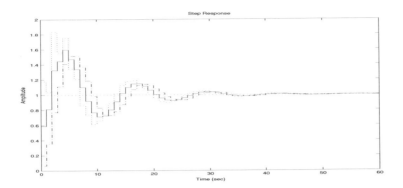

Figure 11.2: Step responses of the transfer functions in (11.31) through (11.33) followed by zero-order holds.

computes its step response for $n = 0 : N = 60$. We then type "hold on" in order to have the step responses of $H_2(z)$ and $H_3(z)$ plotted on the same plot. We name the program f112. It is called an m-file because an extension .m is automatically attached to the file name. Typing in the command window >> f112 will yield in a figure window the plot in Figure 11.2. We see that even though their step responses are all of the form in (11.36), the responses right after the application of the step input are all different. This is due to different sets of k_i. In conclusion, poles dictate the general form of responses; zeros affect only the coefficients k_i. Thus we conclude that zeros play a lesser role than poles in determining responses of systems.

To conclude this section, we mention that the response generated by step is not computed using the transfer function ni/d for the same reasons discussed in Section 9.3.2. As in the CT case, the program first transforms the transfer function into an ss equation, as we will discuss in a later section, and then uses the latter to carry out the computation.

11.4.2   *s*-plane and *z*-plane

Consider a CT pole at s_0 or $1/(s - s_0)$ whose time response is $e^{s_0 t}$, for $t \geq 0$. The sampling of $e^{s_0 t}$ with sampling period $T = 1$ yields the DT signal $e^{s_0 n}$, for $n \geq 0$. The z-transform of $e^{s_0 n}$ is $z/(z - e^{s_0})$ which has a pole at e^{s_0}. Thus the pole $s = s_0$ in CT systems is transformed into the pole $z = e^{s_0}$ in DT systems. Consequently the Laplace-transform variable s and the z-transform variable z can be related by

$$z = e^s$$

We show that $z = e^s$ maps the $j\omega$-axis on the s-plane into the unit circle on the z-plane and the left-half s-plane into the interior of the unit circle. Let $s = \sigma + j\omega$. Then we have

$$z = s^{\sigma + j\omega} = e^{\sigma} e^{j\omega}$$

If $\sigma = 0$ (the imaginary axis of the s-plane), then $e^{\sigma} = 1$ and $|z| = |e^{j\omega}| = 1$, for all ω (the unit circle of the z-plane). Thus the $j\omega$-axis is mapped into the unit circle.

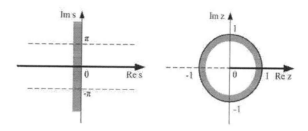

Figure 11.3: Mapping between the s-plane and z-plane by $z = e^s$.

Note that the mapping is not one-to-one. For example, all $s = 0, \pm j2\pi, \pm j4\pi, \ldots$ are mapped into $z = 1$. If $\sigma < 0$ (left half s-plane), then

$$|z| = |e^\sigma||e^{j\omega}| = |e^\sigma| < 1$$

(interior of the unit circle on the z-plane). Thus the left half s-plane is mapped into the interior of the unit circle on the z-plane as shown in Figure 11.3.

If $s = \sigma$, with $-\infty \le \sigma \le \infty$, then we have $0 \le z = e^\sigma \le \infty$. Thus the entire real axis denoted by the solid line on the s-plane is mapped into the positive real axis denoted by the solid line on the z-plane. Note that the entire negative real axis $[-\infty, 0]$ on the s-plane is mapped into $[0, 1]$ on the z-plane and the positive real axis $[0, \infty)$ on the s-plane is mapped into $[1, \infty)$ on the z-plane. If $s = \sigma \pm j\pi$, with $-\infty \le \sigma \le \infty$, then we have $-\infty \le z = e^{\sigma \pm j\pi} = -e^\sigma \le 0$. Thus the entire dashed lines passing through $\pm\pi$ shown in Figure 11.3(a) are mapped into the negative real axis on the z-plane as shown in Figure 11.3(b). Thus the meanings of the two real axes are different. In the CT case or on the s-plane, the real axis denotes zero frequency. In the DT case or on the z-plane, the positive real axis denotes zero frequency; whereas the negative real axis denotes the highest frequency π.

In the CT case, the s-plane is divided into three parts: the left-half plane (LHP), right-half plane (RHP), and the $j\omega$-axis. In view of the mapping in Figure 11.3, we will now divide the z-plane into three parts: the unit circle, its interior and its exterior.

11.4.3 Responses of Poles as $n \to \infty$

Let $p = \alpha + j\beta = re^{j\theta}$, where $r \ge 0$ and θ are real, be a simple pole. The pole lies inside the unit circle if $r < 1$, on the unit circle if $r = 1$, and outside the unit circle if $r > 1$. The pole will contribute the time response $p^n = r^n e^{jn\theta}$. If $\theta = 0$, that is, the pole is on the positive real axis, then its response r^n, as $n \to \infty$, will grow unbounded if $r > 1$, will be a step sequence if $r = 1$, and will vanish if $r < 1$.

If $0 < \theta < \pi$, then the pole $re^{j\theta}$ is a complex pole and its complex conjugate $re^{-j\theta}$ will also be a pole. The pair of complex-conjugate poles will generate the response

$$k_1 r^n \sin(\theta n + k_2)$$

for some real constants k_1 and k_2. Note that the magnitude r governs the envelope of the response because

$$|k_1 r^n \sin(\theta n + k_2)| = |k_1 r^n| |\sin(\theta n + k_2)| \le |k_1 r^n|$$

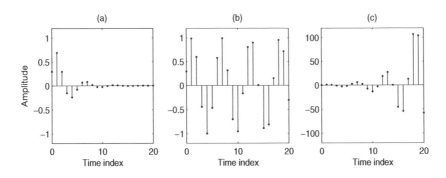

Figure 11.4: (a) $0.7^n \sin(1.1n + 0.3)$. (b) $\sin(1.1n + 0.3)$. (c) $1.3^n \sin(1.1n + 0.3)$.

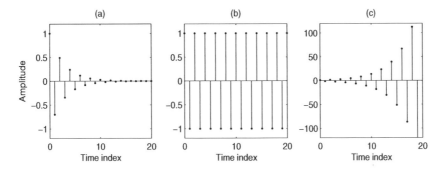

Figure 11.5: (a) $(-0.7)^n$, for $n = 0 : 20$. (b) $(-1)^n$. (c) $(-1.3)^n$.

and the phase θ governs the frequency of oscillation as shown in Figure 11.4. If $r < 1$ or the complex-conjugate poles lie inside the unit circle, then the response vanishes oscillatorily as shown in Figure 11.4(a). If $r = 1$, or the poles lie on the unit circle, then the response is a pure sinusoid which will not vanish nor grow unbounded as shown in Figure 11.4(b). If $r > 1$ or the poles lie outside the unit circle, then the response will grow to $\pm\infty$ as $n \to \infty$ as shown in Figure 11.4(c).

If $\theta = \pi$, then the pole p becomes $re^{j\pi} = -r$ and is located on the negative real axis. Its response is $(-r)^n$. We plot in Figures 11.5(a), (b), and (c) for $r = 0.7, 1, 1.3$ respectively. Note that we have assumed that the sampling period T is 1 and its Nyquist frequency range is $(-\pi/T, \pi/T] = (-\pi, \pi]$. Thus $\theta = \pi$ is the highest possible frequency. In this case, its response will change sign every sample (positive to negative or vice versa) as shown in Figure 11.5. Note that if $\theta = 0$ or the frequency is zero, then the response will never change sign.

We next discuss repeated poles. For repeated poles, we have

$$\mathcal{Z}^{-1}\left[\frac{b_2 z^2 + b_1 z + b_0}{(z - r)^2}\right] = k_0 \delta_d[n] + k_1 r^n + k_2 n r^n$$

$$\mathcal{Z}^{-1}\left[\frac{b_4 z^4 + b_3 z^3 + b_2 z^2 + b_1 z + b_0}{(z - re^{j\theta})^2 (z - re^{-j\theta})^2}\right] = k_0 \delta_d[n] + k_1 r^n \sin(\theta n + k_2) + k_3 n r^n \sin(\theta n + k_4)$$

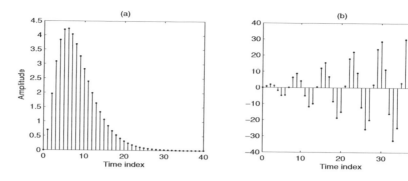

Figure 11.6: (a) $n^2 \times 0.7^n$. (b) $n\sin(0.9n)$.

If $r > 1$, then r^n and nr^n approach ∞ as $n \to \infty$. Thus if a pole, simple or repeated, real or complex, lying outside the unit circle, then its time response approaches infinite as $n \to \infty$.

The response of a repeated pole contains nr^n. It is a product of ∞ and 0 as $n \to \infty$ for $r < 1$. However, because

$$\lim_{n \to \infty} \frac{(n+1)r^{n+1}}{nr^n} = r < 1$$

the response approaches zero as $n \to \infty$ following Cauchy's ratio test. See Reference [P5]. Similarly, we can show that, for any $r < 1$ and any positive integer k, $n^k r^n$ approaches 0 as $n \to \infty$. Thus we conclude that the time response of a pole, simple or repeated, real or complex, lying inside the unit circle, approach 0 as $n \to \infty$. This is indeed the case as shown in Figure 11.6(a) for $n^2(0.7)^n$.

The situation for poles with $r = 1$, or on the unit circle, is more complex. The time response of a simple pole at $z = 1$ is a constant for all n. The time response of a simple pair of complex-conjugate poles on the unit circle is $k_1 \sin(\theta n + k_2)$ which is a sustained oscillation for all n as shown in Figure 11.4(b). However the time response of a repeated pole at $z = 1$ is n which approaches ∞ as $n \to \infty$. The time response of a repeated pair of complex-conjugate poles on the unit circle contains $k_3 n \sin(\theta n + k_4)$ which approaches ∞ or $-\infty$ oscillatorily as $n \to \infty$ as shown in Figure 11.6(b). We summarize the preceding discussion in the following:

Pole Location	Response as $n \to \infty$
Interior of the unit circle, simple or repeated	0
Exterior of the unit circle, simple or repeated	∞ or $-\infty$
Unit circle, simple	constant or sustained oscillation
Unit circle, repeated	∞ or $-\infty$

In conclusion, the response of a pole approaches zero as $n \to \infty$ *if and only if* the pole lies inside the unit circle. In view of the mapping in Figure 11.3, this is consistent with the CT case where the response of a pole approaches zero as $t \to \infty$ if and only if the pole lies inside the LHP.

11.5   Stability

This section introduces the concept of stability for DT systems. If a DT system is not stable, its response excited by any input generally will grow unbounded or overflow. Thus every DT system designed to process signals must be stable. Let us give a formal definition.

Definition 11.1 A DT system is BIBO (bounded-input bounded-output) stable or, simply, stable, if *every* bounded input sequence excites a bounded output sequence. Otherwise, the system is said to be unstable.□

A signal is bounded if it does not grow to ∞ or $-\infty$. In other words, a signal $u[n]$ is bounded if there exists a constant M_1 such that $|u[n]| \leq M_1 < \infty$ for all $n \geq 0$. As in the CT case, Definition 11.1 cannot be used to conclude the stability of a system because there are infinitely many bounded inputs to be checked. Fortunately, stability is a property of a system and can be determined from its mathematical descriptions.

Theorem 11.1 A DT LTI system with impulse response $h[n]$ is BIBO stable if and only if $h[n]$ is absolutely summable in $[0, \infty)$, that is,

$$\sum_{n=0}^{\infty} |h[n]| \leq M < \infty$$

for some constant M.□

Proof The input $u[n]$ and output $y[n]$ of any DT LTI system which is initially relaxed at $n_0 = 0$ can be described by (11.9) or

$$y[n] = \sum_{k=0}^{\infty} h[n-k]u[k]$$

where $h[n]$ is the impulse response of the system and has the property $h[n] = 0$, for all $n < 0$ as we study only causal systems. We use the equation to prove the theorem.

We first show that the system is BIBO stable if $h[n]$ is absolutely summable. Let $u[k]$ be any bounded input. Then there exists an M_1 such that $|u[n]| \leq M_1$ for all $n \geq 0$. We compute

$$|y[n]| = \left| \sum_{k=0}^{\infty} h[n-k]u[k] \right| \leq \sum_{k=0}^{\infty} |h[n-k]||u[k]| \leq M_1 \sum_{k=0}^{\infty} |h[n-k]|$$

Note that $h[n-k]u[k]$ can be positive or negative and their summation can cancel out. This will not happen in the second summation because $|h[n-k]||uk|$ are all positive. Thus we have the first inequality. The second inequality follows from $|u[k]| \leq M_1$, for all $k \geq 0$. Let us introduce the new integer variable $\bar{k} := n - k$ and use the causality condition $h[\bar{k}] = 0$ for all $\bar{k} < 0$. Then the preceding equation becomes

$$|y[n]| \quad \leq \quad M_1 \sum_{\bar{k}=n}^{-\infty} |h[\bar{k}]| = M_1 \sum_{\bar{k}=-\infty}^{n} |h[\bar{k}]| = M_1 \sum_{\bar{k}=0}^{n} |h[\bar{k}]|$$

$$\leq \quad M_1 \sum_{\bar{k}=0}^{\infty} |h[\bar{k}]| \leq M_1 M$$

for all $n > 0$. This shows that if $h[n]$ is absolutely summable, every bounded input will excite a bounded output. Thus the system is BIBO stable.

Next we show that if $h[n]$ is not absolutely summable, then there exists a bounded input that will excite an output whose magnitude approaches ∞ as $n \to \infty$. Because ∞ is not a number, the way to show $|y[n]| \to \infty$ is to show that no matter how large M_2 is, there exists an n_1 such that $|y[n_1]| > M_2$. If $h[n]$ is not absolutely summable, then for any arbitrarily large M_2, there exists an n_1 such that

$$\sum_{k=0}^{n_1} |h[k]| > M_2$$

Let us select an input $u[n]$ as, for $n = 0 : n_1$,

$$u[n] = \begin{cases} 1 & \text{if } h[n_1 - n] \geq 0 \\ -1 & \text{if } h[n_1 - n] < 0 \end{cases}$$

For this bounded input, the output $y[n]$ at $n = n_1$ is

$$y[n_1] = \sum_{k=0}^{\infty} h[n_1 - k]u[k] = \sum_{k=0}^{\infty} |h[n_1 - k]| = \sum_{\bar{k}=n_1}^{-\infty} |h[\bar{k}]| = \sum_{\bar{k}=n_1}^{0} |h[\bar{k}]| > M_2$$

where we have introduced $\bar{k} := n_1 - k$ and used the causality condition $h[\bar{k}] = 0$ for all $\bar{k} < 0$. This shows that if $h[n]$ is not absolutely summable, then there exists a bounded input that will excite an output with an arbitrarily large magnitude. Thus the system is not stable. This establishes the theorem.□

A direct consequence of this theorem is that every DT FIR system is BIBO stable. Indeed, a DT FIR system has only a finite number of nonzero entries in $h[n]$. Thus its impulse response is absolutely summable. We next use the theorem to develop the next theorem.

Theorem 11.2 A DT LTI lumped system with proper rational transfer function $H(z)$ is stable if and only if every pole of $H(z)$ has a magnitude less than 1 or, equivalently, all poles of $H(z)$ lie inside the unit circle on the z-plane. □

Proof: If $H(z)$ has one or more poles lying outside the unit circle, then its impulse response grows unbounded and is not absolutely summable. If it has poles on the unit circle, then its impulse response will not approach 0 and is not absolutely summable. In conclusion if $H(z)$ has one or more poles on or outside the unit circle, then the system is not stable.

Next we argue that if all poles lie inside the unit circle, then the system is stable. To simplify the discussion, we assume that $H(z)$ has only simple poles and can be expanded as

$$H(z) = \sum_i \frac{k_i z}{z - p_i} + k_0$$

where the poles p_i can be real or complex. Then its inverse z-transform or the impulse response of the system is

$$h[n] = \sum_i k_i p_i^n + k_0 \delta_d[n]$$

for $n \geq 0$. We compute

$$\sum_{n=0}^{\infty} |h[n]| = \sum_{n=0}^{\infty} \left| \sum_i k_i p_i^n + k_0 \delta_d[n] \right| \leq \sum_i \left(\sum_{n=0}^{\infty} |k_i p_i^n| \right) + |k_0| \qquad (11.37)$$

If $p_i = r_i e^{j\theta_i}$ and if $0 \leq r_i < 1$, then

$$\sum_{n=0}^{\infty} \left| k_i r_i^n e^{jn\theta_i} \right| = \sum_{n=0}^{\infty} |k_i r_i^n| = \frac{|k_i|}{1 - r_i} =: M_i < \infty$$

where we have used $\left| e^{j\theta_i n} \right| = 1$, for every θ_i and for all n. Thus if every pole of $H(z)$ has a magnitude less than 1, then every term in (11.37) is absolutely summable. Consequently $h[n]$ is absolutely summable and the system is BIBO stable. This establishes Theorem 11.2.□

The stability of a system with a proper rational transfer function depends only on the poles. A system can be stable with its zeros lying anywhere: inside, outside, or on the unit circle on the z-plane. We study only LTI lumped systems, thus we use mostly Theorem 11.2.

Example 11.5.1 Consider a DT system with transfer function

$$H(z) = \frac{(z + 2)(z - 10)}{(z - 0.9)(z + 0.95)(z + 0.7 + j0.7)(z + 0.7 - j0.7)}$$

Its two real poles 0.9 and -0.95 have magnitudes less than 1. We compute the magnitude of the complex poles:

$$\sqrt{(0.7)^2 + (0.7)^2} = \sqrt{0.49 + 0.49} = \sqrt{0.98} = 0.9899$$

It is less than 1. Thus the system is stable. Note that $H(z)$ has two zeros outside the unit circle on the z-plane. □

Example 11.5.2 Consider an FIR system with impulse response $h[n]$, for $n = 0 : N$, with $h[N] \neq 0$ and $h[n] = 0$, for all $n > N$. Its transfer function is

$$H(z) = \sum_{n=0}^{N} h[n] z^{-n} = \frac{h[0]z^N + h[1]z^{N-1} + \cdots + h[N-1]z + h[N]}{z^N}$$

All its N poles are located at $z = 0.$[5] They all lie inside the unit circle. Thus it is stable.□

---

[5]If $h[N] = 0$, then $H(z)$ has at most $N - 1$ number of poles at $z = 0$.

Corollary 11.2 A DT LTI lumped system with impulse response $h[n]$ is BIBO stable if and only if $h[n] \to 0$ as $n \to \infty$. $\square$

As discussed in Subsection 11.4.3, the response of a pole approaches 0 as $n \to \infty$ if and only if the pole lies inside the unit circle on the z-plane. Thus this corollary follows directly from Theorem 11.2. This corollary has a very important practical implication. If a DT system is known to be LTI and lumped, then its stability can easily be checked by measurement. We apply $u[0] = 1$ and $u[n] = 0$ for $n > 0$ (an impulse sequence), then the output is the impulse response. Thus the system is stable if and only if the response approaches zero as $n \to \infty$.

11.5.1 What holds for lumped systems may not hold for distributed systems

We mention that the results for LTI lumped systems may not be applicable to LTI distributed systems. For example, consider a DT LTI system with impulse response

$$h[n] = \frac{1}{n}$$

for $n \geq 1$. It approaches 0 as $n \to \infty$. Can we conclude that the system is stable? Or is Corollary 11.2 applicable? To answer this, we must check whether the system is lumped or not. The transfer function or the z-transform of $h[n] = 1/n$ can be computed as $-\ln(1 - z^{-1})$. It is not a rational function of z. Thus the system is not lumped and Corollary 11.2 is not applicable.

Let us compute

$$S := \sum_{n=0}^{\infty} |h[n]| = \sum_{n=1}^{\infty} \frac{1}{n} = 1 + \frac{1}{2} + \left(\frac{1}{3} + \frac{1}{4}\right) + \left(\frac{1}{5} + \frac{1}{6} + \frac{1}{7} + \frac{1}{8}\right)$$
$$+ \left(\frac{1}{9} + \frac{1}{10} + \cdots + \frac{1}{15} + \frac{1}{16}\right) + \cdots$$

The first pair of parentheses contains two terms, each term is $1/4$ or larger, thus the sum is larger than $2/4 = 1/2$. The second pair of parentheses contains four terms, each term is $1/8$ or larger. Thus the sum is larger than $4/8 = 1/2$. The third pair of parentheses contains eight terms, each term is $1/16$ or larger. Thus the sum is larger than $8/16 = 1/2$. Proceeding forward, we have

$$S > 1 + \frac{1}{2} + \frac{1}{2} + \frac{1}{2} + \cdots = \infty$$

and the sequence is not absolutely summable. See also Problem 3.19. Thus the DT system with impulse response $1/n$, for $n \geq 1$, is not stable according to Theorem 11.1. In conclusion, Theorem 11.1 is applicable to lumped and distributed systems; whereas Theorem 11.2 and its corollary are applicable only to lumped systems.

11.5.2   The Jury test

Consider a system with proper rational transfer function $H(z) = N(z)/D(z)$. We assume that $N(s)$ and $D(s)$ are coprime. Then the poles of $H(z)$ are the roots of $D(z)$. If $D(z)$ has degree three or higher, computing its roots by hand is not simple. This, however, can be carried out using a computer. For example, consider

$$D(z) = 2z^3 - 0.2z^2 - 0.24z - 0.08$$

Typing

```
>> d=[2 -0.2 -0.24 -0.08];
>> roots(d)
```

in MATLAB will yield $-0.5, -0.2 \pm j0.2$. It has one real root and one pair of complex conjugate roots. They all have magnitudes less than 1. Thus any transfer function with $D(z)$ as its denominator is BIBO stable.

A polynomial is defined to be *DT stable* if all its roots have magnitudes less than 1. We discuss a method of checking whether a polynomial is DT stable or not without computing its roots.[6] We use the following $D(z)$ to illustrate the procedure:

$$D(z) = a_0 z^5 + a_1 z^4 + a_2 z^3 + a_3 z^2 + a_4 z + a_5 \quad \text{with } a_0 > 0 \qquad (11.38)$$

We call a_0 the leading coefficient. If $a_0 < 0$, we apply the procedure to $-D(z)$. Because $D(z)$ and $-D(z)$ have the same set of roots, if $-D(z)$ is DT stable, so is $D(z)$. The polynomial $D(z)$ has degree 5 and six coefficients a_i, $i = 0 : 5$. We form Table 11.2, called the *Jury table*. The first row is simply the coefficients of $D(z)$ arranged in the descending power of z. The second row is the reversal of the first row. We compute $k_1 = a_5/a_0$, the ratio of the last entries of the first two rows. The first b_i row is obtained by subtracting from the first a_i row the product of the second a_i row and k_1. Note that the last entry of the first b_i row is automatically zero and is discarded in the subsequent discussion. We then reverse the order of b_i to form the second b_i row and compute $k_2 = b_4/b_0$. The first b_i row subtracting the product of the second b_i row and k_2 yields the first c_i row. We repeat the process until the table is completed as shown. We call b_0, c_0, d_0, e_0, and f_0 the *subsequent leading coefficients*. If $D(z)$ has degree N, then the table has N subsequent leading coefficients.

Theorem 11.3 A polynomial with a positive leading coefficient is DT stable if and only if every subsequent leading coefficient is positive. If any subsequent leading coefficient is 0 or negative, then the polynomial is not DT stable.□

The proof of this theorem can be found in Reference [2nd ed. of C6]. We discuss only its employment.

Example 11.5.3 Consider

$$D(z) = z^3 - 2z^2 - 0.8$$

---

[6]The remainder of this subsection may be skipped without loss of continuity.

a_0	a_1	a_2	a_3	a_4	a_5	
a_5	a_4	a_3	a_2	a_1	a_0	$k_1 = a_5/a_0$
b_0	b_1	b_2	b_3	b_4	0	(1st a_i row) $-k_1$(2nd a_i row)
b_4	b_3	b_2	b_1	b_0		$k_2 = b_4/b_0$
c_0	c_1	c_2	c_3	0		(1st b_i row) $-k_2$(2nd b_i row)
c_3	c_2	c_1	c_0			$k_3 = c_3/c_0$
d_0	d_1	d_2	0			(1st c_i row) $-k_3$(2nd c_i row)
d_2	d_1	d_0				$k_4 = d_2/d_0$
e_0	e_1	0				(1st d_i row) $-k_4$(2nd d_i row)
e_1	e_0					$k_5 = e_1/e_0$
f_0	0					(1st e_i row) $-k_5$(2nd e_i row)

Table 11.2: The Jury Table

Note that the polynomial has a missing term. We form

1	-2	0	-0.8	
-0.8	0	-2	1	$k_1 = -0.8/1 = -0.8$
0.36	2	-1.6	0	
-1.6	-2	0.36		$k_2 = -1.6/0.36 = -4.44$
-6.74				

A negative subsequent leading coefficient appears. Thus the polynomial is not DT stable.$\square$

Example 11.5.4 Consider

$$D(z) = 2z^3 - 0.2z^2 - 0.24z - 0.08$$

We form

2	-0.2	-0.24	-0.08	
-0.08	-0.24	-0.2	2	$k_1 = -0.08/2 = -0.04$
1.9968	-0.2096	-0.248	0	
-0.248	-0.2096	1.9968		$k_2 = -0.248/1.9968 = -0.124$
1.966	-0.236	0		
-0.236	1.966			$k_3 = -0.236/1.966 = -0.12$
1.938	0			

The three subsequent leading coefficients are all positive. Thus the polynomial is DT stable.$\square$

We defined in Section 9.5.3 a polynomial to be CT stable if all its roots have negative real parts. A CT stable polynomial cannot have missing terms or negative coefficients. This is not so for DT stable polynomials. For example, the polynomial in the preceding Example has negative coefficients and is still DT stable. Thus the conditions for a polynomial to be CT stable or DT stable are different and are independent of each other.

11.6   Steady-state and transient responses

In this section, we introduce the concepts of steady-state and transient responses. Consider a system with transfer function $H(z)$. Let $y[n]$ be the output excited by some input $u[n]$. We call

$$y_{ss}[n] = \lim_{n \to \infty} y[n] \tag{11.39}$$

the *stead-state (ss)* response and

$$y_{tr}[n] = y[n] - y_{ss}[n] \tag{11.40}$$

the *transient (tr)* response of the system excited by the input $u[n]$. For example, consider the system in Example 11.4.1 with transfer function

$$H(z) = \frac{2z^2 + 10}{(z - 2)(z + 0.8)}$$

Its step response was computed in (11.25) as

$$y[n] = 6.43 \cdot 2^n + 2.24 \cdot (-0.8)^n - 6.67 \cdot 1^n = 6.43 \cdot 2^n + 2.24 \cdot (-0.8)^n - 6.67 \tag{11.41}$$

for $n \geq 0$. Because $(-0.8)^n$ approaches 0 as $n \to \infty$, the steady-state response of (11.41) is

$$y_{ss}[n] = 6.43 \cdot 2^n - 6.67$$

and the transient response is

$$y_{tr}[n] = 2.24 \cdot (-0.8)^n$$

for $n \geq 0$. We see that the steady-state response consists of the time responses due to the pole of the input and a pole of the system. Because the pole $z = 2$ is outside the unit circle, we have $y_{ss}[n] \to \infty$ as $n \to \infty$. Note that $H(z)$ is not stable, and its steady-state response generally approaches infinite no matter what input is applied. □

Example 11.6.1 Consider a system with transfer function

$$
\begin{aligned}
H(z) &= \frac{z + 2}{z^3 + 0.4z^2 + 0.36z - 0.405} = \frac{z + 2}{(z - 0.5)(z + 0.45 - j0.78)(z + 0.45 + j0.78)} \\
&= \frac{z + 2}{(z - 0.5)(z - 0.9e^{j2.1})(z - 0.9e^{-j2.1})} \tag{11.42}
\end{aligned}
$$

It has a real pole at 0.5 and a pair of complex-conjugate poles at $0.9e^{\pm j2.1}$. They all lie inside the unit circle, thus the system is stable.

The step response of (11.42) in the transform domain is

$$Y(z) = H(z)U(z) = \frac{z + 2}{(z - 0.5)(z - 0.9e^{j2.1})(z - 0.9e^{-j2.1})} \cdot \frac{z}{z - 1}$$

and its time response is of the form, for $n \geq 0$,

$$y[n] = H(1) \times 1^n + k_1 \times 0.5^n + k_2 \times 0.9^n \sin(2.1n + k_3) \tag{11.43}$$

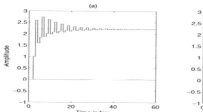 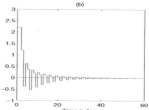

Figure 11.7: (a) Step response of (11.42) followed by zero-order hold. (b) Its transient response.

with, as in (11.35),

$$H(1) = \frac{1+2}{1+0.4+0.36-0.405} = \frac{3}{1.355} = 2.21$$

and some real constants k_i.

Because the poles of $H(z)$ all have magnitudes less than 1, their responses all approach zero as $n \to \infty$. Thus the steady-state response of (11.43) is

$$y_{ss}[n] = H(1) \cdot 1^n = H(1) = 2.21 \qquad (11.44)$$

and the transient response is

$$y_{tr}[n] = k_1 \times 0.5^n + k_2 \times 0.9^n \sin(\theta n + k_3) \qquad (11.45)$$

We see that the steady-state response is determined by the pole of the input alone and the transient response is due to all poles of the transfer function in (11.42). $\square$

Systems are designed to process signals. The most common processions are amplification and filtering. The processed signals or the outputs of the systems clearly should be dictated by the signals to be processed. If a system is not stable, its output will generally grow unbounded. Clearly such a system cannot be used. On the other hand, if a system is stable, after its transient response dies out, the processed signal is dictated by the applied signal. Thus the transient response is an important criterion in designing a system. The faster the transient response vanishes, the faster the response reaches steady state.

11.6.1   Time constant and response time of stable systems

Consider the transfer function in (11.42). Its step response is computed in (11.43) and its steady-state and transient responses are given in (11.44) and (11.45) respectively. We plot in Figure 11.7(a) its step response followed by a zero-order hold and in Figure 11.7(b) its transient response followed by a zero-order hold.[7] We see that the transient response approaches zero and the total response approaches the steady state $y_{ss}[n] = H(1) = 2.21$ as $n \to \infty$.

---

[7]If we plot the responses directly, they will be more difficult to visualize. In Figure 11.7(b), we actually plot $y_{tr}[n]$ for easier viewing.

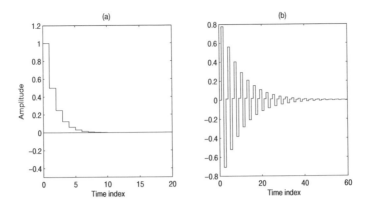

Figure 11.8: (a) The sequence 0.5^n followed by zero-order hold. (b) The sequence $0.9^n \sin(2.1n)$ followed by zero-order hold.

How fast or how much time will the transient response take to approach zero? Recall from Subsection 2.8.1 that the real exponential sequence r^n with $r < 1$ is defined to have a time constant of $n_c := -1/\ln r$ and takes $5n_c$ in samples to have its magnitude decreases to less than 1% of its peak value. Indeed, for $r = 0.5$, we have $n_c = 1.44$ and $5n_c = 7.2$.[8] Thus it takes 0.5^n roughly 7 samples to reach zero as shown in Figure 11.8(a). Because $\mathcal{Z}[r^n] = z/(z - r)$, we may define the time constant of a real pole r, with $0 \le |r| < 1$, as $n_c := -1/\ln|r|$. Note that if $|r| \ge 1$, its response will not approach zero and its time constant is not defined.

For a pair of complex-conjugate poles $re^{\pm j\theta}$ with $0 \le r < 1$, we will define its time constant as $-1/\ln r$. This definition is justified because

$$|r^n \sin(\theta n)| = |r^n||\sin(\theta n)| \le |r^n|$$

For example, the time constant of $0.9e^{\pm j2.1}$ is $n_c = -1/\ln 0.9 = 9.49$, and the response due to the pair of poles vanishes in $5n_c = 47.45$ samples as shown in Figure 11.8(b).

With the preceding discussion, we may define the *time constant* of a stable $H(z)$ as

$$n_c = \frac{-1}{\ln \text{ (largest magnitude of all poles)}} \tag{11.46}$$

If $H(z)$ has many poles, the smaller the pole in magnitude, the faster its time response approaches zero. Thus the pole with the largest magnitude dictates the time for the transient response to approach zero. For example, the transfer function in (11.42) has three poles 0.5 and $0.9e^{\pm j2.1}$. The largest magnitude of all poles is 0.9. Thus the time constant of (11.42) is $n_c = -1/\ln 0.9 = 9.49$.

The response time of a DT stable system will be defined as the time for its step response to reach steady-state or its transient response to become zero. With the preceding discussion, the response time of a stable transfer function can be considered

---

[8]Strictly speaking, the number of samples must be an integer. We disregard this fact because we are interested in only an estimate.

to equal five time constants. The transfer function in (11.42) has time constant 9.49 in samples and, consequently, response time 47.45 in samples as shown in Figure 11.7(a). The transfer functions in (11.31) through (11.33) also have time constant 9.49 and response time 47.45 in samples as shown in Figure 11.2. Note that the rule of five time constants should be considered as a guide rather than a strict rule. For example, the transfer function $0.7z(z + 0.7)/(z - 0.7)^3$ has time constant $n_c = -1/\ln 0.7 = 2.8$ and response time 14 in samples. But its transient response takes more than 30 samples to become zero as shown in Figure 11.6(a). In general, if the poles of a stable transfer function are not clustered in a small region as in the preceding example, the rule of five time constants is applicable.

The time constant of a stable transfer function is defined from its poles. Its zeros do not play any role. For example, the transfer functions from (11.31) through (11.33) have the same set of poles but different zeros. Although their transient responses are different, their responses all approach steady state in roughly 47 samples or 47 seconds. Note that the sampling period is assumed to be 1. Thus the time constant and response time depend mainly on the poles of a transfer function.

11.7   Frequency Responses

In this section, we introduce the concept of frequency responses. Consider a DT transfer function $H(z)$. Its values along the unit circle on the z-plane — that is, $H(e^{j\omega})$, for all ω — is called the *frequency response*. In general, $H(e^{j\omega})$ is complex-valued and can be expressed in polar form as

$$H(e^{j\omega}) = A(\omega)e^{j\theta(\omega)} \tag{11.47}$$

where $A(\omega)$ and $\theta(\omega)$ are real-valued functions of ω and $A(\omega) \geq 0$. We call $A(\omega)$ the *magnitude response* and $\theta(\omega)$ the *phase response*. We first give an example.

Example 11.7.1 Consider the DT transfer function

$$H(z) = \frac{z + 1}{10z - 8} \tag{11.48}$$

Its frequency response is

$$H(e^{j\omega}) = \frac{e^{j\omega} + 1}{10e^{j\omega} - 8}$$

We compute

$$\omega = 0 : \quad H(1) = \frac{1 + 1}{10 - 8} = 1 = 1 \cdot e^{j0}$$

$$\omega = \pi/2 = 1.57 : \quad H(j1) = \frac{j1 + 1}{j10 - 8} = \frac{1.4e^{j\pi/4}}{12.8e^{j2.245}} = 0.11e^{-j1.46}$$

$$\omega = \pi = 3.14 : \quad H(-1) = \frac{-1 + 1}{-10 - 8} = 0$$

Other than these ω, its computation is complicated. Fortunately, the MATLAB function freqz, where the last character z stands for the z-transform, carries out the

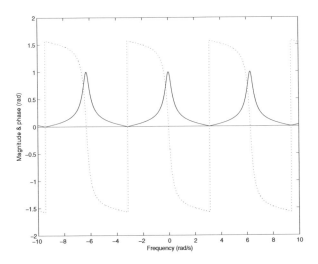

Figure 11.9: Magnitude response (solid line) and phase response (dotted line) of (11.48).

computation. To compute the frequency response of (11.48) from $\omega = -10$ to 10 with increment 0.01, we type

```
n=[1 1];d=[10 -8];
w=-10:0.01:10;
H=freqz(n,d,w);
plot(w,abs(H),w,angle(H),':')
```

The result is shown in Figure 11.9. We see that the frequency response is periodic with period 2π, that is

$$H(e^{j\omega}) = H(e^{j(\omega+k2\pi)})$$

for all ω and all integer k. This is in fact a general property and follows directly from $e^{j(\omega+k2\pi)} = e^{j\omega}e^{jk2\pi} = e^{j\omega}$. Thus we need to plot $H(e^{j\omega})$ for ω only in a frequency interval of 2π. As in plotting frequency spectra of DT signals, we plot frequency responses of DT systems only in the Nyquist frequency range $(-\pi/T, \pi/T]$ with $T = 1$, or in $(-\pi, \pi]$. See Section 4.8.

If all coefficients of $H(z)$ are real, then we have $[H(e^{j\omega})]^* = H(e^{-j\omega})$, which implies

$$[A(\omega)e^{j\theta(\omega)}]^* = A(\omega)e^{-j\theta(\omega)} = A(-\omega)e^{j\theta(-\omega)}$$

Thus we have

$$A(\omega) = A(-\omega) \quad \text{(even)} \tag{11.49}$$

and

$$\theta(-\omega) = -\theta(\omega) \quad \text{(odd)} \tag{11.50}$$

In other words, if $H(z)$ has only real coefficients, then its magnitude response is even and its phase response is odd. Thus we often plot frequency responses only in the

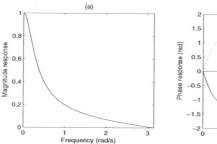
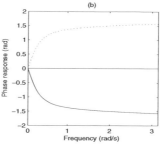

Figure 11.10: (a) Magnitude responses of (11.48) (solid line) and (11.51) (dotted line). (b) Phase responses of (11.48) (solid line) and (11.51) (dotted line).

positive frequency range $[0, \pi]$. If we do not specify ω in using `freqz(n,d)`, then `freqz` selects automatically 512 points in $[0, \pi]$.[9] Thus typing

```
n=[1 1];d=[10 -8];
[H,w]=freqz(n,d);
subplot(1,2,1)
plot(w,abs(H)),axis square
subplot(1,2,2)
plot(w,angle(H)),axis square
```

generates the magnitude and phase responses of (11.47) in Figures 11.10(a) and (b) with solid lines. Note that we did not specify the frequencies in the preceding program.

Next we consider the DT transfer function

$$H_1(z) = \frac{z+1}{-8z+10} \tag{11.51}$$

Its pole is $10/8 = 1.25$ which lies outside the unit circle. Thus the system is not stable. Replacing `d=[10 -8]` by `d=[-8 10]`, the preceding program generates the magnitude and phase responses of (11.51) in Figures 11.10(a) and (b) with dotted lines. We see that the magnitude responses of (11.48) and (11.51) are identical, but their phase responses are different. See Problem 11.18.

We next discuss the physical meaning of frequency responses under the stability condition. Consider a DT system with transfer function $H(z)$. Let us apply to it the input $u[n] = ae^{j\omega_0 n}$. Note that $u[n]$ is not a real-valued sequence. However its employment will simplify the derivation. The z-transform of $u[n] = ae^{j\omega_0 n}$ is $az/(z - e^{j\omega_0})$. Thus the output of $H(z)$ is

$$Y(z) = H(z)U(z) = H(z)\frac{az}{z - e^{j\omega_0}}$$

---

[9]The function `freqz` is based on FFT, thus its number of frequencies is selected to be a power of 2 or $2^9 = 512$. However, the function `freqs` for computing frequency responses of CT systems has nothing to do with FFT, and its default uses 200 frequencies.

We expand $Y(z)/z$ as

$$\frac{Y(z)}{z} = H(z)\frac{a}{z - e^{j\omega_0}} = \frac{k_1}{z - e^{j\omega_0}} + \text{terms due to poles of } H(z)$$

with

$$k_1 = aH(z)|_{z=e^{j\omega_0}} = aH(e^{j\omega_0})$$

Thus we have

$$Y(z) = aH(e^{j\omega_0})\frac{z}{z - e^{j\omega_0}} + z \times [\text{terms due to poles of } H(z)]$$

which implies

$$y[n] = aH(e^{j\omega_0})e^{j\omega_0 n} + \text{responses due to poles of } H(z)$$

If $H(z)$ is stable, then all responses due to its poles approach 0 as $n \to \infty$. Thus we have

$$y_{ss}[n] := \lim_{n\to\infty} y[n] = aH(e^{j\omega_0})e^{j\omega_0 n} \tag{11.52}$$

It is the steady-state response of $H(z)$ excited by $u[n] = ae^{j\omega_0 n}$. It is the DT counterpart of (9.50). We write it as, using (11.47),

$$\begin{aligned}
y_{ss}[n] &= aA(\omega_0)e^{j\theta(\omega_0)}e^{j\omega_0 n} = aA(\omega_0)e^{j[\omega_0 n + \theta(\omega_0)]} \\
&= aA(\omega_0)\left[\cos(\omega_0 n + \theta(\omega_0)) + j\sin(\omega_0 n + \theta(\omega_0))\right] \tag{11.53}
\end{aligned}$$

We list some special cases of (11.52) and (11.53) as a theorem.

Theorem 11.4 Consider a DT system with proper rational transfer function $H(z)$. If the system is stable, then

$$\begin{aligned}
u[n] = a \quad &\text{for } n \geq 0 \quad \to \quad y_{ss}[n] = aH(1) \\
u[n] = a\sin\omega_0 n \quad &\text{for } n \geq 0 \quad \to \quad y_{ss}[n] = a|H(e^{j\omega_0})|\sin(\omega_0 n + \angle H(e^{j\omega_0})) \\
u[n] = a\cos\omega_0 n \quad &\text{for } n \geq 0 \quad \to \quad y_{ss}[n] = a|H(e^{j\omega_0})|\cos(\omega_0 n + \angle H(e^{j\omega_0}))
\end{aligned}$$

□

The steady-state responses in (11.52) and (11.53) are excited by the input $u[n] = ae^{j\omega_0 n}$. If $\omega_0 = 0$, the input is a step sequence with amplitude a, and the output approaches $aH(e^{j0})e^{j0\cdot n} = aH(1)$, a step sequence with amplitude $aH(1)$. If $u[n] = a\sin\omega_0 n = \text{Im } ae^{j\omega_0 n}$, where Im stands for the imaginary part, then the output approaches the imaginary part of (11.53) or $aA(\omega_0)\sin(\omega_0 n + \theta(\omega_0))$. Using the real part of $e^{j\omega_0 n}$, we will obtain the third equation. In conclusion, if we apply a sinusoidal input to a system, then the output approaches a sinusoidal output with the same frequency, but its amplitude will be modified by $A(\omega_0) = |H(e^{j\omega_0})|$ and its phase by $\angle H(e^{j\omega_0}) = \theta(\omega_0)$. We give an example.

Example 11.7.2 Consider a system with transfer function $H(z) = (z + 1)/(10z + 8)$. The system has pole $-8/10 = -0.8$ which has a magnitude less than 1, thus it is stable. We compute the steady-state response of the system excited by

$$u[n] = 2 + \sin 6.38n + 0.2\cos 3n$$

Is there any problem with the expression of $u[n]$? Recall that we have assumed $T = 1$. Thus the Nyquist frequency range is $(-\pi, \pi] = (-3.14, 3.14]$ in rad/s. The frequency 6.38 in $\sin 6.38n$ is outside the range and must be shifted inside the range by subtracting $2\pi = 6.28$. Thus the input in principal form is

$$u[n] = 2 + \sin(6.38 - 6.28)n + 0.2\cos 3n = 2 + \sin 0.1n + 0.2\cos 3n \qquad (11.54)$$

See Section 4.6.2. In the positive Nyquist frequency range $[0, \pi] = [0, 3.14]$, $2 = 2\cos(0 \cdot n)$ and $\sin 0.1n$ are low-frequency components and $\cos 3n$ is a high-frequency component.

In order to apply Theorem 11.4, we read from Figures 11.10(a) and (b) $H(1) = 1$, $H(e^{j0.1}) = 0.9e^{-j0.4}$, and $H(e^{j3}) = 0.008e^{-j1.6}$. Clearly, the reading cannot be very accurate. Then Theorem 11.4 implies

$$y_{ss}[n] = 2 \times 1 + 0.9\sin(0.1n - 0.4) + 0.2 \times 0.008\cos(3n - 1.6) \qquad (11.55)$$

We see that $\cos 3n$ is greatly attenuated, whereas the dc component 2 goes through without any attenuation and $\sin 0.1n$ is only slightly attenuated. Thus the system passes the low-frequency signals and stops the high-frequency signal and is therefore called a lowpass filter. □

We mention that the condition of stability is essential in Theorem 11.4. If a system is not stable, then the theorem is not applicable as demonstrated in the next example.

Example 11.7.3 Consider a system with transfer function $H_1(z) = (z+1)/(-8z+10)$. It has pole at 1.25 and is not stable. Its magnitude and phase responses are shown in Figure 11.10(a) and (b) with dotted lines. We can also read out $H_1(e^{j0.1}) = 0.9e^{j0.4}$. If we apply the input $u[n] = \sin 0.1n$, the output can be computed as

$$y[n] = -0.374(1.25)^n + 0.9\sin(0.1n + 0.4)$$

for $n \geq 0$. Although the output contains $|H_1(e^{j0.1})|\sin(0.1n + \measuredangle H_1(e^{j0.1}))$, the term is buried by $-0.374(1.25)^n$ as $n \to \infty$. Thus the output grows unbounded and Theorem 11.4 is not applicable. Furthermore, the frequency response $H_1(e^{j\omega})$ has no physical meaning. □

In view of Theorem 11.4, if we can design a stable system with magnitude response as shown in Figure 11.11(a) with a solid line and phase response with a dotted line, then the system will pass sinusoids with frequency $|\omega| < \omega_c$ and stop sinusoids with frequency $|\omega| > \omega_c$. We require the phase response to be linear to avoid distortion as we will discuss in a later section. We call such a system an *ideal lowpass filter* with cutoff frequency ω_c. The frequency range $[0, \omega_c)$ is called the *passband*, and $[\omega_c, \pi]$, the *stopband*. Figures 11.11(b) and (c) show the characteristics of ideal bandpass and highpass filters. The ideal bandpass filter will pass sinusoids with frequencies lying inside the range $[\omega_l, \omega_u]$, where ω_l and ω_u are the lower and upper cutoff frequencies, respectively. The ideal highpass filter will pass sinusoids with frequencies lying inside $[\omega_c, \pi]$. They are called *frequency selective filters*. Note that Figure 11.11 is essentially

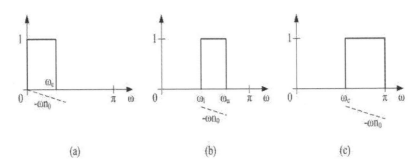

Figure 11.11: (a) Ideal lowpass filter with cutoff frequency ω_c. (b) Ideal bandpass filter with upper and lower cutoff frequencies ω_u and ω_l. (c) Ideal highpass filter with cutoff frequency ω_c.

the same as Figure 9.11. The only difference is that the frequency range in Figure 9.11 is $[0, \infty)$ and the frequency range in Figure 11.11 is $[0, \pi]$ because of the assumption $T = 1$.

The impulse response $h[n]$ of the ideal lowpass filter with linear phase $-\omega n_0$, where n_0 is a positive integer can be computed as

$$h[n] = \frac{\sin[\omega_c(n - n_0)]}{\pi(n - n_0)}$$

for all n in $(-\infty, \infty)$. See Problem 11.20. It is not zero for all $n < 0$, thus the ideal lowpass filter is not causal and cannot be built in the real world. In practice, we modify the magnitude responses in Figure 11.11 to the ones shown in Figure 11.12. We insert a *transition band* between the passband and stopband. Furthermore, we specify a *passband tolerance* and *stopband tolerance* as shown with shaded areas and require the magnitude response to lie inside the areas. The transition band is generally not specified and is the "don't care" region. We also introduce the *group delay* defined as

$$\text{Group delay} = \tau(\omega) = -\frac{d\theta(\omega)}{d\omega} \qquad (11.56)$$

For an ideal filter with linear phase such as $\theta(\omega) = -n_0\omega$, its group delay is n_0, a constant. Thus instead of specifying the phase response to be a linear function of ω, we specify the group delay in the passband to be roughly constant or $n_0 - \epsilon < \tau(\omega) < n_0 + \epsilon$, for some constants n_0 and ϵ. Even with these more relaxed specifications on the magnitude and phase responses, if we specify both of them, it is still difficult to design causal filters to meet both specifications. Thus in practice, we often specify only magnitude responses as shown in Figure 11.12. The design problem is then to find a proper stable rational function $H(z)$ of a degree as small as possible to have its magnitude response lying inside the specified region. See, for example, Reference [C7].

11.8   Frequency responses and frequency spectra

We showed in the preceding section that the output of a stable system with transfer function $H(z)$ excited by $u[n] = ae^{j\omega_0 n}$ approaches $aH(e^{j\omega_0})e^{j\omega_0 n}$ as $n \to \infty$. Real-

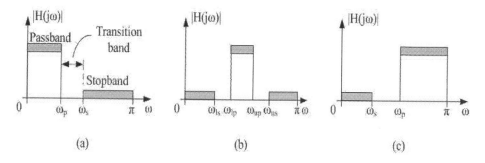

Figure 11.12: Specifications of practical (a) lowpass filter, (b) bandpass filter, and (c) highpass filter.

world signals are rarely pure sinusoids. We now extend the result to the general case.

Consider a DT positive-time signal $x[n]$ with sampling period 1. If it is absolutely summable, its frequency spectrum is defined in (4.65), with $T = 1$, as

$$X_d(\omega) = \sum_{n=0}^{\infty} x[n]e^{-jn\omega}$$

Let $X(z)$ be the z-transform of $x[n]$, that is,

$$X(z) = \sum_{n=0}^{\infty} x[n]z^{-n}$$

We see that replacing $z = e^{j\omega}$, the z-transform of $x[n]$ becomes the frequency spectrum of $x[n]$, that is,

$$X_d(\omega) = \mathcal{Z}[x[n]]|_{z=e^{j\omega}} = X(e^{j\omega}) \qquad (11.57)$$

This is similar to the CT case where the frequency spectrum of a positive-time and absolutely integrable $x(t)$ equals its Laplace transform with s replaced by $j\omega$. It is important to mention that (11.57) holds only if $x[n]$ is positive time and absolutely summable. For example, Consider $x[n] = 1.2^n$, for $n \geq 0$. Its z-transform is $X(z) = z/(z-1.2)$. Its frequency spectrum however is not defined and cannot equal $X(e^{j\omega}) = e^{j\omega}/(e^{j\omega} - 1.2)$.

The input and output of a DT system with transfer function $H(z)$ are related by

$$Y(z) = H(z)U(z) \qquad (11.58)$$

The equation is applicable whether the system is stable or not and whether the frequency spectrum of the input signal is defined or not. For example, consider $H(z) = z/(z-2)$, which is not stable, and $u[n] = 1.2^n$, which grows unbounded and its frequency spectrum is not defined. The output of the system excited by $u[n]$ is

$$Y(z) = H(z)U(z) = \frac{z}{z-2}\frac{z}{z-1.2} = \frac{2.5z}{z-2} - \frac{1.5z}{z-1.2}$$

which implies $y[n] = 2.5 \times 2^n - 1.5 \times 1.2^n$, for $n \geq 0$. The output grows unbounded and its frequency spectrum is not defined.

Before proceeding, we show that if $H(z)$ is stable (its impulse response $h[n]$ is absolutely summable) and if the input $u[n]$ is absolutely summable, then so is the output $y[n]$. Indeed, we have

$$y[n] = \sum_{k=0}^{\infty} h[n-k]u[k]$$

which implies

$$|y[n]| \leq \sum_{k=0}^{\infty} |h[n-k]||u[k]|$$

Thus we have

$$\sum_{n=0}^{\infty} |y[n]| \leq \sum_{n=0}^{\infty} \left(\sum_{k=0}^{\infty} |h[n-k]||u[k]| \right) = \sum_{k=0}^{\infty} \left(\sum_{n=0}^{\infty} |h[n-k]| \right) |u[k]|$$

$$= \sum_{k=0}^{\infty} \left(\sum_{\bar{n}=-k}^{\infty} |h(\bar{n})| \right) |u[k]| = \left(\sum_{\bar{n}=0}^{\infty} |h[\bar{n}]| \right) \left(\sum_{k=0}^{\infty} |u[k]| \right)$$

where we have interchanged the order of summations, introduced a new index $\bar{n} = n-k$ (where k is fixed), and used the causality condition $h[\bar{n}] = 0$ for all $\bar{n} < 0$. Thus if $h[n]$ and $u[n]$ are absolutely summable, so is $y[n]$.

Let us substitute $z = e^{j\omega}$ into (11.58) to yield

$$Y(e^{j\omega}) = H(e^{j\omega})U(e^{j\omega}) \tag{11.59}$$

The equation is meaningless if the system is not stable or if the input frequency spectrum is not defined. However, if the system is stable and if the input is absolutely summable, then the output is absolutely summable and consequently its frequency spectrum is well defined and equals the product of the system's frequency response $H(e^{j\omega})$ and the input's frequency spectrum $U(e^{j\omega})$. Equation (11.59) is the basis of digital filter design and is the DT counterpart of (9.80).

Let us consider the magnitude and phase responses of an ideal lowpass filter with cutoff frequency ω_c shown in Figure 11.11(a) or

$$H(e^{j\omega}) = \begin{cases} 1 \cdot e^{-j\omega n_0} & \text{for } |\omega| \leq \omega_c \\ 0 & \text{for } \omega_c < |\omega| \leq \pi \end{cases} \tag{11.60}$$

where n_0 is a positive integer. Now if $u[n] = u_1[n] + u_2[n]$ and if the magnitude spectra of $u_1[n]$ and $u_2[n]$ are as shown in Figure 11.13, then the output frequency spectrum is given by

$$Y(e^{j\omega}) = H(e^{j\omega})U(e^{j\omega}) = H(e^{j\omega})[U_1(e^{j\omega}) + U_2(e^{j\omega})] = U_1(e^{j\omega})e^{-j\omega n_0} \tag{11.61}$$

where $H(e^{j\omega})U_2(e^{j\omega}) = 0$, for all ω in $(-\pi, \pi]$ because their nonzero parts do not overlap. If the z-transform of $u_1[n]$ is $U_1(z)$, then the z-transform of $u_1[n - n_0]$ is

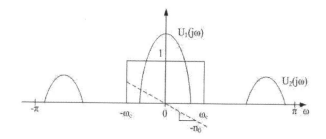

Figure 11.13: Spectra of $u_1[n]$ and $u_2[n]$.

$z^{-n_0}U_1(z)$ as derived in (7.30). Thus the spectrum in (11.61) is the spectrum of $u_1[n - n_0]$. In other words, the output of the ideal lowpass filter is

$$y[n] = u_1[n - n_0] \tag{11.62}$$

That is, the filter stops completely the signal $u_2[n]$ and passes $u_1[n]$ with only a delay of n_0 samples. This is called a *distortionless transmission* of $u_1[n]$.

We stress once again that the equation in (11.58) is more general than the equation in (11.59). Equation (11.59) is applicable only if the system is stable and the input frequency spectrum is defined.

11.9 Realizations – State-space equations

Although DT transfer functions are used to carry out design of digital filters, they are not used, as discussed in Subsection 9.3.2, in computer computation and real-time processing. Thus we discuss the realization problem. That is, given a proper rational transfer function $H(z)$, find a DT state-space (ss) equation of the form

$$\mathbf{x}[n+1] = \mathbf{A}\mathbf{x}[n] + \mathbf{b}u[n] \tag{11.63}$$
$$y[n] = \mathbf{c}\mathbf{x}[n] + du[n] \tag{11.64}$$

that has $H(z)$ as its transfer function. We call the ss equation a realization of $H(z)$. Once a realization is available, we can use it, as discussed in Subsection 7.6.1, for computer computation and real-time processing.

The realization procedure is almost identical to the one in Section 8.9 and its discussion will be brief. Consider the DT proper rational transfer function

$$H(z) = \frac{Y(z)}{U(z)} = \frac{\bar{b}_1 z^4 + \bar{b}_2 z^3 + \bar{b}_3 z^2 + \bar{b}_4 z + \bar{b}_5}{\bar{a}_1 z^4 + \bar{a}_2 z^3 + \bar{a}_3 z^2 + \bar{a}_4 z + \bar{a}_5} \tag{11.65}$$

with $\bar{a}_1 \neq 0$. We call $\bar{a}_1$ the leading coefficient. The rest of the coefficients can be zero or nonzero. The transfer function is proper and describes a causal system. The first step in realization is to write (11.65) as

$$H(z) - \frac{N(z)}{D(z)} + d = \frac{b_1 z^3 + b_2 z^2 + b_3 z + b_4}{z^4 + a_2 z^3 + a_3 z^2 + a_4 z + a_5} + d \tag{11.66}$$

with $D(z) = z^4 + a_2 z^3 + a_3 z^2 + a_4 z + a_5$ and $N(z)/D(z)$ is strictly proper. This can be achieved by dividing the numerator and denominator of (11.65) by $\bar{a}_1$ and then carrying out a direct division. The procedure is identical to the one in Example 8.9.1 and will not be repeated.

Now we claim that the following ss equation realizes (11.66):

$$\mathbf{x}[n+1] \;=\; \begin{bmatrix} -a_2 & -a_3 & -a_4 & -a_5 \\ 1 & 0 & 0 & 0 \\ 0 & 1 & 0 & 0 \\ 0 & 0 & 1 & 0 \end{bmatrix} \mathbf{x}[n] + \begin{bmatrix} 1 \\ 0 \\ 0 \\ 0 \end{bmatrix} u[n] \qquad (11.67)$$

$$y[n] \;=\; [b_1 \quad b_2 \quad b_3 \quad b_4]\mathbf{x}[n] + du[n]$$

with $\mathbf{x}[n] = [x_1[n] \ x_2[n] \ x_3[n] \ x_4[n]]'$. This ss equation can be obtained directly from the coefficients in (11.66). We place the denominator's coefficients, except its leading coefficient 1, with sign reversed in the first row of $\mathbf{A}$, and the numerator's coefficients, without changing sign, directly as $\mathbf{c}$. The constant d in (11.66) is the direct transmission part in (11.67). The rest of the ss equation have fixed patterns. The second row of $\mathbf{A}$ is $[1 \ 0 \ 0 \cdots]$. The third row of $\mathbf{A}$ is $[0 \ 1 \ 0 \cdots]$ and so forth. The column vector $\mathbf{b}$ is all zero except its first entry which is 1.

To show that (11.67) is a realization of (11.66), we must compute its transfer function. Following the procedure in Section 8.9, we write (11.67) explicitly as

$$\begin{aligned}
x_1[n+1] &= -a_2 x_1[n] - a_3 x_2[n] - a_4 x_3[n] - a_5 x_4[n] + u[n] \\
x_2[n+1] &= x_1[n] \\
x_3[n+1] &= x_2[n] \\
x_4[n+1] &= x_3[n]
\end{aligned}$$

Applying the z-transform, using (11.18), and assuming zero initial conditions yield

$$\begin{aligned}
zX_1(z) &= -a_2 X_1(z) - a_3 X_2(z) - a_4 X_3(z) - a_5 X_4(z) + U(z) \\
zX_2(z) &= X_1(z) \\
zX_3(z) &= X_2(z) \\
zX_4(z) &= X_3(z)
\end{aligned} \qquad (11.68)$$

From the second to the last equation, we can obtain

$$X_2(z) = \frac{X_1(z)}{z}, \quad X_3(z) = \frac{X_2(z)}{z} = \frac{X_1(z)}{z^2}, \quad X_4(z) = \frac{X_1(z)}{z^3} \qquad (11.69)$$

Substituting these into the first equation of (11.68) yields

$$\left(z + a_2 + \frac{a_3}{z} + \frac{a_4}{z^2} + \frac{a_5}{z^3} \right) X_1(z) = U(z)$$

or

$$\left[\frac{z^4 + a_2 z^3 + a_3 z^2 + a_4 z + a_5}{z^3} \right] X_1(z) = U(z)$$

which implies

$$X_1(z) = \frac{z^3}{z^4 + a_2 z^3 + a_3 z^2 + a_4 z + a_5} U(z) =: \frac{z^3}{D(z)} U(z) \qquad (11.70)$$

where $D(z) = z^4 + a_2 z^3 + a_3 z^2 + a_4 z + a_5$. Substituting (11.70) into (11.69) and then into the following z-transform of the output equation in (11.67) yields

$$\begin{aligned} Y(z) &= b_1 X_1(z) + b_2 X_2(z) + b_3 X_3(z) + b_4 X_4(z) + dU(z) \\ &= \left(\frac{b_1 z^3}{D(z)} + \frac{b_2 z^3}{D(z)} + \frac{b_3 z}{D(z)} + \frac{b_4}{D(z)} \right) U(z) + dU(z) \\ &= \left(\frac{b_1 z^3 + b_2 z^2 + b_3 z + b_4}{D(z)} + d \right) U(z) \end{aligned} \qquad (11.71)$$

This shows that the transfer function of (11.67) equals (11.66). Thus (11.67) is a realization of (11.66) or (11.65). The realization in (11.67) is said to be in the *controllable form*. We first give an example.

Example 11.9.1 Consider the transfer function in (7.28) or

$$H(z) = \frac{3z^3 - 2z^2 + 5}{z^3} = 3 + \frac{-2z^2 + 0 \cdot z + 5}{z^3 + 0 \cdot z^2 + 0 \cdot z + 0} \qquad (11.72)$$

It is in the form of (11.66) and its realization can be obtained form (11.67) as

$$\begin{aligned} \mathbf{x}[n+1] &= \begin{bmatrix} 0 & 0 & 0 \\ 1 & 0 & 0 \\ 0 & 1 & 0 \end{bmatrix} \mathbf{x}[n] + \begin{bmatrix} 1 \\ 0 \\ 0 \end{bmatrix} u[n] \\ y[n] &= \begin{bmatrix} -2 & 0 & 5 \end{bmatrix} \mathbf{x}[n] + 3u[n] \end{aligned} \qquad (11.73)$$

where $\mathbf{x}[n] = [x_1[n]\ x_2[n]\ x_3[n]]'$. This is the ss equation in (7.18) and (7.19). □

Example 11.9.2 Find a realization for the transfer function

$$H(z) = \frac{Y(z)}{U(z)} = \frac{3z^4 + 5z^3 + 24z^2 + 23z - 5}{2z^4 + 6z^3 + 15z^2 + 12z + 5} \qquad (11.74)$$

Dividing all coefficients by 2 and then carrying out a direct division as in Example 8.9.1, we can obtain

$$H(z) = \frac{-2z^3 + 0.75z^2 + 2.5z - 6.25}{z^4 + 3z^3 + 7.5z^2 + 6z + 2.5} + 1.5 \qquad (11.75)$$

Thus its controllable-form realization is, using (11.66),

$$\begin{aligned} \mathbf{x}[n+1] &= \begin{bmatrix} -3 & -7.5 & -6 & -2.5 \\ 1 & 0 & 0 & 0 \\ 0 & 1 & 0 & 0 \\ 0 & 0 & 1 & 0 \end{bmatrix} \mathbf{x}[n] + \begin{bmatrix} 1 \\ 0 \\ 0 \\ 0 \end{bmatrix} u[n] \\ y[n] &= \begin{bmatrix} -2 & 0.75 & 2.5 & -6.25 \end{bmatrix} \mathbf{x}[n] + 1.5u[n] \end{aligned} \qquad (11.76)$$

We see that the realization can be read out from the coefficients of the transfer function. Note that the $\{\mathbf{A}, \mathbf{b}, \mathbf{c}, d\}$ of this realization is identical to its CT counterpart in Example 8.9.2.□

The MATLAB function tf2ss, an acronym for transfer function to ss equation, carries out realizations. For the transfer function in (11.74), typing in the command window

```
>> n=[3 5 24 23 -5];de=[2 6 15 12 5];
>> [a,b,c,d]=tf2ss(n,de)
```

will yield

$$
a \;=\; \begin{array}{cccc}
-3.0000 & -7.5000 & -6.0000 & -2.5000 \\
1.0000 & 0 & 0 & 0 \\
0 & 1.0000 & 0 & 0 \\
0 & 0 & 1.0000 & 0
\end{array}
$$

$$
b \;=\; \begin{array}{c}
1 \\
0 \\
0 \\
0
\end{array}
$$

$$
c \;=\; \begin{array}{cccc} -2.0000 & 0.7500 & 2.5000 & -6.2500 \end{array}
$$

$$
d \;=\; 1.5000
$$

This is the controllable-form realization in (11.76). In using tf2ss, there is no need to normalize the leading coefficient and to carry out direct division. Thus its use is simple and straightforward.

Whenever we use a transfer function, the system must be initially relaxed or all its initial conditions must be zero. Even though an ss equation is developed from a transfer function, it is applicable even if the initial conditions are different from zero. As discussed in Section 7.6.1, in order to compute the output $y[n]$, for $n \geq 0$, excited by the input $u[n]$, for $n \geq 0$, we must also specify the initial state $\mathbf{x}[0]$. For the FIR system with memory of three samples in Example 11.9.1, the initial state is, as discussed in Chapter 7, $\mathbf{x}[0] = [u[-1]\; u[-2]\; u[-3]]'$. For the IIR system in Example 11.9.2, the relationship between the initial state $\mathbf{x}[0]$ and $u[n]$ and $y[n]$, for $n < 0$, is complicated. Fortunately, the knowledge of the relationship is not needed in using the ss equation. Moreover, in most practical applications, we assume $\mathbf{x}[0] = \mathbf{0}$.

All discussion in Chapter 8 regarding transfer functions and ss equations for the CT case is directly applicable to the DT case. For example, if $N(z)$ and $D(z)$ are coprime, then the realization of the DT transfer function $H(z) = N(z)/D(z)$ will be minimum. Even though a transfer function is an external description and applicable only to zero initial conditions and an ss equation is an internal description and applicable to nonzero initial conditions, the two descriptions are equivalent if the degree of $H(z)$ equals the dimension of the ss equation. Under this condition, either one can be used in analysis and design. In practice, we use transfer functions to carry out design. Once a satisfactory transfer function is designed, we use its minimal realization to carry out

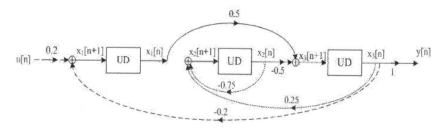

Figure 11.14: Basic block diagram of (11.77) and (11.78).

computer computation and real-time processing. The realization can also be used in specialized hardware implementation as we discuss next.

11.9.1 Basic block diagrams

We developed in Section 7.6 an ss equation from a basic block diagram. We now show that every ss equation can be built using the basic elements shown in Figure 7.2. We use an example to illustrate the procedure. Consider the ss equation

$$\begin{bmatrix} x_1[n+1] \\ x_2[n+1] \\ x_3[n+1] \end{bmatrix} = \begin{bmatrix} 0 & 0 & -0.2 \\ 0 & -0.75 & 0.25 \\ 0.5 & -0.5 & 0 \end{bmatrix} \begin{bmatrix} x_1[n] \\ x_2[n] \\ x_3[n] \end{bmatrix} + \begin{bmatrix} 0.2 \\ 0 \\ 0 \end{bmatrix} u[n] \quad (11.77)$$

$$y[n] = \begin{bmatrix} 0 & 0 & 1 \end{bmatrix} \begin{bmatrix} x_1[n] \\ x_2[n] \\ x_3[n] \end{bmatrix} + 0 \times u[n] \quad (11.78)$$

The ss equation is the CT counterpart in (8.14) and (8.15). It has dimension three and needs three unit-delay elements as shown in Figure 11.14. We assign the output of each unit-delay element as a state variable $x_i[n]$. Then its input is $x_i[n+1]$. To carry out simulation, we write (11.77) explicitly as

$$\begin{aligned} x_1[n+1] &= -0.2x_3[n] + 0.2u[n] \\ x_2[n+1] &= -0.75x_2[n] + 0.25x_3[n] \\ x_3[n+1] &= 0.5x_1[n] - 0.5x_2[n] \end{aligned}$$

We use the first equation to generate $x_1[n+1]$ in Figure 11.14 using dashed lines. We use the second equation to generate $x_2[n+1]$ using dotted lines. Using the last equation and the output equation in (11.78), we can readily complete the connections in Figure 11.14. This is the DT counterpart of Figure 8.8. Replacing every integrator by a unit-delay element, Figure 8.8 becomes Figure 11.14.

Additions and multiplications are basic operations in every PC, DSP processor, or specialized hardware. A unit-delay element is simply a memory location. We store a number in the location and then fetch it in the next sampling instant. Thus a basic block diagram will provide a schematic diagram for specialized hardware implementation of the ss equation. See, for example, Reference [C7]. In conclusion, ss equations

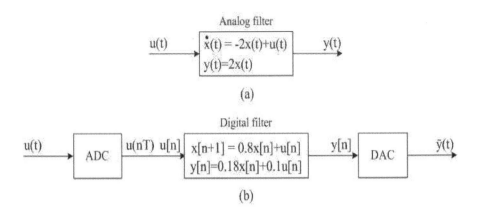

Figure 11.15: (a) Analog procession of CT signal. (b) Digital processing of CT signal.

are used in computer computation, real-time processing, and specialized hardware implementation. Even so, ss equations are not used, as discussed in Section 9.10, in design.

11.10   Digital processing of CT signals

Because of many advantages of DT techniques, CT signals are now widely processed using DT systems. We use an example to illustrate how this is carried out. Consider the CT signal

$$u(t) = \cos 0.2t + 0.2 \sin 25t \qquad (11.79)$$

It consists of two sinusoids. Suppose $\cos 0.2t$ is the desired signal and $0.2 \sin 25t$ is noise. We first use a CT or analog filter and then a DT or digital filter, as shown in Figures 11.15(a) and (b), to pass the desired signal and to stop the noise.

Consider $H(s) = 2/(s + 2)$. It is a lowpass filter with 3-dB bandwidth 2 rad/s. Because

$$H(j0.2) = \frac{2}{2 + j0.2} \approx \frac{2}{2} = 1$$

and

$$H(j25) = \frac{2}{j25 + 2} \approx \frac{2}{j25} = 0.08e^{-j\pi/2}$$

the steady-state response of the filter excited by $u(t)$ is

$$y_{ss}(t) = \cos 0.2t + 0.2 \times 0.08 \sin(25t - \pi/2)$$

Thus the filter passes the desired signal and attenuates greatly the noise. We plot in in Figures 11.16(a) and (b) the input and output of the analog filter. Indeed the noise $0.2 \cos 20t$ is essentially eliminated in Figure 11.16(b). It is important to mention that in computer computation or real-time processing, we use the realization of $H(s) =$

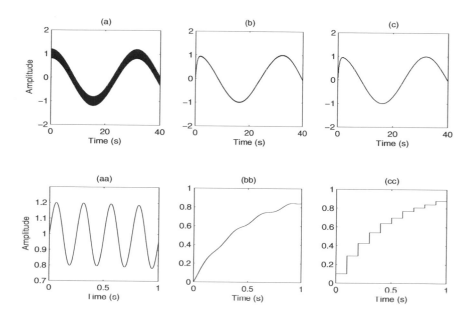

Figure 11.16: (a) The CT signal in (11.79). (b) The output $y(t)$ of the analog filter $2/(s+2)$ in Figure 11.15(a). (c) The output $\bar{y}(t)$ in Figure 11.15(b), that is, the output $y[n]$ of the digital filter $(z+1)/(10z-8)$ passing through a zero-order hold. (aa) Small segment of (a). (bb) Small segment of (b). (cc) Small segment of (c)

$2/(s+2)$ or

$$\begin{aligned}
\dot{x}(t) &= -2x(t) + u(t) \\
y(t) &= 2x(t)
\end{aligned}$$

as shown in Figure 11.15(a).

We now discuss digital processing of the CT signal as shown in Figure 11.15(b). In order to process $u(t)$ digitally, we first select a sampling period. The Nyquist sampling theorem requires $T < \pi/25 = 0.1257$. Let us select $T = 0.1$. Then the output of the ADC in Figure 11.15(b) is

$$u(nT) = \cos 0.2nT + 0.2 \sin 25nT \qquad (11.80)$$

with $T = 0.1$ and $n \geq 0$. Its positive Nyquist frequency range is $[0, \pi/T = 31.4]$ in rad/s.

In designing DT filters, we first normalize the sampling period to 1. This can be achieved by multiplying all frequencies by $T = 0.1$. Thus the two frequencies 0.2 and 25 in the positive Nyquist frequency range $[0, \pi/T = 31.4]$ become 0.02 and 2.5 in the normalized positive Nyquist frequency range $[0, \pi = 3.14]$. Thus we design a DT filter to pass sinusoids with frequency 0.02, and to stop sinusoid with frequency 2.5. The DT filter $H(z) = (z+1)/(10z-8)$, whose magnitude response is shown in Figure 11.10(a), can achieve this. Note that the filter is designed using $T = 1$. But the resulting filter can be used for any $T > 0$. Let us type in an edit window the following

```
%Program 11.2 (f1116.m)
n=0:400;T=0.1;
uT=sin(0.2*n*T)+0.2.*sin(25*n*T);
dsys=tf([1 1],[10 -8],T);
yT=lsim(dsys,uT)
stairs(n*T,yT)
```

The first line is the number of samples to be used and the selected sampling period. The second line is the sampled input or, equivalently, the output of the ADC in Figure 11.15(b). The third line uses transfer function (tf) to define the DT system by including the sampling period T. We mention that the T used in defining the DT filter must be the same as the sampling period used in the input signal. The function lsim, an acronym for linear simulation, computes the output of the digital filter $(z+1)/(10z-8)$. It is the sequence of numbers $y[n]$ shown in Figure 11.15(b). The MATLAB function stairs carries out zero-order hold to yield the output $\bar{y}(t)$. It is the output of the DAC shown in Figure 11.15(b). We name and save the program as f1116. Typing in the command window of MATLAB >> f1116 will yield Figure 11.16(c) in a figure window. The result is comparable to the one in Figure 11.16(b) obtained using an analog filter. Note that even though the responses in Figures 11.16(b) and (c) are indistinguishable, the former is a continuous function of t as shown in Figure 11.16(bb), the latter is a staircase function as shown in Figure 11.16(cc).

In computer computation or specialized hardware implementation, the digital filter $H(z) = (z+1)/(10z-8)$ must be realized as an ss equation. We write $H(z)$ as

$$H(z) = \frac{z+1}{10z-8} = \frac{0.1z+0.1}{z-0.8} = 0.1 + \frac{0.18}{z-0.8}$$

then its controllable-form realization is

$$\begin{aligned} x[n+1] &= 0.8x[n] + u[n] \\ y[n] &= 0.18x[n] + 0.1u[n] \end{aligned}$$

This is the equation in Figure 11.15(b). Note that although we use the tf model in Program 11.2, the transfer function is first transformed into the preceding ss equation by calling the MATLAB function tf2ss in actual computation.

To conclude this section, we mention that the response time of a system defined for a step input is also applicable to sinusoidal inputs and general inputs. For example, the CT filter $H(s) = 2/(s+2)$ has time constant $1/2 = 0.5$ and response time 2.5 seconds. The output of $H(s)$ shown in Figure 11.16(b) takes roughly 2.5 seconds to reach steady state as shown. The DT filter $H(z) = (0.1z+0.1)/(z-0.8)$ has time constant $n_c = -1/\ln 0.8 = 4.48$ in samples and response response $5n_c = 22.4$ in samples or $5n_c T = 2.24$ in seconds. This is roughly the case as shown in Figure 11.16(c). These response times are small and are often ignored. If a response time is not small as shown in Figure 10.15(b), then it must be taken into consideration in design.

11.10.1   Filtering the sound of a piano's middle C

Consider the transduced signal shown in Figure 1.3. It is the sound of the middle C of a piano. According to Wikipedia, it has frequency 261 Hz. By this, it is implicitly

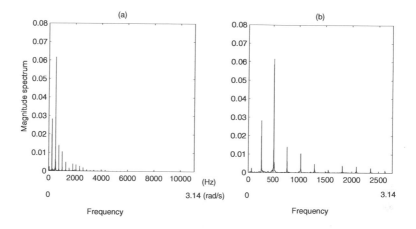

Figure 11.17: (a) Magnitude spectrum in $[0, 11025]$ (Hz) for $f_s = 22050$ and in normalized positive Nyquist frequency range $[0, \pi]$ (rad/s). (b) Magnitude spectrum in $[0, 2756]$ (Hz) for $\bar{f}_s = f_s/4$ and in normalized positive Nyquist frequency range $[0, \pi]$ (rad/s).

assumed that the signal is a pure sinusoid with frequency 261 Hz. It is not the case at all as we can see from the waveform shown in Figure 1.3. Indeed from its magnitude spectrum shown in Figure 5.11, we see that the signal has its energy centered around $f_c = 261$ Hz and its harmonics $2f_c$, $3f_c$, ... Thus spectra of real-world signals are generally complicated.

In this subsection, we will design a digital filter to pass the part of the waveform in Figure 1.3 whose spectrum centered around $2f_c = 522$ and stop all other frequencies. Recall that the signal was obtained using the sampling frequency $f_s = 22050$ Hz or sampling period $T = 1/f_s = 1/22050$. Thus its positive Nyquist frequency range is $[0, 0.5f_s] = [0, 11025]$ in Hz as shown in Figure 11.17(a). To design a digital filter, the positive Nyquist frequency range must be normalized to $[0, \pi] = [0, 3.14]$ (rad/s) as shown in the lower horizontal scale. Because the frequency spikes in Figure 11.17(a) are closely clustered, it is difficult to design a filter to pass one spike and stop the rest.

The magnitude spectrum of the middle-C signal is practically zero for frequencies larger than 2500 Hz as shown in Figure 11.17(a). Thus we may consider it to be bandlimited to $f_{max} = 2500$ and the sampling frequency can be selected as $\bar{f}_s > 2f_{max} = 5000$. Clearly the sampling frequency $f_s = 22050$ used is unnecessarily large or its sampling period $T = 1/f_s$ is unnecessarily small. As discussed in Subsection 5.5.2, in order to utilize the recorded data, we select $\bar{f}_s = f_s/4 = 5512.5$ or the new sampling period as $\bar{T} = 4T$. For this $\bar{f}_s$, the positive Nyquist frequency range becomes $[0, 2756]$ in Hz and the computed magnitude spectrum is shown in Figure 5.11 and repeated in Figure 11.17(b). We see that the spikes are less clustered and it will be simpler to design a digital filter to pass one spike and stop the rest.

In order to design a digital filter to pass the part of the waveform with frequency centered around 522 Hz in the Nyquist frequency range $[0, 2756]$ in Hz, we must nor-

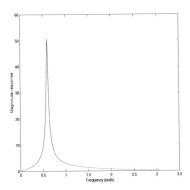

Figure 11.18: Magnitude response of (11.81).

malize 522 to $522 \times \pi/2756 = 0.595$ in the normalized Nyquist frequency range $[0, \pi]$ in rad/s. Consider the digital filter

$$H(z) = \frac{(z+1)(z-1)}{(z - 0.98e^{j0.595})(z - 0.98e^{-j0.595})} = \frac{z^2 - 1}{z^2 - 0.98(2\cos 0.595)z + 0.98^2} \quad (11.81)$$

Its magnitude response is plotted in Figure 11.18. It is obtained in MATLAB by typing

```
n=[1 0 -1];d=[1 -1.9*cos(0.595) 0.95*0.95];
[H,w]=freqz(n,d);plot(w,abs(H))
```

Recall that the function freqz automatically select 512 frequencies in $[0, \pi]$ and then compute its frequency response. We plot only its magnitude response in Figure 11.18. We see that the filter in (11.81) has a narrow spike around 0.595 rad/s in its magnitude response, thus it will pass the part of the signal whose spectrum centered around 0.595 rad/s and attenuate greatly the rest. The filter is called a *resonator*. See Reference [C7].

Although the filter in (11.81) is designed using $T = 1$, it is applicable for any $T > 0$. Because the peak magnitude of $H(z)$ is roughly 50 as shown in Figure 11.18, the transfer function $0.02H(z)$ has its peak magnitude roughly 1 and will carry out only filtering without amplification. We plot in Figure 11.19(a) the middle-C signal, for t from 0 to 0.12 second, obtained using the sampling period $\bar{T} = 4T = 4/22050$ and using linear interpolation. Applying this signal to the filter $0.02H(z)$ yields the output shown in Figure 11.19(b). It is obtained using the program

```
n=0.02*[1 0 -1];d=[1 -1.96*cos(0.595) 0.98*0.98];
dog=tf(n,d,4/22050);
y=lsim(dog,u)
```

where u is the signal in Figure 11.19(a). We see from Figure 11.19(b) that there is some transient response. We plot in Figure 11.19(c) a small segment from 0.05 to

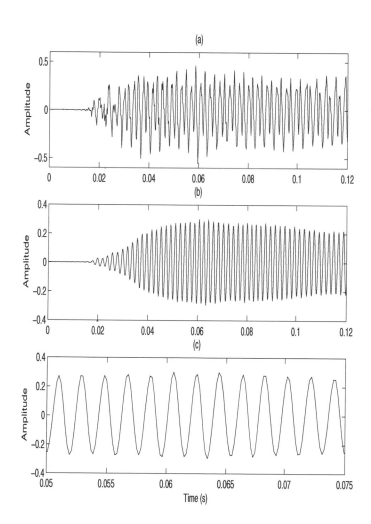

Figure 11.19: (a) Middle-C signal. (b) The output of $0.02H(z)$. (c) Its small segment.

0.075 second of Figure 11.19(b). It is roughly a sinusoid. It has 13 cycles in 0.025 second and thus has frequency $13/0.025 = 520$ Hz.

The number of samples in Figure 11.19 is 5500. The non-real time procession y=lsim(dog, u) takes 0.016 second to complete. This elapsed time is obtained using tic and toc.

Problems

11.1 The impulse response of the savings account studied in Example 7.3.2 was computed as $h[n] = (1.00015)^n$, for $n \geq 0$. What is its transfer function? Is the result the same as the one obtained in Problem 7.22.

11.2 Use the z-transform to solve Problem 7.8. Are the results the same?

11.3 The 20-point moving average studied in Example 7.4.1 can be described by the convolution or nonrecursive difference equation in (7.12) and the recursive difference equation in (7.13). Compute the transfer functions from (7.12) and (7.13). Are they the same?

11.4 Find a difference equation to describe a system with transfer function

$$H(z) = \frac{Y(z)}{U(z)} = \frac{2z^2 + 5z + 3}{4z^2 + 3z + 1}$$

11.5 Find the poles and zeros for each of the following transfer functions.

1. $H_1(z) = \frac{3z+6}{2z^2+2z+1}$
2. $H_2(z) = \frac{z^{-1}-z^{-2}-6z^{-3}}{1+2z^{-1}+z^{-2}}$

11.6 Compute the poles of the transfer functions of (7.12) and (7.13) computed in Problem 11.3. Do they have the same set of poles?

11.7 Find the impulse and step responses of a DT system with transfer function

$$H(z) = \frac{0.9z}{(z+1)(z-0.8)}$$

11.8 Verify that the output of $H(z) = (z+1)/(-8z+10)$ excited by $u[n] = \sin 0.1n$ is given by

$$y[n] = -0.374(1.25)^n + 0.91\sin(0.1n + 0.42)$$

for $n \geq 0$.

11.9 Consider (11.28) or

$$\mathcal{Z}^{-1}\left[\frac{b_1 z + b_0}{z - p}\right] = k_0 \delta_d[n] + k_1 p^n$$

for $n \geq 0$. Verify $k_0 = -b_0/p$ and $k_1 = (b_1 p + b_0)/p$.

11.10 What is the general form of the response of

$$H(z) = \frac{z^2 + 2z + 1}{(z-1)(z-0.5+j0.6)(z-0.5-j0.6)}$$

excited by a step sequence?

11.11 Consider the DT transfer function

$$H(z) = \frac{N(z)}{(z+0.6)^3(z-0.5)(z^2+z+0.61)}$$

where $N(z)$ is a polynomial of degree 4 with leading coefficient 2 and $H(1) = 10$. What is the general form $y[n]$ of its step response? What is $y[n]$ as $n \to \infty$? What are $y[n]$, for $n = 0 : 2$?

11.12 Find polynomials of degree 1 which are, respectively, CT stable and DT stable, CT stable but not DT stable, DT stable but not CT stable, and neither CT stable nor DT stable.

11.13 Determine the DT stability of the following systems

1. $\frac{z+1}{(z-0.6)^2(z+0.8+j0.6)(z+0.8-j0.6)}$

2. $\frac{3z-6}{(z-2)(z+0.2)(z-0.6+j0.7)(z-0.6-j0.7)}$

3. $\frac{z-10}{z^2(z+0.95)}$

11.14 Use Jury's test to check the DT stability of the polynomials

1. $z^3 + 4z^2 + 2$
2. $z^3 - z^2 + 2z - 0.7$
3. $2z^4 + 1.6z^3 + 1.92z^2 + 0.64z + 0.32$

11.15 A polynomial cannot be CT stable if it has a missing term or a negative coefficient. Is this also true for a polynomial to be DT stable? If not, give a counterexample.

11.16 Compute the values of

$$H(z) = \frac{z-1}{z+0.8}$$

at $z = e^{j\omega}$ with $\omega = 0, \pi/4, \pi/2$. Use the MATLAB function `freqz` to compute and to plot the magnitude and phase responses of $H(z)$ for ω in $[0, \pi]$. Compare your values with the computer-generated ones. Is it a lowpass or highpass filter?

11.17 Find the steady-state response of the DT system in Problem 11.16 excited by

$$u[n] = 2 + \sin 0.1n + \cos 3n$$

Read the needed values from the plots in Problem 11.16. How many samples will it take to reach steady state?

11.18 Verify analytically that the magnitude responses of

$$H_1(z) = \frac{z+1}{10z-8} \quad \text{and} \quad H_2(z) = \frac{z+1}{-8z+10}$$

are identical. Do they have the same phase responses?

11.19 Consider a system with transfer function

$$H(z) = \frac{1}{z+1}$$

Is the system stable? What is the form of the steady-state response of the system excited by $u[n] = \sin 0.6n$? Does the response contain any sinusoid with frequency other than 0.6 rad/s? Does Theorem 11.4 hold?

11.20 The DT Fourier transform of $x[n]$ with $T = 1$ is defined in (4.65) as

$$X_d(\omega) = \sum_{n=-\infty}^{\infty} x[n]e^{-j\omega n}$$

The inverse DT Fourier transform of $X_d(\omega)$ is given by

$$x[n] = \frac{1}{2\pi} \int_{\omega=-\pi}^{\pi} X_d(\omega)e^{j\omega n}d\omega$$

for all n in $(-\infty, \infty)$. Verify that for the ideal lowpass filter defined in (11.60), its inverse DT Fourier transform is

$$h[n] = \frac{\sin[\omega_c(n-n_0)]}{\pi(n-n_0)}$$

for all integers n in $(-\infty, \infty)$. Because $h[n] \neq 0$, for all $n < 0$, the ideal filter is not causal and cannot be built in the real world.

11.21 Find a state-space equation realization of the transfer function in Problem 11.4.

11.22 Find a realization for the transfer function in Problem 11.1. Is the result the same as the one in Problem 7.19.

11.23 Find a one-dimensional realization of $H(z) = z/(z-0.8)$. Find a two-dimensional realization of $H(z)$ by introducing the factor $z-0.5$ to its numerator and denominator.

11.24 Draw a basic block diagram for the system in Problem 11.4.

11.25 Let $X(z)$ be the z-transform of $x[n]$. Show that if all poles, except possibly a simple pole at $z = 1$, of $X(z)$ have magnitudes less than 1, then $x[n]$ approaches a constant (zero or nonzero) as $n \to \infty$ and

$$\lim_{n\to\infty} x[n] = \lim_{z\to 1}(z-1)X(z)$$

This is called the *final-value theorem* of the z-transform. This is the DT counterpart of the one in Problem 9.27. Before using the equation, we first must check essentially the stability of $X(z)$. The theorem is not as useful as Theorem 11.4.

References

(Brackets with an asterisk denote books on Signals and Systems)

[A1] Anderson, B.D.O., and J.B. Moore, *Optimal Control - Linear Quadratic Methods*, Upper Saddle River, NJ: Prentice Hall, 1990.

[B1]* Baraniuk, R., *Signals and systems*, Online text, Connexions.

[B2] Burton, D., *The History of mathematics*, 5/e, New York: McGraw Hill, 2003.

[C0]* Cadzow, J.A. and H.F. Van Landingham, *Signals and Systems*, Prentice-Hall, 1985.

[C1]* Carlson, G.E., *Signal and Linear System Analysis*, 2/e, New York: Wiley, 1998.

[C2] Carslaw, H.S. and J.C. Jaeger, *Operational methods in Applied Mathematics*, 2/e, London: Oxford University press, 1943.

[C3]* Cha, P.D. and J.I. Molinder, *Fundamentals of Signals and Systems: A Building Block Approach*, Cambridge University Press, 2006.

[C4]* Chaparro, L., *Signals and Systems using MATLAB*, NewYork: Academic Press, 2010.

[C5] Chen, C.T., *Analog and Digital Control System Design: Transfer-Function, State-Space, and Algebraic Methods*, New York: Oxford University Press, 1993.

[C6] Chen, C.T., *Linear System Theory and Design*, 3/e, New York: Oxford University Press, 1999.

[C7] Chen, C.T., *Digital Signal Processing: Spectral Computation and Filter Design*, New York: Oxford University Press, 2001.

[C8]* Chen, C.T., *Signals and Systems*, 3/e, New York: Oxford University Press, 2004.

[C9] Chua, L.O., C.A. Desoer, and E.S. Kuo, *Linear and Nonlinear Circuits*, New York: McGraw-Hill, 1987.

[C10] Churchill, R.V., *Modern Operational Mathematics in Engineering*, New York: McGraw-Hill, 1944.

[C11] Couch, L.W. II, *Digital and Analog Communication Systems*, 7/e, Upper Saddle River, NJ: Prentice Hall, 2007.

[C12] Crease, R.P., *The Great Equations*, New York: W.W. Norton & Co., 2008.

[D0] Doetsch, G., *Laplace-transformation*, New York: Dover, 1943.

[D1] Dorf, R.C. and R.H. Bishop, *Modern Control Systems*, 11/e, Upper Saddle River, NJ: Prentice Hall, 2008.

[D2] Dunsheath, P., *A History of Electrical Power Engineering*, Cambridge, Mass.: MIT press, 1962.

[E1] Eccles, P.J., *An Introduction to Mathematical Reasoning*, Cambridge University Press, 1997.

[E2] Edwards, C.H. and Oenney, D.E., *Differential Equations and Boundary Value Problems*, 3/e, Upper Saddle River, NJ: Pearson Education Inc., 2004.

[E3]* ElAli, T.S. and M.A. Karim, *Continuous Signals and Systems with MATLAB*, 2/e, Boca Raton, FL: CRC Press, 2008.

[E4] Epp, S.S., *Discrete Mathematics with Applications*, 2nd ed., Boston: PWS Publishing Co., 1995.

[F1] Fraden, J., *Handbook of modern sensors: Physics, design, and applications*, Woodbury: American Institute of Physics, 1997.

[G1]* Gabel,R.A. and R.A. Roberts, *Signals and Linear Systems*, 3/e, New York: John Wiley, 1987.

[G2] Gardner, M.F. and J.L. Barnes, *Transients in Linear Systems: Studied by the Laplace Transformation, Volume 1*, New York: John Wiley, 1942.

[G3] Giancoli, D.C., *Physics: Principles with Applications*, 5th ed., Upper Saddle River, NJ: Prentice Hall, 1998.

[G4]* Girod, B., R. Rabenstein, and A. Stenger, *Signals and Systems*, New York: Wiley, 2003

[G5]* Gopalan, K., *Introduction to Signal and System Analysis*, Toronto: Cengage Learning, 2009.

[H1]* Haykin, S. and B. Van Veen, *Signals and Systems*, 2/e, New York: John Wiley, 2003.

[H2] Haykin, S., *Communiation Systems*, 4/e, New York: John Wiley, 2001.

[H3] Hayt, W.H., J.E. Kemmerly, and S.M.Durbin, *Engineering Circuit Analysis*, 6th ed., New York: McGraw Hill, 2002.

[H4] Horenstein, M.N., *Microelectronic Circuits and Devices*, 2/e, Upper Saddle River, NJ: Prentice Hall, 1996.

[H5]* Houts, R.C., *Signal Analysis in Linear Systems*, New York: Saunders, 1991.

[I1] Irwin, J.D. and R.M. Nelms, *Basic Engineering Circuit Analysis*, 8th ed., New York: John Wiley, 2005.

[I2] Isaacson, W., *Einstein: His Life and Universe*, New York: Simon & Schuster, 2007.

[J1]* Jackson, L.B., *Signals, Systems, and Transforms*, Boston: Addison Wesley, 1991.

[J2] Johnson, C.D., *Process Control Instrumentation Technology*, New York: John Wiley, 1977.

[K1] Kailath, T., *Linear Systems*, Upper Saddle River, NJ: Prentice Hall, 1980.

[K2]* Kamen, E.W. and B.S. Heck, *Fundamentals of Signals and Systems*, 3/e, Upper Saddle River, NJ: Prentice Hall, 2007.

[K3] Kammler, D.W., *A First Course in Fourier Analysis*, Upper Saddle River, NJ: Prentice Hall, 2000.

[K4] Kingsford, P.W., *Electrical Engineering: A History of the Men and the Ideas*, New York: St. Martin's Press, 1969.

[K5] Körner, T.W., *Fourier Analysis*, Cambridge, U.K.: Cambridge Univ. Press, 1988.

[K6]* Kudeki, E. and D.C. Munson Jr., *Analog Signals and Systems*, Upper Saddle River, NJ: Prentice Hall, 2009.

[K7]* Kwakernaak, H. and R. Sivan, *Modern Signals and Systems*, Upper Saddle River, NJ: Prentice Hall, 1991.

[L1]* Lathi, B. P. *Linear Systems and Signals*, New York: Oxford Univ. Press, 2002.

[L2] Lathi, B. P. *Modern Digital and Analog Communication Systems*, 3/e, New York: Oxford Univ. Press, 1998.

[L3]* Lee, E.A., and P. Varaiya, *Structure and Interpretation of Signals and Systems*, Boston, Addison Wesley, 2003.

[L4] Lindley, D., *The end of Physics*, New York: BasicBooks, 1993.

[L5]* Lindner, D.K., *Introduction to Signals and Systems*, New York: McGraw-Hill, 1999.

[L6]* Liu, C.L., and Jane W.S. Liu, *Linear Systems Analysis*, New York: McGraw-Hill, 1975.

[L7] Love, A. and J.S. Childers, *Listen to Leaders in Engineering*, New York: David McKay Co., 1965.

[M1]* Mandal, M., and A. Asif, *Continuous and Discrete Time Signals and Systems*, New York: Cambridge University Press, 2009.

[M2] Manley, J.M., " The concept of frequency in linear system analysis", *IEEE Communication Magazine*, vol. 20, issue 1, pp. 26-35, 1982.

[M3]* McGillem, C.D., and G.R. Cooper, *Continuous and Discrete Signal and System Analysis*, 3/e, New York: Oxford University Press, 1983.

[M4]* McClellan, J.H., R.W. Schafer, and M.A. Yoder, *Signal Processing First*, Upper Saddle River, NJ: Prentice Hall, 2003.

[**M5**] McMahon, A.M., *The Making of a Profession: A Century of Electrical Engineering in America*, New York: IEEE Press, 1984.

[**M6**]* McMahon, D., *Signals and Systems Demystified*, New York: McGraw-Hill, 2007.

[**N1**] Nahin, P.J., *Oliver Heaviside: Sage in Solitude*, New York: IEEE press, 1988.

[**O1**]* Oppenheim, A.V., and A.S. Willsky, *Signals and Systems*, Upper Saddle River, NJ: Prentice Hall, 1983. Its 2nd edition, with S.H. Nawab, 1997.

[**P1**] Papoulis, A., *Signal Analysis*, New York: McGraw-Hill, 1977.

[**P2**] Petroski, H., *The essential engineer: Why science alone will not solve our global problems* , New York: Alfred A. Knopf, 2010.

[**P3**] Petroski, H., *Small things considered: Why there is no perfect design*, New York: Alfred A. Knopf, 2003.

[**P4**]* Phillips, C.L., J.M. Parr, and E.A. Riskin, *Signals, Systems, and Transforms*, 3/e, Upper Saddle River, NJ: Prentice Hall, 2003.

[**P5**] Pipes, L.A. and L.R. Harville, *Applied Mathematics for Engineers and Physicists*, 3/e, New York: McGraw-Hill, 1970.

[**P6**] Pohlmann, K.C., *Principles of Digital Audio*, 4/e, New York: McGraw-Hill, 2000.

[**P7**]* Poularikas, A., and S. Seely, *Signals and Systems*, Boston: PWS-Kent, 1994.

[**P8**]* Poularikas, A., *Signals and Systems Primer with MATLAB*, Boca Raton, FL: CRC Press, 2007.

[**R1**]* Robert, M.J., *Signals and Systems: Analysis using transform methods and MATLAB*, New York: McGraw-Hill, 2012.

[**S1**] Schroeder, M.R., *Computer Speech*, 2/e, New York: Springer, 2004.

[**S2**] Sedra, A.S., and K.C. Smith, *Microelectronic Circuits*, 5/e, New York: Oxford University Press, 2003.

[**S3**]* Sherrick, J.D., *Concepts in Systems and Signals*, 2/e, Upper Saddle River, NJ: Prentice Hall, 2004.

[**S4**] Singh, B., *Fermat's Enigma*, New York: Anchor Books, 1997.

[**S5**] Singh, B., *Big Bang*, New York: Harper Collins Publishers, Inc., 2004.

[**S6**]* Siebert, W.M., *Circuits, Signals, and Systems*, New York: McGraw-Hill, 1986.

[**S7**] Slepian, D., " On Bandwidth", *Proceedings of the IEEE*, vol. 64, no. 3, pp. 292-300, 1976.

[**S8**]* Soliman, S., and M. Srinath, *Continuous and Discrete Signal and Systems*, Upper Saddle River, NJ: Prentice Hall, 1990.

[S9] Sony (UK) Ltd, *Digital Audio and Compact disc technology*, 1988.

[S10]* Stuller, J.A., *An Introduction to Signals and Systems*, Ontario, Canada: Thomson, 2008.

[S11]* Sundararajan, D., *A Practical Approach to Signals and Systems*, New York: John Wiley, 2009.

[T1] Terman, F.E., "A brief history of electrical engineering education", *Proc. of the IEEE*, vol. 64, pp. 1399-1404, 1976. Reprinted in vol. 86, pp. 1792-1800, 1998.

[U1] Uyemura, J.P., *A First Course in Digital Systems Design*, New York: Brooks/Cole, 2000.

[V1]* Varaiya, L., *Structure and Implementation of Signals and Systems*, Boston, MA: Addison-Wesley, 2003.

[W1] Wakerly, J.F., *Digital Design: Principles & Practices*, 3/e, Upper Saddle River, NJ: Prentice Hall, 2000.

[W2] *Wikipedia*, web-based encyclopedia at wikipedia.org

[Z1] Zemanian, A.H., *Distribution Theory and Transform Analysis*, New York: McGraw-Hill, 1965.

[Z2]* Ziemer, R.E., W.H. Tranter, and D.R. Fannin, *Signals and Systems*, 4/e, Upper Saddle River, NJ: Prentice Hall, 1998.

Index

Made in the USA
Lexington, KY
27 February 2012